Mathematics and Scientific Representation

OXFORD STUDIES IN PHILOSOPHICAL SCIENCE

General Editor:
Paul Humphreys, University of Virginia

Advisory Board
Anouk Barberousse (European Editor)
Robert Batterman
Jeremy Butterfield
Peter Galison
Philip Kitcher
Margaret Morrison
James Woodward

The Book of Evidence
Peter Achinstein

Science, Truth, and Democracy
Philip Kitcher

Inconsistency, Asymmetry, and Non-Locality
A Philosophical Investigation of Classical Electrodynamics
Mathias Frisch

The Devil in the Details: Asymptotic Reasoning in Explanation, Reduction, and Emergence
Robert W. Batterman

Science and Partial Truth: A Unitary Approach to Models and Scientific Reasoning
Newton C. A. da Costa and Steven French

Inventing Temperature: Measurement and Scientific Progress
Hasok Chang

The Reign of Relativity: Philosophy in Physics 1915–1925
Thomas Ryckman

Making Thing Happen: A Theory of Causal Explanation
James Woodward

Mathematics and Scientific Representation
Christopher Pincock

Simulation and Similarity: Using Models to Understand the World
Michael Weisberg

Systematicity: The Nature of Science
Paul Hoyningen-Huene

Causation and Its Basis in Fundamental Physics
Douglas Kutach

Reconstructing Reality: Models, Mathematics, and Simulations
Margaret Morrison

The Ant Trap: Rebuilding the Foundations of the Social Sciences
Brian Epstein

Mathematics and Scientific Representation

Christopher Pincock

UNIVERSITY PRESS

Oxford University Press is a department of the University of Oxford.
It furthers the University's objective of excellence in research, scholarship,
and education by publishing worldwide.

Oxford   New York
Auckland   Cape Town   Dar es Salaam   Hong Kong   Karachi
Kuala Lumpur   Madrid   Melbourne   Mexico City   Nairobi
New Delhi   Shanghai   Taipei   Toronto

With offices in
Argentina   Austria   Brazil   Chile   Czech Republic   France   Greece
Guatemala   Hungary   Italy   Japan   Poland   Portugal   Singapore
South Korea   Switzerland   Thailand   Turkey   Ukraine   Vietnam

Oxford is a registered trade mark of Oxford University Press
in the UK and certain other countries.

© Oxford University Press 2012

First issued as an Oxford University Press paperback, 2014.

All rights reserved. No part of this publication may be reproduced, stored in a
retrieval system, or transmitted, in any form or by any means, without the prior
permission in writing of Oxford University Press, or as expressly permitted by law,
by license, or under terms agreed with the appropriate reproduction rights organization.
Inquiries concerning reproduction outside the scope of the above should be sent to the
Rights Department, Oxford University Press, at the address above.

You must not circulate this work in any other form
and you must impose this same condition on any acquirer.

Library of Congress Cataloging-in-Publication Data
Pincock, Christopher.
Mathematics and scientific representation / Christopher Pincock.
p. cm.
ISBN 978-0-19-975710-7 (hardcover : acid-free paper); 978-0-19-020139-5 (paperback : acid-free paper)
1. Science—Mathematics. 2. Science—Methodology. I. Title.
Q175.32.M38P56 2011
501'.51—dc22
2011002551

For my parents

There is no philosophy which is not founded upon knowledge of the phenomena, but to get any profit from this knowledge it is absolutely necessary to be a mathematician.

—Daniel Bernoulli

CONTENTS

List of Figures xi
Preface xiii
Mathematical Notation xv

1. Introduction 3
    1.1. A Problem 3
    1.2. Classifying Contributions 5
    1.3. An Epistemic Solution 8
    1.4. Explanatory Contributions 12
    1.5. Other Approaches 15
    1.6. Interpretive Flexibility 18
    1.7. Key Claims 21

PART I Epistemic Contributions

2. Content and Confirmation 25
    2.1. Concepts 25
    2.2. Basic Contents 27
    2.3. Enriched Contents 29
    2.4. Schematic and Genuine Contents 31
    2.5. Inference 33
    2.6. Core Conceptions 35
    2.7. Intrinsic and Extrinsic 36
    2.8. Confirmation Theory 38
    2.9. Prior Probabilities 41

3. Causes 45
    3.1. Accounts of Causation 45
    3.2. A Causal Representation 48
    3.3. Some Acausal Representations 51
    3.4. The Value of Acausal Representations 60
    3.5. Batterman and Wilson 63

4. Varying Interpretations 66
    4.1. Abstraction as Variation 66
    4.2. Irrotational Fluids and Electrostatics 68
    4.3. Shock Waves 74

4.4. The Value of Varying Interpretations 78
4.5. Varying Interpretations and Discovery 80
4.6. The Toolkit of Applied Mathematics 82

5. Scale Matters 87
    5.1. Scale and Scientific Representation 87
    5.2. Scale Separation 88
    5.3. Scale Similarity 93
    5.4. Scale and Idealization 96
    5.5. Perturbation Theory 104
    5.6. Multiple Scales 105
    5.7. Interpreting Multiscale Representations 113
    5.8. Summary 120

6. Constitutive Frameworks 121
    6.1. A Different Kind of Contribution 121
    6.2. Carnap's Linguistic Frameworks 122
    6.3. Kuhn's Paradigms 126
    6.4. Friedman on the Relative A Priori 131
    6.5. The Need for Constitutive Representations 137
    6.6. The Need for the Absolute A Priori 138

7. Failures 141
    7.1. Mathematics and Scientific Failure 141
    7.2. Completeness and Segmentation Illusions 142
    7.3. The Parameter Illusion 146
    7.4. Illusions of Scale 153
    7.5. Illusions of Traction 155
    7.6. Causal Illusions 161
    7.7. Finding the Scope of a Representation 163

PART II Other Contributions

8. Discovery 169
    8.1. Semantic and Metaphysical Problems 169
    8.2. A Descriptive Problem 175
    8.3. Description and Discovery 179
    8.4. Defending Naturalism 183
    8.5. Natural Kinds 187

9. Indispensability 190
    9.1. Descriptive Contributions and Pure Mathematics 190
    9.2. Quine and Putnam 190
    9.3. Against the Platonist Conclusion 196
    9.4. Colyvan 200

Contents

10. Explanation 203
    10.1. Explanatory Contributions 203
    10.2. Inference to the Best Mathematical Explanation 210
    10.3. Belief and Understanding 217

11. The Rainbow 221
    11.1. Asymptotic Explanation 221
    11.2. Angle and Color 223
    11.3. Explanatory Power 228
    11.4. Supernumerary Bows 229
    11.5. Interpretation and Scope 236
    11.6. Batterman and Belot 239
    11.7. Looking Ahead 242

12. Fictionalism 243
    12.1. Motivations 243
    12.2. Literary Fiction 244
    12.3. Mathematics 250
    12.4. Models 256
    12.5. Understanding and Truth 261

13. Facades 264
    13.1. Physical and Mathematical Concepts 264
    13.2. Against Semantic Finality 265
    13.3. Developing and Connecting Patches 268
    13.4. A New Approach to Content 275
    13.5. Azzouni and Rayo 278

14. Conclusion: Pure Mathematics 279
    14.1. Taking Stock 279
    14.2. Metaphysics 280
    14.3. Structuralism 284
    14.4. Epistemology 285
    14.5. Peacocke and Jenkins 290
    14.6. Historical Extensions 293
    14.7. Nonconceptual Justification 295
    14.8. Past and Future 297

Appendix A: Method of Characteristics 301
Appendix B: Black-Scholes Model 303
Appendix C: Speed of Sound 305
Appendix D: Two Proofs of Euler's Formula 307
Bibliography 309
Index 317

LIST OF FIGURES

3.1	The Bridges of Königsberg	52
3.2	A Non-Eulerian Graph	52
3.3	An Eulerian Graph	52
3.4	Velocity and Flux	56
4.1	Source	72
4.2	Flow Around an Object	73
4.3	Characteristic Base Curves	76
5.1	Normal versus ENSO Conditions	90
5.2	Temperature versus Distance	93
5.3	Self-Similar Coordinates	94
5.4	Surface Waves	98
5.5	Wave Dispersion	100
5.6	Boundary Layer Flow	107
5.7	Ideal Fluid Flow	110
5.8	The Failure of a Single Scale	115
5.9	The Success of Two Scales	116
5.10	Bénard Cells	116
7.1	Column Storage	143
7.2	Two-Support System	143
7.3	Three-Support System	144
7.4	The Failure of the Three-Support System	144
7.5	LTCM Return on Investment	149
7.6	A Fold Catastrophe	157
7.7	A Cusp Catastrophe	158
7.8	Divergent Trajectories	160
11.1	A Rainbow	224
11.2	Ray Path	225
11.3	Rainbow with Supernumerary Bows	230
11.4	Impact Parameter	233
11.5	Critical Points	234
11.6	Comparing Predictions	235
13.1	Polar Coordinates	272
13.2	A Riemann Surface	274

PREFACE

This book discusses the contributions that mathematics makes to the success of science. This topic falls between the philosophy of mathematics and the philosophy of science as they are currently practiced, and this may account for its relative neglect in contemporary philosophy. Despite this situation, I believe that the presence of mathematics in much of science has significant philosophical implications for our understanding of both mathematics and science. Though many readers will find much to disagree with in this book, I hope to place the topic of applications on the agenda of philosophy of science and mathematics. For those who reject the particular approach I have taken to this problem, I look forward to other proposals that my efforts might generate.

This project began as my dissertation, which was completed in the spring of 2002 under the supervision of Paolo Mancosu. I am extremely grateful to him for his continuing encouragement and critical suggestions over the many years since then. I have been strongly influenced by his many articles and his overall approach to the philosophy of mathematics, as reflected in the anthology he recently edited, *The Philosophy of Mathematical Practice* (Mancosu 2008c). I hope that my book will be a step in the right direction toward reconnecting philosophers of mathematics with the philosophy of science and scientific practice.

Shortly after 2002 I became convinced that issues connected with applications of mathematics are best pursued by focusing on the epistemic contributions which mathematics makes. This approach is first reflected in Pincock (2007). The philosophers most responsible for this shift in my thinking are Robert Batterman and Mark Wilson. A reader of their work will find their influence throughout this book, from the choice of examples to the selection of the issues most deserving of extended discussion. While I expect that both Batterman and Wilson will find much to criticize in this book, I would like to express my gratitude to both for their work on these topics and for their many discussions and expressions of support.

The writing of this book began in earnest while I was a visiting fellow at the Center for the Philosophy of Science at the University of Pittsburgh in the 2008–2009 academic year. I thank John Norton, the director of that center, along with the other fellows, Hanne Andersen, Claus Beisbart, Erik Curiel, Laura Felline, Ulrich Krohs, Flavia Padovani, Daniel Parker and Federica Russo for a stimulating and engaging atmosphere. Additional support for this project came in the form of a National Endowment for the Humanities fellowship. I am thankful to the NEH for supporting a project of this kind. Finally, Purdue University provided generous leave and sabbatical support, and I am thankful to the Philosophy Department and my colleagues for their help in completing this book manuscript.

Many friends and colleagues beyond those mentioned made important contributions to this book as chapters were being drafted or presented at conferences and colloquia. I especially thank André Ariew, Jody Azzouni, Alan Baker, Sorin Bangu, Otávio Bueno, Elisabeth Camp, Anjan Chakravarrty, Mark Colyvan, Gabriele Contessa, William Demopoulos, James Franklin, Steven French, Michael Friedman, Roman Frigg, Paul Humphreys, Ken Manders, Stathis Psillos, Joshua Schechter, Mark Steiner, Susan Vineberg, and Michael Weisberg. Of course, none of these people are to blame for any errors in this book. I am also very much indebted to Peter Ohlin and Jenny Wolkowicki at Oxford University Press for their support in completing this project and seeing it through to publication. Two referees for Oxford made many valuable suggestions for how an earlier version could be improved.

My greatest debt is to my wife, Tansel Yilmazer, who has suffered through the highs and lows of composing and revising fourteen chapters over a two-and-a-half-year period. Whatever light is shed on the problems discussed here, I owe to her.

Mississauga, Ontario, Canada
December 31, 2010

PREVIOUSLY PUBLISHED

Sections from several chapters are adapted from articles.

Some of the material on confirmation in chapter 2 derives from "Mathematics, Science and Confirmation Theory," *Philosophy of Science* (Symposium Proceedings), 77 (2010): 959–970.

Sections 4.6 and 5.6 overlap with "Towards a Philosophy of Applied Mathematics," in O. Bueno and Ø. Linnebo (eds.), *New Waves in Philosophy of Mathematics* (Palgrave Macmillan, 2009), 173–194.

Section 5.4 is related to an unpublished paper, "How to Avoid Inconsistent Idealizations."

The discussion of indispensability and explanation is pursued in an unpublished paper, "Mathematical Contributions to Scientific Explanation."

Chapter 11 is based on "Mathematical Explanations of the Rainbow," *Studies in History and Philosophy of Modern Physics*, 42 (2011): 13–22, with permission from Elsevier.

MATHEMATICAL NOTATION

A book on the contributions that mathematics makes to the success of science must include some mathematics. Because not all readers will be equally comfortable with the notation deployed in the cases I discuss, it seemed prudent to collect the symbols used.

Functions from trigonometry like *sine*, *cosine*, and *tangent* appear in some case studies.

For a function of one variable $f(t)$, the following symbols are used for the *derivative*: $\frac{d}{dt}f(t), \dot{f}$, and $f'(t)$. Each of these can be repeated to indicate a second-order derivative: $\frac{d^2}{dt^2}f(t), \ddot{f}, f''(t)$.

For a function of two or more variables $f(x, t)$, the following symbols are used for the *partial derivative*: $\frac{\partial}{\partial x}f(x, t)$ and f_x. Each of these can be repeated to indicate a second-order partial derivative: $\frac{\partial^2}{\partial x^2}f(x, t)$ and f_{xx}.

The symbol ∇ (nabla) has various uses in connection with scalar and vector fields. When ϕ is a scalar field, meaning that it assigns a number to each point in a domain, $\nabla \phi$ is the *gradient* of the field. However, if F is a vector field, meaning that it assigns a vector to each point in a domain, $\nabla \cdot F$ is the *divergence* of the field. The Laplacian of a scalar field ϕ is expressed by $\nabla^2 \phi$. Finally, $\nabla \times F$ is the *curl* of the vector field F.

The italicized words are defined and discussed in introductory calculus textbooks such as Stewart (2007b).

On a few occasions functions with a complex variable $f(z)$ appear. An introductory textbook in this area is Brown and Churchill (1996).

Mathematics and Scientific Representation

1

Introduction

1.1 A PROBLEM

The success of science is undeniable, but the nature of that success remains opaque. To make a start on characterizing the success of science, I will stipulate that science is in the business of producing representations of the physical world. These representations will be accurate just in case some region of the world is a certain way. When approached from this angle, it seems obvious that the success of science should be characterized by saying that our scientific representations are, for the most part, accurate. That is, they present this or that region to be a certain way, and in fact the world is that way.

This natural form of scientific realism is opposed on many fronts. In the history and sociology of science, many would argue that the success of science is not best characterized by the accuracy of its representations. Instead, the success is defined in terms of the agreement of the community of scientists. The main question, for a given historical or sociological episode, is how a representation came to be widely accepted by a given community, and there is no easy inference from widespread agreement to realistic correctness. Call this the *social conception* of the success of science. Within the philosophy of science, the social conception is largely ignored, but there are still several apparently viable alternatives to realism. Each agrees that the success of science has something to do with the accuracy of scientific representations, but qualifies the respects in which a given representation has to be accurate for it to be successful. Instrumentalism and constructive empiricism in very different ways restrict the success of science to what is observable. Others focus on the existence of entities or the structural relations between entities. Cartwright offers an even more qualified picture of the success of science by restricting the scope of the regularities offered by our scientific representations (Cartwright 1999).

All of these differing characterizations of the success of science need to face the problem that I articulate here. Put most generally, the problem is to account for the presence of mathematics in our scientific representations. The central issue is to see to what extent the central place of mathematics in science contributes to the success of science. More carefully, for a given physical situation, context, and mathematical representation of the situation, (1) what does the mathematics contribute to the representation, (2) how does it make this contribution, and (3) what must be in place for this contribution to occur? The parameters of this more careful formulation of the problem make clear that the way a given piece of mathematics contributes to the success of science in one representation can differ from the contribution from some other piece in some other context.

Advocates of the social conception of the success of science are likely to emphasize the respects in which mathematics contributes to the convergence of scientists on a given representation. Porter's *Trust in Numbers* is an especially thorough exploration of this approach (Porter 1995). Although this strategy for tackling our problem cannot be dismissed out of hand, it surely would be a disappointing result for the realist, or anybody else who thinks scientific representations aim to capture some aspect of the real world, if this social conception of the contribution of mathematics was the full story. For if this was the whole account of the contributions of mathematics, then it would be hard to resist the social conception of science more generally.

Still, it is my experience that philosophers otherwise resistant to the social conception of science are tempted to take this way out when it comes to the role of mathematics in science. One reason for this seems to be that several exaggerated claims have been made on behalf of the role of mathematics in the success of science, and the only alternative to these claims seems to be a social conception. The position I have in mind here might be called the *metaphysical conception* of the way mathematics contributes to the success of scientific representations. According to the metaphysical conception, mathematics contributes to accurate scientific representations because the physical world is itself essentially mathematical. An inventory of the genuine constituents of the physical world reveals that these constituents include mathematical entities. So it is no surprise that our best scientific representations include a lot of mathematics. For this mathematics is needed to characterize this or that part of reality accurately.[1] A similar attitude has been held for scientific representations that include causation. A metaphysical account of the causal character of these representations is simply that the world itself contains causal relationships and these should be reflected in our most accurate representations.

As with the social conception, it is hard to deny that there are certain cases that fit the metaphysical conception of the contribution of mathematics to the success of science. At the same time, there is good reason to think that this cannot be the whole story. In the vast majority of cases in which we find mathematics being deployed, there is little inclination for the scientist to take the mathematics to reflect the underlying reality of the phenomenon in question. We see this, for example, whenever a representation employs a mathematical idealization. An idealized representation, for my purposes, is a representation that involves assumptions that the agent deploying the representation takes to be false. When a fluid is represented as continuous, even though the agent working with the representation believes that the system is discrete, then we have some useful mathematics that resists any simple metaphysical interpretation. Precisely these sorts of cases tend to provoke a slide into a social conception of their contribution. For if the mathematics does not reflect

---

1. This appears to be the conclusion of Franklin (1989) and Franklin (2008), but it is not clear how widely he intends to extend this metaphysical approach.

Introduction

any underlying physical constituents of the system, then what other contribution could it be making besides helping generate agreement among scientists?

1.2 CLASSIFYING CONTRIBUTIONS

To move beyond this unsatisfying dilemma, we need a more careful inventory of the contributions that mathematics seems to be making in our successful scientific representations. With this in hand, we can see if there are any other general strategies for answering our question. I start by presenting four different dimensions that we can consider for a given representation. For each dimension, mathematics seems to be doing something different as we move from one extreme to the other. Though somewhat elaborate, this classification scheme will allow us to isolate more or less pure cases of how mathematics contributes in a given way. Other, more difficult cases can then be understood as impure mixtures of these different sorts of contributions.

Before we explore these four dimensions, we should first distinguish between the mathematical content of the representation itself and the mathematical techniques that were used to arrive at that representation. We dub the mathematics of a representation *intrinsic* when it figures in the content of the representation. But there are usually additional *extrinsic* mathematical elements that contribute to the representation beyond those that figure in the content. Suppose, for example, we have a representation of the dynamics of a particular physical system like a cannonball. Here the intrinsic mathematics of the representation includes the differential equations with which the trajectory of the cannonball is specified. These equations are used directly to fix how the world must be for the representation to be correct. At the same time, there may very well be different elements of mathematics relevant to the representation, in both its derivation from other representations and its further application in allowing the derivation of other representations. Understanding the contribution of this extrinsic mathematics is often just as important as understanding what the intrinsic mathematics is doing.

Our first two dimensions relate to the degree to which a given representation is abstract. One sense of abstractness I pursue is the degree to which a given representation purports to capture the genuine causal structure underlying a given phenomenon. A dynamic representation aims to include the causal details of a series of events, where causal effects propagate along the series in accordance with the mechanisms deployed in the representation. Mathematics can play a central role here in allowing the different parts of the representation to track the genuine causal structure. For example, we may aim to represent the causal structure of the development of a traffic jam. The mathematical representation could accomplish this by having a series of real numbers standing for positions of the road. The representation would also include functions and mathematical rules or equations governing those functions corresponding to the causal laws about the propagation of cars along the road. If everything worked properly, these mathematical transformations would mimic the causal effects of the cars on one another, for example, how the slowing down of one car would change the velocity of the cars behind it. We would then have a *concrete causal* representation of the system.

Here we have the sort of role for mathematics envisaged by the metaphysical conception. Its general viability depends in the end on how generous we are in populating the world with genuine causal relationships. Even the traffic case could be questioned by someone who restricted causes to relationships between fundamental physical entities. On this sparse approach to causes, a given car slowing down is not the cause of its neighbor slowing down, and the real cause of this event is a complicated chain of events involving the fundamental particles making up the cars and the drivers. There is little motivation to restrict causal relationships so severely, and the most popular accounts of causes are much more liberal. Still, even on the most profligate picture of causes, there are mathematical representations where causes are absent. What I have in mind are representations that aim to represent the system at some designated equilibrium point or the respects in which the system is stable around such a point. I call such cases *abstract acasual* because they abstract away from the causal relationships found in concrete causal representations. In our traffic case, we might imagine a representation of a given system of cars that simply presented what the system would look like after all the cars had achieved some fixed velocity for a given density of cars. Here we have two respects in which this abstract acasual representation could be mathematical corresponding to our intrinsic/extrinsic distinction. The specification of the velocity for the given density exhausts the intrinsic mathematics of the representation. But there are additional extrinsic mathematical elements that contribute to the representation such as the mathematical transformations that produce the abstract acasual representation. These might involve a study of the concrete causal representation of the traffic, but there is no general assurance that an abstract acasual representation must result from an associated concrete causal representation.

This is one sense of abstractness. A second dimension also involves the abstractness of the representation, but in a different sense of "abstract." It sometimes turns out that there is a family of representations with a constant core of intrinsic mathematics, but where each representation in the family aims at a different kind of physical system. Such *abstract varying* representations can be contrasted with the *concrete fixed* representations that aim to capture the features of just one kind of physical system. With an abstract varying representation we can bring together a collection of representations in virtue of the mathematics that they share, and it can turn out that the success of this family of representations is partly due to this mathematical similarity. As an example of this we can take the mathematics associated with the harmonic oscillator model. The physical systems studied using this model range from simple systems like a pendulum and a spring through to domains like acoustic systems, electromagnetic fields and even nuclear phenomena (Marion and Thornton 1995, §3.7). What unites our representations of these different systems is the constant mathematical contribution. Exactly how this mathematics might provide for the success of this sort of representation is not initially clear. But we should not make the mistake of assuming that mathematics helps with abstract varying cases in the same way that it helps with abstract acasual cases.

A third dimension along which we can consider the contribution of mathematics is *scale*. The scale of a representation is given by the size of the magnitudes involved in the representation. Spatial representations vary in scale from the very small, as

Introduction

with many quantum mechanical representations, through to the extremely large, as with much of general relativity. Time and energy are other magnitudes that can be represented at a wide range of different scales. The intriguing aspects of the scale of a representation often arise when we have successful representations of a given phenomenon at different scales and the features of these representations fail to fit together in a transparent way. For example, at the microscale we can represent a substance like water or air as a collection of molecules flying around and hitting each other. A macroscale representation of the same substance may treat it very differently, that is, as a fluid that occupies space continuously. Even some of the magnitudes that the representations assign to the substance can differ. Although both will assign a mass to the water, the macroscopic representation may also include the fluid's temperature and pressure distribution. Neither magnitude is a fundamental feature of the substance in the microscopic representation. A difference in scale, then, can involve a host of other changes in the content of what is represented.

In some sense, the opposite sort of case where scale is important is where we represent a phenomenon as "self-similar" or scale invariant. We can use small-scale model ships, for example, to decide how to design large-scale ships because we take the representation of the relevant aspects of the motion of the ship through water to be scale invariant. This is a special case of a family of abstract varying representations united by scale-invariant mathematics. But the success of this sort of abstract varying representation need not have any immediate implications for other sorts of abstract varying representations.

The relationship between representations of different scale is often specifiable using mathematics. Here, then, we have a contribution of extrinsic mathematics to the success of this or that representation. It can also happen that a single representation combines two or more representations where the links between these representations are mathematically significant. As we will see in chapter 5, in the boundary layer representation of fluid flow around an object like an airplane wing, one set of equations is used for most of the flow, and another set of equations is used to represent the flow in the boundary layer close to the wing. The success of this hybrid representation depends on delicate considerations of scale. In effect, the two sets of equations are derived by different rescalings of a more realistic representation of the fluid flow. An understanding of scale and these rescaling techniques is essential, then, to seeing how the mathematics is successful in these sorts of cases. Exactly what the system has to be like for these manipulations to be successful is a delicate topic that I return to.

Another sort of hybrid representation that involves considerations of scale is where two or more representations at different scales are tied together so that they mutually inform each other about the represented system's features. An intuitive example of this multiscale representation is a representation of the Earth's climate. Climatologists are interested in the year-to-year behavior of the Earth's climate and the large-scale patterns of things like precipitation and temperature. Still, there is a complex interaction between these large-scale phenomena and small-scale phenomena like vegetation and hurricanes. These interactions find their place via sophisticated representations that incorporate effects between these phenomena at different scales. Again, mathematical aspects of these representations are often central to their success, so here is another place where we need to consider what the mathematics is doing and how it is doing it.

The final dimension along which to consider the contribution of mathematics to the success of a scientific representation is its constitutive versus derivative character. Several philosophers of science, from Carnap and Kuhn to Michael Friedman, have argued for the need for a *constitutive framework* as a backdrop against which a given area of science with its *derivative* representations is possible. There is no clear agreement on exactly why these frameworks are needed or what role they perform, but the frameworks often reserve a prominent place for mathematics. For example, general relativity theory operates under the assumption that space-time can be accurately represented by the mathematics of differential geometry. Friedman presents this framework as a necessary presupposition for the rest of general relativity theory, most notably the more particular claims about the relationship between mass and the geometry of space-time encapsulated in Einstein's field equations (Friedman 2001). The claims of the constitutive framework, then, have a different status from the more derivative representations that are made from within the framework. We should be open to the possibility that the contribution of mathematics to the framework is different in kind from what we find with the more usual contributions to ordinary representations.

1.3 AN EPISTEMIC SOLUTION

For a given mathematical representation we can start by asking, then, whether the mathematics in question is intrinsic or extrinsic. Then it is possible to determine how the mathematics contributes to (1) concrete causal or abstract acausal content, (2) a concrete fixed or abstract varying content, (3) issues of scale, or (4) a constitutive or derivative content. Some cases will combine different features from each of these choices, whereas others will be best understand as more pure instances of a single element of our classification scheme. The hope is that we can say something informative about what mathematics is doing as we move through these components of our framework and that understanding each of these components will undermine the unsatisfactory social and metaphysical conceptions. The main conclusion that I argue for in this book is that mathematics makes an *epistemic* contribution to the success of our scientific representations. Epistemic contributions include aiding in the confirmation of the accuracy of a given representation through prediction and experimentation. But they extend further into considerations of calibrating the content of a given representation to the evidence available, making an otherwise irresolvable problem tractable and offering crucial insights into the nature of physical systems. So even when mathematics is not playing the metaphysical role of isolating fundamentally mathematical structures inherent in the physical world, it can still be making an essential contribution to the success of science. For part of this success is the fact that we take our evidence to confirm the accuracy of our best scientific representations. And it is here that the mathematical character of these representations makes its decisive mark.

Introduction

The vindication of this proposal depends on the careful consideration of a range of examples which fit into the various parts of the classification scheme developed so far. Still, a prima facie case for this epistemic approach can be made by considering a few kinds of cases and what their epistemic contributions might be. To start, I return to the abstract acausal representations. Recall that in one instance of this sort of case the mathematics involves a representation of the system at some equilibrium point and may also include claims about the stability properties of such a point. In the traffic case, we consider the representation of the cars after they have reached some fixed velocity for a given density. There are two initially clear epistemic aspects of this acausal representation. Most obviously, if the acausal representation has been derived from a causal representation, then we can get some information about the system that was not explicit in the causal representation. We might learn, for example, at what density the maximum velocity is 50 km/h. The mathematics of the acausal representation allows us to know something about the traffic system that we might otherwise be ignorant of.

A second epistemic benefit of the acausal representation is somewhat less obvious. This is the highly restricted content of the acausal representation when compared to a causal representation of the same system. At an intuitive level, the acausal representation says less, and so takes fewer risks when representing the system. This impoverished content has an immediate benefit when we think about how we might confirm the acausal representation. Because it says less about the system, it will be easier to confirm. This aspect comes through clearly in our traffic case. If we had doubts about the accuracy of the acausal representation, we could remedy these doubts by carrying out the following experiment: set up a traffic system so that it corresponded to features of the causal model and see how the system stabilized as more cars were added to the road. Agreement on the density required to produce a velocity of 50 km/h is then good evidence that the acausal representation is correct.

It might seem like the same point would hold for a causal representation of the traffic system. Why would the same experiment not confirm the causal representation to the same degree that it confirmed the acausal representation? A detailed answer would depend on an account of the causal claims, but it is not too controversial to assume that claims about causal chains involve more than stable features of the eventual result of the causal chain. Many accounts of causation emphasize counterfactual considerations, that is, what would have happened had the system been different in this or that respect. Simply examining an actual traffic system does little to confirm these features of the causal representation, and so to this extent it will be less well confirmed by the sort of experiment I have described.

There is another kind of case that also illustrates the relative ease of confirmation of acausal over causal representations. This is the case where we lack a causal representation of the development of the system, but are still able to formulate an acausal representation of what the system might look like at equilibrium. In a trivial sense, this sort of situation is one where the acausal representation can be confirmed and the causal representation cannot. Though it might seem that such cases would be rare, they are easily found in the practice of science. Typically, causal representations of a given situation cannot be formulated because of their daunting complexity. And even when they can be presented, they may be untestable because the mathematics associated with the complex causal interactions proves intractable. As a result, the scientist or applied mathematician shifts to an acausal representation. It need not be a wholly acausal representation, as abstracting away from one family of causes need not involve a complete abstraction from all causal representation. An example of this that we will return to is the representation of steady flow in fluid mechanics. Steady flow involves functions that do not vary over time. As a result, steady flow representations are acausal with respect to the causal features of the flow that depend on time. In many cases, we can confirm aspects of the steady flow representation of some fluid flow situation, and we remain totally unable to formulate—let alone confirm—a representation of the time-dependent causes of the phenomena in question.

A related kind of case has received some discussion recently by philosophers under the name of "robustness analysis."[2] Here we have a causal model, but we do not know how to set its parameters to produce an accurate representation. A second acausal model is then developed based on the features of the original model that are preserved under variations in the unknown parameters. This can lead to greater confidence in the acausal model than in any of the various causal models.

Mathematics can contribute to the epistemic goals of science, then, by contributing to the formulation of abstract acausal representations. The same result can be seen for the contributions of mathematics to abstract varying representations. Recall that an abstract varying representation was a family of representations of different kinds of physical systems with a common mathematical core. An important difference with the earlier abstract acausal representation is that an abstract varying representation has more content than an associated concrete fixed representation. This is because the varying representation includes representations of a wide variety of physical systems, whereas the fixed representation targets just one. Nevertheless, the varying representation can have epistemic advantages over the fixed representation and these advantages can be traced back to the mathematical character of the varying representation.

One sort of case involves a transfer of evidential support from one member of the family to another where the transfer is mediated by the mathematical similarities between the parts of the abstract varying representation. Suppose, for example, that we have thoroughly explored the simple harmonic oscillator representation of the pendulum. The core features of this representation are that the pendulum has an equilibrium point, when the bob is pointing straight down, and the force restoring it to this point is proportional to the displacement from equilibrium. These features allow us to classify different kinds of pendulums and accurately predict the motion of the bobs across a wide variety of changes in mass, initial velocity, and angular displacement. Using experiments, we can confirm the accuracy of this representation in a wide variety of respects. Suppose further that we have arrived at a representation of a certain kind of spring where the spring representation is mathematically similar in the relevant aspects to the successful pendulum representation. That is, there is an equilibrium point and a linear restoring force. With the spring the equilibrium

---

2. Weisberg (2006) and Weisberg (2008).

and the linear restoring force are determined by the materials making up the spring. Still, the mathematical similarities between the representations allow us to transfer our confidence in the pendulum representation over to the spring representation. Confirmation flows, then, within the family of representations making up an abstract varying representation. We can come to be more confident in the members of the family in virtue of their mathematical similarities than would have been possible via the consideration of the available data and each representation individually.

This process of abstraction and widening of the scope of a family of representations can be extended in a much more radical direction. So far we have only considered the relatively modest move of varying the restoring force from the force restoring the pendulum to equilibrium to the force restoring the spring to equilibrium. An even more abstract perspective involves considering the family of representations where the restoring force is linear, no matter what its physical basis is. This fully abstract representation is typically described as the simple harmonic oscillator model. Once it has been mathematically isolated, it can be investigated in a rigorous fashion completely independently of its physical manifestations. That is, we can develop a wholly mathematical theory of this model in its own right. This provides us with a host of claims that we can reasonably believe describe any physical system once we determine, either experimentally or otherwise, that it meets the minimal conditions to be a member of the family of representations. A model, mathematically identified, can lead to a huge range of well-confirmed representations of vastly different physical systems. This abstract unifying power allows the harmonic oscillator to encompass the wide range of physical systems mentioned earlier.[3]

Issues connected to the scale of a representation are more difficult to survey as the way scale is deployed can vary so much from case to case. As we have already seen, sometimes a representation is of a single scale, while in other cases the representation mixes scales or purports to be scale invariant. The main point to emphasize at this stage is that mathematics can play a central role in both the intrinsic specification of the scale or scales of the representation and in the extrinsic derivation of a representation where scale is significant. As with the two sorts of abstract representations already considered, epistemic benefits may result. In the simplest case, the mathematics may lead to a representation that is highly restricted in content because it represents the system only at a single scale. When this occurs, we have the same benefits that we saw for the abstract acausal representation. Because the content is so restricted, it is easier to confirm that the representation is accurate. At the other extreme, we may derive a mathematical representation that purports to be scale invariant. Successful scale-invariant representations clarify under which conditions the phenomenon will be unaffected by scale. As a result, scale-invariant representations take on all the positive features of the abstract varying representations just considered. We can transfer the evidence for one representation in the family to another representation in the family. This is, after all, just what happens

---

3. The value of this approach to modeling has also been emphasized by Weisberg at, for example, Weisberg (2007, §4).

when we decide how to design a ship based on experiments with a small-scale ship. What varies in such a family is not the interpretation of the elements of the mathematical representation, but instead the scale at which the magnitudes associated with the elements are interrelated. Other multiscale representations involve further epistemic benefits, which we will return to later.

Finally, we come to the constitutive versus derivative dimension of a mathematical representation. As with scale considerations, here the issues are complex. It is not even clear if a different kind of framework representation is necessary in addition to ordinary representations; even if such representations are needed, there is no consensus on why they are needed. I argue, though, that there are decisive reasons in favor of requiring constitutive frameworks and that these reasons involve issues of the confirmation of derivative representations. This strand is prominent in Kuhn and Friedman, but not always as clearly articulated as one might like. Still, here is one summary of the position I will be defending: "A constitutive framework ... defines a space of empirical possibilities ... and the procedure of empirical testing against the background of such a framework then functions to indicate which empirical possibilities are actually realized" (Friedman 2001, p. 84). Very roughly, the mathematics deployed in a framework for a given domain has the effect of representing a determinate range of states of affairs as physically possible. Taking this framework for granted, more derivative representations can then be confirmed by comparing data against these physical possibilities. A central example for Friedman is the framework representation of the general theory of relativity according to which the structure of space-time is identified with a member of the family of a certain kind of manifold. Delicate issues remain, though, about how mathematical these frameworks really need to be, and also the basis for the confirmation of the frameworks themselves.

Although none of the discussion so far is conclusive, it at least holds out the hope of understanding how a given mathematical representation contributes to the overall success of science. The general outlines of the account make clear that mathematics has a more important role than just facilitating the consensus of scientists, as the social conception suggests. It is also consistent with a more modest place for mathematics than the metaphysical conception requires. Even when the mathematics fails to track fundamental features of the world, it still affords us access to successful scientific representations. The contribution of the mathematics is in helping us formulate and investigate representations that we can actually confirm. This epistemic role puts mathematics at the center of science without the appeal to dubious metaphysical assumptions.

1.4 EXPLANATORY CONTRIBUTIONS

There is a growing debate on the existence of mathematical explanations of physical phenomena. Advocates of the explanatory power of mathematics in science often present their examples as part of the so-called indispensability argument for mathematical platonism. Critics object either that the alleged explanations are not especially mathematical, or else that nonplatonists about mathematics can also account for the explanatory power in the examples. The issue is pressing for the account I aim to develop because of the complex interactions between issues of explanation and confirmation. After surveying some of these links, I turn to the place of explanatory issues in the epistemic account of the contributions of mathematics to scientific representation. In the next section I consider the links between this project and the indispensability argument.

Accounts of explanation can be divided into pragmatic and objectivist approaches. A pragmatic approach conceives of explanation as something that we do with representations that we have already elected to accept based on other virtues. As a result, explanatory power is not itself a consideration that should be used to decide which representation to accept. The most prominent example of a pragmatic approach is van Fraassen (1980). He argues that explanations are best conceived as answers to why questions. Various different why questions receive different scientific answers using the representations we have already adopted. But just because a representation can contribute to an answer to a why question, we should not conclude that it is likely to be true (or more likely to be true than its competitors). Van Fraassen's pragmatic conception of explanation, then, undercuts inference to the best explanation.

Opposed to these pragmatic accounts, we find objectivists about explanation. They place additional constraints on an acceptable explanation and so are able to draw stronger conclusions than van Fraassen from the existence of an explanation of a phenomenon. The most ambitious versions of an objectivist approach insist that our best available explanation is likely to be true. This allows one to start with more or less uncontroversial cases of successful explanations and conclude that the ingredients of the explanation receive a boost in confirmation simply in virtue of appearing in the best explanation available. Two strategies for reaching this more substantial conception of explanation are especially popular. The first insists that a good explanation must give the cause of what is being explained, whereas the second requires that a good explanation be part of the best scheme of unifying a wide range of phenomena. Causal and unification approaches to explanation consequently disagree on which examples are cases of genuine explanation, but both vindicate some form of inference to the best explanation for their respective explanations.

The three different approaches to explanation surveyed here will have quite different reactions to the classification scheme developed in the previous sections. A pragmatic approach can concede that mathematics contributes to successful explanations, but insist that this is only because mathematics appears in our best representations as determined by criteria besides explanatory power. There is no deep connection between the mathematics and the explanatory power on the pragmatic approach in line with the more general resistance to draw any substantial consequences from explanatory power. Advocates of the causal and unification approach will likely to be tempted to draw more interesting conclusions. If we insist that all explanations involve causes and grant that there are cases where the mathematics contributes to explanation, then it seems inevitable that we conclude that there are mathematical causes of physical phenomena. From this perspective, we have independent confirmation of the metaphysical conception. That is, the physical world is essentially mathematical, and so when we explain phenomena, we need to invoke the mathematical entities that make up, in part, the system in question.

The puzzling interpretation of this kind of conclusion has prompted most advocates of mathematical explanation to go beyond causal explanation. One strategy is to emphasize the unifying powers of mathematics and claim that mathematics can contribute to explanation precisely because it can unify diverse scientific phenomena.[4] We can place this position in our scheme by aligning it with the abstract varying representations such as the simple harmonic oscillator model. These representations bring together representations of vastly different physical systems. In virtue of linking these representations, the abstract varying representation is able to represent the similarities between these systems. Presumably, it is these similarities that form the core appeal of the claim that unification is the key to explanation. We have also seen in outline how an abstract varying representation can mediate the confirmation of its different components. If this picture can be sustained, then we can understand the alleged link between explanatory unification and a boost in confirmation. This would allow the justification of a limited form of inference to the best explanation.

There are two problems with this happy reconciliation, however. First, there seem to be mathematical explanations that resist incorporation into the unification framework. I turn to these shortly. Second, there are prima facie convincing cases of unified mathematical representations that fail to be explanatory (Morrison 2000). The suspicion here is that bringing together a family of representations via some shared mathematics may take us away from the relevant details of the different systems, and so actually decrease the overall explanatory power of the different representations. This may be because all explanation is really causal, or else because the mathematics unifies the phenomena in a misleading way. Either way, if these examples stand up to scrutiny, it will not be possible to trace the explanatory power of the mathematics in all cases to its unifying power.

What sorts of cases resist the unification framework? One example is the topological explanation for why there is no nonoverlapping return path across the bridges of Königsberg.[5] It is quite artificial to force this explanation into the unification model, but at the same time it seems fairly clear that there is something mathematical and noncausal going on. Here we can appeal to another slot of our classification scheme, namely, abstract acausal representations. With the bridges, we have abstracted away from the material causes of the bridges being in the shape that they are in, and represented them as they are independently of the details of their construction. This can be thought of as a representation of the system at equilibrium in an extended sense where the structure of the system is stable under changes in its material construction. As with the abstract varying cases, we have seen how abstract acausal representations can lead to an increase in confirmation. If that story can be applied to these explanatory cases as well, then we have a different way to vindicate inference

---

4. See Kitcher (1989), discussed in Mancosu (2008a).

5. This example is discussed later in chapter 3.

Introduction

to the best explanation. Abstract acausal representations can figure in explanations and at the same time receive a boost in confirmation.

There are, then, some prospects for linking the picture of mathematical representation that I have sketched to discussions of mathematical explanation and perhaps even using this approach to ground some form of inference to the best explanation. In what follows, though, I proceed somewhat cautiously. As there is no agreement on cases or accounts of explanation, I do not assume that mathematics contributes explanatory power in this or that example. Instead, my focus will be on confirmation. Here there is also disagreement, but the question of confirmation seems more tractable and amenable to a naturalistic treatment via a careful study of the actual practice of scientists. If, as I hope will be the case, there are substantial links that can be drawn between boosts in confirmation and apparent mathematical explanations, then we will have a basis for further discussion about the explanatory contributions of mathematics. Still, if it turns out that there are no such links, the point about the contribution of mathematics to confirmation can be independently defended.

1.5 OTHER APPROACHES

There are a variety of other ways to approach the problem of mathematics in science, and it is worth setting out how my strategy promises to link up with these different approaches. The most prominent alternative approach focuses on the correct metaphysical interpretation of pure mathematics. Is pure mathematics best thought of as involving abstract objects, as the platonist argues? Many nominalist accounts of mathematics aim to interpret pure mathematics so that it does not require the existence of any abstract objects. Most nominalist interpretations of mathematics preserve the intuitive objective truth-values associated with mathematical statements, but there are some more recent nonstandard approaches that I group under the label of "fictionalist" that do not. Fictionalists provide an account of pure mathematics that preserves some features of ordinary mathematical practice, but give up the standard truth-values we would assign to some statements. That is, they are not truth-value realists, while the standard nominalists remain realists about truth-values, even if they are not realists about abstract objects. Platonists are realists about both truth-values and abstract objects.

The indispensability argument for platonism argues that we should accept the existence of abstract objects based on their essential or indispensable contribution to our best scientific theories (Colyvan 2001). This contribution can be characterized in different ways. Initially with Quine and Putnam, the contribution was specified in terms of the formulation of our best scientific theories. That is, it was not possible to formulate acceptable versions of these theories that did not involve mathematical terms. More recently, the focus has shifted to the explanatory power of mathematics. That is, there are superior explanations available from within our mathematical scientific theories. Either way, the indispensable contribution of mathematics is supposed to require that we believe in the existence of abstract mathematical objects.

From the perspective of this project, the problem with the positions of both advocates and critics of these indispensability arguments is that they have not considered the prior question that I have put at center stage: what is the contribution that mathematics makes to this or that scientific representation? A proper evaluation of the role of mathematics in science in determining the interpretation of pure mathematics turns on delicate considerations of the content of mathematical scientific representations. The platonist picture seems to be that it is only by interpreting mathematics in terms of abstract objects that we can make sense of these contents. Unfortunately, the platonist advocates of indispensability have only recently turned to a clarification of what these contents are (Bueno and Colyvan 2011). Critics of indispensability have done little better. There was initially a laudable attempt to provide apparently nonmathematical formulations of some of our best scientific theories (Field 1980). Although most critics of indispensability arguments would grant that these attempts failed, there is little attempt to diagnose why they failed. Instead, many antiplatonists have gone on to insist that they do not need to reformulate these theories, but can accept them without giving any account of the content of the claims of the theories. In certain cases, some discussion of the contents has been given, but only for simple examples with no indication of how to extend this to more interesting cases.[6]

The benefit of formulating the issue as I have presented it, then, is that the content of these scientific representations takes center stage. We have already seen in broad outline an argument that the contribution of mathematics is epistemic based on the different ways it contributes to the content of our representations. As I argue in the next chapter, the content can be further characterized in terms of broadly structural relationships between physical situations and mathematical structures. Although there are many details to work out, this picture will support some criticisms of the indispensability argument. For if the content can be understood in this structural way, the accuracy of these representations will not require much from the mathematical objects. In particular, applications will turn on extrinsic features of the mathematical entities, and so they will be unable to pin down the intrinsic features of these entities, including whether they are abstract. Such a conclusion does not support all forms of antiplatonism, however. For I will argue that fictionalists are not able to recover the contents of our most successful scientific representations, at least on the account of their content that I develop in chapter 2. Until they respond to these concerns about content, then, we have a strong reason to reject fictionalist interpretations of pure mathematics. It remains possible that some nonfictionalist, antiplatonist interpretation can be sustained. Whether or not this is possible seems to me to turn largely on questions internal to pure mathematics and its epistemology. I conclude this book with a brief discussion of these questions and their independence from reflection on applications of mathematics.

Beyond the indispensability argument, the most well-known philosophical debate surrounding mathematical scientific representations concerns what is

---

6. See, for example, Balaguer (1998). I return to this issue in chapter 12.

deemed the unreasonable effectiveness of the contribution of mathematics to science. Wigner famously concluded from a survey of impressive scientific discoveries involving mathematics that "the miracle of the appropriateness of the language of mathematics for the formulation of the laws of physics is a wonderful gift which we neither understand nor deserve" (Wigner 1960, p. 14). It is not clear exactly what Wigner finds mysterious when it comes to the interactions between mathematics and physics, but the examples he emphasizes turn on the development of sophisticated mathematics, which only later proved crucial in the formulation of successful scientific theories. Unlike the calculus, for example, which was discovered as part of attempts to understand physical phenomena like motion, the twentieth century saw the application of domains of mathematics that were developed based on internal mathematical considerations. A simple example of this is the theory of complex numbers. They were introduced into mathematics based on reflections on the real numbers. Rounding out the real numbers to include complex numbers contributed to a superior mathematical theory. The complex numbers lacked any intuitive physical significance, but in spite of this they provide a crucial part of the mathematics deployed in science. The mystery, then, is that mathematics developed independently of applications should prove so effective when it is eventually applied.

Part of the problem with Wigner's conclusions and the resulting debate is that it is often not clear exactly what sort of contribution is supposed to be surprising or unreasonable. Steiner has rectified many of the flaws of Wigner's original discussion by focusing on the role of mathematics in suggesting how to formulate new scientific theories (Steiner 1998). The conclusion of Steiner's argument is correspondingly much more definite and controversial than any of the vague suggestions by Wigner. Steiner concludes that this role for mathematics undermines certain forms of naturalism. Naturalism, in Steiner's argument, is the view that the universe is not constituted in a way that fits with our cognitive faculties. Antinaturalists, then, include not only theists who believe that the human mind has been designed to understand the natural world but also those whom Steiner calls "Pythagoreans," who insist that the universe is formed in line with the mathematics that our cognitive faculties find of interest.

The main premises of Steiner's antinaturalist argument involve historical cases where mathematical considerations were crucial in getting a scientist to formulate a theory she deemed worth testing. These mathematical considerations, in turn, are linked to species-specific aesthetic judgments. The combination of these two premises entails for Steiner that the scientist acted as if species-specific judgments would give good insight into the way the physical world works. This challenges the scientists' own claims to be naturalists. More seriously, Steiner argues that the overall success of this strategy for discovering new scientific theories should undermine our own confidence in naturalism.

On a first pass, there is little connection between the question that I have asked about the contributions of mathematics to successful scientific representations and Steiner's discovery problem. Most obviously, the discovery issue concerns how or why a scientist came to formulate a representation that she judged worth testing. As with other questions of innovation and discovery, we often have little understanding of how this occurs, and there is no reason to think that the ultimate story of a given discovery would not rely on subjective psychological considerations like intuition or "gut feeling." There should be no mystery about why a scientist employed methods of discovery without a rational foundation or how these methods were successful in some cases. By contrast, my focus has been on the way mathematics helps a scientist confirm a representation through ordinary experimental testing. This confirmation or justification is independent of the origins of the representation or why the scientist decided to test it.

Still, this division between discovery and justification is hard to enforce when we turn to the history of science and actual scientific practice. It may turn out that reflection on how mathematics makes its epistemic contribution to science helps us understand the sorts of cases discussed by Steiner. To see how this might work, recall the discussion of abstract varying representations like the simple harmonic oscillator. Once it is validated for certain pendulum systems, we can transfer the confirmation from this domain to a new domain like spring systems based on the claimed mathematical similarities. Now, independently of the actual historical details of the scientific understanding of pendulum and spring systems, it certainly could have happened that a scientist sought to represent springs as harmonic oscillators based on this kind of analogical reasoning. If we can understand how confirmation can be transferred in this sort of way, this discovery procedure becomes far less puzzling. For the scientist is in a position to recognize that if this way of representing the spring is successful, then the family of well-confirmed harmonic oscillator systems will be expanded.

Although this is just a quick sketch of how the epistemic conception might bear on Steiner's argument, it is enough to show that an answer to my representation question might provide a rationale for some of the discovery procedures that Steiner considers. If this is right, and the epistemic approach does not violate naturalism, then there is something wrong with Steiner's argument. The crucial flaw, I argue, is the link between mathematical theories and species-specific aesthetic judgments. If this link were correct, then mathematics would be ill-suited to contribute to successful scientific representations. The connection between mathematics and confirmation established by the epistemic conception requires that mathematics be objective, and further reflection on the practice of pure mathematics reinforces this objectivity. So we can block Steiner's antinaturalist conclusion by focusing on mathematical representations and clarifying the link between pure mathematics and aesthetic judgment.

1.6 INTERPRETIVE FLEXIBILITY

Discussion of Wigner's and Steiner's worries about the success of appealing to mathematics leads one naturally to wonder just how successful mathematics has really been in aiding science. In fact, it might seem like the whole way in which I have formulated my main topic has been biased toward the positive role of mathematics in science. Perhaps in addition to the contributions that mathematics makes to the success of science, there are also many ways in which mathematics hinders the development of accurate scientific representations. We have seen this worry already in a specific case, where it seems that unifying diverse representations using some single mathematical framework can undermine the usefulness of the overall representation. The thought is that mathematics is a tool that sometimes works and sometimes does not, and to understand its contribution in any realistic way we must take the bad along with the good. The failures of mathematics to help science are many, and even though these failures are not always celebrated in textbooks and histories along with the successes, a philosophical understanding of the issue cannot ignore them.

I argue that these failures are to be expected. The features of mathematics that make it possible to make a positive contribution also allow negative contributions. Above all, the *interpretive flexibility* of mathematical representations creates the opportunities exploited by the epistemic conception as well as the sorts of failures that result from an overly confident reliance on mathematics to guide our scientific representations. This can happen in many ways, but for now notice only that the abstract nature of these representations (in both the acausal and varying senses discussed so far) permit these representations to be interpreted in ways that outstrip their well-confirmed core. There is a risk in taking the well-confirmed aspects of a given representation and extrapolating our confidence to other aspects that are completely unconfirmed. To consider two simple examples, we might start with a causal concrete representation and obtain an associated abstract acausal representation. If the abstract acausal representation is confirmed, then there is a temptation to take this as evidence of the correctness of the causal concrete representation. Here the mathematical link between the acausal and the causal encourages us to carry over the confirmation even if the link makes it clear that the acausal representation is easier to confirm. A second kind of simple case involves concrete fixed and abstract varying representations like a pendulum and the harmonic oscillator. When we transfer our confidence to the harmonic oscillator representation from the pendulum to the spring, we must be careful not to transfer too much of the pendulum interpretation over to the spring representation. For example, we cannot assume that the force restoring the bob to the pendulum's equilibrium is the same kind of force as that restoring the spring to its equilibrium. A scientist that understands both systems well will of course not make this mistake, but in other sorts of cases the error is easy to make. We overinterpret the members of the family of the representation and wind up thinking that the systems have greater physical similarities underlying their mathematical similarities.

Mathematics contributes flexibility, then, and this flexibility creates the opportunity for scientific failures. Often these failures will escape detection for some time because they involve features that are remote from our experimental access. It follows that there is no simple path to well-confirmed scientific representations using mathematics. As with everything else, it all turns on considerations arising from eventual experimental testing, and the realist's hope must be that no matter what interpretive errors have been made at a given stage of scientific theorizing, there will be a later stage when this error can be isolated and rectified.

A final caveat concerning the project undertaken here concerns the approach to confirmation that I appeal to. There can be a whiff of paradox in asserting that mathematics contributes to the confirmation of our best scientific theories, just as we saw there was some mystery with the role of mathematics in discovery with Wigner and Steiner. On the discovery side, the puzzle came from the way mathematicians allegedly anticipated the needs of scientists. With confirmation, it might seem impossible for mathematical beliefs to play any positive role in supporting scientific claims. For mathematical beliefs are not about the physical world, and it is commonly assumed that mathematical truths are necessary, that is, true no matter how the physical world happens to be. If this is right, then how can mathematical truths have any relevance to the truth about the physical world, and how can believing some mathematics help tell us which way our evidence bears on a given physical possibility?

As we will see in some detail in chapter 2, the problem is not remedied by drawing on any of the main contenders for a comprehensive account of confirmation. These theories try to explain when boosts in confirmation occur, but they are ill-equipped to factor in a role for mathematical beliefs in this process. To take a simple case, a scientist may assign some prior probability to a mathematical scientific representation T in advance of experimental testing. But suppose that the scientist must know another mathematical theory M to extract any predictions from this representation. There is little to no discussion in the confirmation literature about how coming to believe M can allow the scientist to determine the predictions and actually test the theory. The main culprit here is the assumption of logical omniscience, that is, that scientists should assign the same probability to logically equivalent propositions. As mathematical truths are metaphysically necessary, it seems to follow that they are all assigned the probabilities of logical truths. But this obscures an important place where mathematics is making its mark, namely, in helping us see what implies what. Questions of logical entailment need to be distinguished from the mathematical analysis of the content of representations.

A new approach is needed, then, to understand how mathematics can help us confirm scientific representations. My approach does not involve a new comprehensive theory of scientific confirmation. Instead, I rely on various principles that relate the content of a representation and the evidence in different sorts of cases. These principles have some prima facie plausibility and will be illustrated in the case studies I develop in this book. Some accord well with certain extant theories of confirmation, but when there is a conflict with these theories, I argue that it is the theories that are wrong, not the principles. Only in this way can we isolate when mathematics makes its contribution to confirmation and come to understand what philosophers of science seem to have overlooked in their reflections on scientific confirmation to date.

1.7 KEY CLAIMS

The goal of this introductory chapter has been to set out the main claims that I argue for in the rest of this book and to situate my project with respect to some other topics associated with mathematics and scientific representation. Here are claims I clarify and defend in the course of this book:

1. A promising way to make sense of the way in which mathematics contributes to the success of science is by distinguishing several different contributions (chapter 1).
2. These contributions can be individuated in terms of the contents of mathematical scientific representations (chapter 2).
3. A list of these contributions should include at least the following five: concrete causal, abstract acausal, abstract varying, scaling, and constitutive (chapters 3–6).
4. For each of these contributions, a case can be made that the contribution to the content provides an epistemic aid to the scientist. These epistemic contributions come in several kinds (chapters 3–6).
5. At the same time, these contributions can lead to scientific failures, thus complicating any general form of scientific realism for representations that deploy mathematics (chapter 7).
6. Mathematics does not play any mysterious role in the discovery of new scientific theories. This point is consistent with a family of abstract varying representations having some limited benefits in suggesting new representations that are worthy of testing (chapter 8).
7. The strongest form of indispensability argument considers the contributions I have emphasized and argues for realism of truth-value for mathematical claims (chapter 9).
8. These contributions can be linked to explanatory power, so we can articulate an explanatory indispensability argument for mathematical realism (chapter 10).
9. However, even such an argument based on explanatory contributions faces the challenge of articulating a plausible form of inference to the best explanation (IBE) which can support mathematical claims (chapter 10).
10. This challenge to IBE for mathematical claims is consistent with mathematics contributing to successful IBE for nonmathematical claims, as in the extended example of the rainbow (chapter 11).
11. Fictionalist approaches to mathematics and scientific models face challenges that undermine their main motivations (chapter 12).
12. The way our physical and mathematical concepts relate to their referents suggests that our representations depend for their contents both on our grasp of concepts and our beliefs (chapter 13).

I end this chapter with an outline of the remaining chapters. In chapter 2, I develop two of the central areas of my project: content and confirmation. The content of a given scientific representation turns on a variety of considerations that need to be clarified before we can investigate the contribution of mathematics. A main aim of this discussion is to sharpen the intrinsic/extrinsic distinction that I have worked with informally so far. Another priority is confirmation. This is a topic of intense discussion, although, as I have just noted, little of it is directly relevant to my main concerns. Still, a survey of some views on confirmation and the necessary principles needed to make sense of mathematics in science must be on the table to vindicate the epistemic conception.

After these preliminaries, I turn to four chapters exploring the dimensions of mathematical contributions. Chapter 3, on the concrete causal/abstract acausal distinction, considers several examples along this spectrum as well as some influential accounts of causation and its representation. In chapter 4 I bring in the other sense of abstractness with the concrete fixed/abstract varying dimension. Again, examples are used to illustrate the differences between these sorts of representations and the epistemic benefits of the mathematics. Chapter 5 continues this theme with a survey of examples that involve scale in different ways. The constitutive/derivative distinction in chapter 6 rounds out this positive survey of how mathematics makes its positive contribution. A cautionary chapter 7 emphasizes the respects in which things can go wrong by discussing several cases where mathematics seems to have contributed to the failure of a given representation. These failures are diagnosed as a consequence of the interpretive flexibility that goes hand in hand with using mathematics in science.

The next six chapters link up the epistemic conception with other discussions of the role of mathematics of science. In chapter 8, I start by surveying the worries about discovery raised by Wigner and Steiner. Chapter 9 turns in earnest to the debates about the indispensability argument for platonism. The more recent focus on mathematical explanations in science and their significance is considered in chapter 10. An extended case study involving IBE makes up chapter 11. Chapter 12 considers the antiplatonist views I group under the label of "fictionalism" and their prospects for adopting the epistemic conception of the contribution of mathematics to science. Finally, in chapter 13 I consider relaxing the assumptions introduced in chapter 2 about how the content of a representation is determined. It turns out that these assumptions are somewhat unrealistic, but I argue that this does not affect the lessons drawn from the examples discussed in earlier chapters. Here I engage with Wilson's picture of scientific representations as facades and conclude that even if Wilson is correct, we can still understand the contribution of mathematics in epistemic terms (Wilson 2006).

The book concludes with chapter 14 on the implications of the previous chapters for our understanding of pure mathematics. It turns out that the role of mathematics in science has little to no bearing on our understanding of pure mathematics, but even this negative conclusion gives indirect support to the approach to pure mathematics known as structuralism. On the epistemic side, I will argue that much of pure mathematics must be a priori if it is to make its contributions to the success of science.

PART I

Epistemic Contributions

2

Content and Confirmation

2.1 CONCEPTS

In the last chapter I introduced a scientific representation as anything that had content. Content possession, in turn, was explained as setting some conditions a target system had to meet for the representation to be correct and other conditions that would make the representation incorrect. This first pass allowed a rough description of the features of mathematical scientific representations, but to make any more progress we must do better. Here I aim to provide an account of where the content of mathematical scientific representations come from. This is a necessary first step in understanding the link between mathematical representations and confirmation. After clarifying my preferred notion of content, the second half of this chapter discusses confirmation. In both cases I am forced to make some controversial assumptions. When I am not in a position to argue for these assumptions, it should be clear that they are part of one package of views that will prove sufficient to make sense of the central role of mathematics in science. Those who reject my assumptions are encouraged to articulate alternative approaches.

The main claim that I try to defend about the content of mathematical scientific representations is that the content is exclusively structural. By this I mean that the conditions of correctness that such representations impose on a system can be explained in terms of a formal network of relations that obtains in the system along with a specification of which physical properties are correlated with which parts of the mathematics. As we will see, this does not preclude an important role for the context in which the representation is produced in fixing the content. But I adopt a clear distinction between what must be in place for a representation with a given content to exist and the content of that representation. A similar distinction is embedded in our commonsense notion of perception. There we concede that we must have brains to accurately represent the world in a perceptual experience. Still, it does not follow that part of the content of the representation is that we have brains.

The contents of our best scientific representations are quite sophisticated, so an explanation of these contents requires some stage setting and terminology. As I will use the terms, *theories*, *models*, and *representations* are different sorts of things. A theory for some domain is a collection of claims. It aims to describe the basic constituents of the domain and how they interact. A model is any entity that is used to represent a target system. Typically we use our knowledge of the model to draw conclusions about the target system. Models may be concrete entities or abstract mathematical structures. Finally, a representation is a model with a content. Contents provide conditions under which the representation is accurate. I distinguish three kinds of content that are prior to the content that I ultimately ascribe to our representations. These are what I call the *basic content*, the *enriched content*, and the *schematic content*. They correspond roughly to increasingly sophisticated kinds of representations, but I do not claim that every representation goes through these stages or that scientists would always be happy making these distinctions. The apparatus is instead a tool to help explain how to find the content in a given case. Crucially, we will see that the derivation of the representation from more basic representations is an essential guide to seeing what the content of these representations actually is. As the term *schematic* suggests, the content of a representation will result from filling in certain gaps in the schematic content of the representation.

I develop my preferred account of content by making two debatable assumptions. First, I assume mathematical platonism. That is, the subject matter of mathematics is a domain of abstract objects like the natural numbers, real numbers, and groups. I take for granted that mathematical language is about these abstract objects. In chapter 10, we will see that this assumption can be considerably weakened, but until then we employ it for the purposes of simplicity of presentation.[1] The second debatable assumption concerns the physical world. I assume that scientists are able to refer to the properties, quantities, and relations that constitute their domain of investigation, at least in a weak sense of "refer." To see what sort of reference I am presupposing, recall the representations of phlogiston. In specifying the contents of these representations, I take the scientist's ability to talk about phlogiston for granted. But sadly, there was no such substance. I do not want it to follow that the phlogiston representations were without content. Instead, I would describe the situation as one where the scientists were referring to phlogiston in the minimal sense that they were coherently discussing a substance that could have existed, and had it existed, some of their phlogiston representations would have been correct. As it stands, such representations all turned out to be incorrect.

In both the mathematical and physical cases, then, I ascribe to agents an ability to refer. A convenient shorthand for these abilities is to say that these agents possess the relevant concepts. It is a delicate issue to determine how we acquire these concepts and what they contribute to the content of our representations. At this stage of our discussion I flag two debatable assumptions. First, I assume that our mathematical concepts contribute all the mathematical content that a representation may have. Among other things, this involves the semantic internalist assumption that the concepts are sufficient to pick out their referents. Second, I assume that our physical concepts pick out properties, but that the features of these properties beyond what is reflected in our concepts can be incorporated into the content of

---

1. Briefly, any account of pure mathematics that agrees with the platonist on the structural relations of the mathematical entities will be adequate. This includes Lewis's megethology and Hellman's modal structuralism, even though these are, strictly speaking, nominalist views. Other nominalist views like Balaguer's and Yablo's fictionalism turn out to be inadequate.

our representations.[2] Semantic externalism, then, is assumed for physical concepts. Both assumptions are considered in more detail in chapter 13 when we consider Wilson's views on mathematical and physical concepts (Wilson 2006).

2.2 BASIC CONTENTS

A mathematical scientific representation will have as its content the existence of a certain kind of relation between a mathematical structure and the arrangement of some properties and quantities in a given scientific domain. I want to insist that the relations at issue here are all what I call "structural":

> A *structural relation* is one that obtains between systems S_1 and S_2 solely in virtue of the formal network of the relations that obtains between the constituents of S_1 and the formal network of the relations that obtains between the constituents of S_2.

A formal network is a network that can be correctly described without mentioning the specific relations that make up the network. One way to visualize this is to imagine substituting the constituents and the relations of one of the systems, while preserving the overall pattern. If this substitution fails to undermine the relation between the two systems, then that relation is structural in my sense.

So for any given mathematical scientific representation, we have found its content if we can answer three questions: (1) What mathematical entities and relations are in question? (2) What concrete entities and relations are in question? (3) What structural relation must obtain between the two systems for the representation to be correct? Unsurprisingly, we can call this approach to content the "structuralist account".[3] We will see that the main thing that needs to be clarified for this structural account to work is the range of acceptable structural relations. After explaining the most basic case, I allow more intricate sorts of structural relations, including those whose specification requires mathematics. In this respect, my structuralist approach to representational content is more open-ended than some of its competitors.

The simplest kind of structural relation is an isomorphism. Two structures $S = <D, R_1, R_2, \ldots>$ and $S^* = <D^*, R_1^*, R_2^*, \ldots>$ are isomorphic just in case there is a function f from the domain D of S to the domain D^* of S^* that is total, one-one, and onto and such that for any relation R_i in S and R_i^* in S^*, and for all $x_1, \ldots, x_n$ in D, $R_i(x_1, \ldots, x_n)$ iff $R_i^*(f(x_1), \ldots, f(x_n))$. Another simple kind of structural relation is a homomorphism. The test for being homomorphic is the same as being isomorphic except we do not require that the function be one-one. Suppose, for example, that our concrete system is a group of five people and the only relation involved is order

---

2. This assumption plays a role in chapter 6.

3. In other places this basic idea is called "the mapping account". See, for example, Pincock (2004), Batterman (2010) and Bueno and Colyvan (2011).

of age. Such a physical system is isomorphic to the natural numbers 1 through 5, with the less than relation, on the assumption that no two individuals share the same age. If two people have the same age, then the function that tracks the order of ages will fail to be an isomorphism because two individuals will be mapped to the same natural number (and so the function will not be one-one). In this case, though, we will still have a homomorphism between the physical system and the natural numbers from 1 to 4 with the less than relation.

In both cases, we can see how the substitution test already sketched would be met. In the first case, replacing the people with their age relation with the numbers 6 to 10 in the less than relation preserves the original isomorphism. To see how nonstructural relations fail this sort of test, suppose we start with the five people in the age relation and the natural numbers 1 through 5, with the less than relation. Suppose, though, that the intersystem relation corresponding to the function f is that a person's age in years is to be mapped to that natural number. That is, the youngest person is 1 years old, the next oldest is 2 years old, and so on. Now substituting these people for other people need not preserve this relation, even if each of the five new people have different ages. This shows that the relation corresponding to the function f is nonstructural. I claim that no relation of this sort is relevant to the content of mathematical scientific representations.

On this structuralist account, offering a mathematical scientific representation can be schematically summarized as claiming that the concrete system S stands in the structural relation M to the mathematical system S^*. If both systems exist and the structural relation obtains, then the representation is correct. Otherwise, the representation is incorrect. Even at this level of abstraction, we can see the need for the assumptions introduced in the last section. First, the only way for an agent to adopt a belief with this content is if they have some referential access to the mathematical system S^*. Similarly, we have taken for granted the ability to specify the relevant constituents of the physical system S. Without these assumptions, the whole picture risks breaking down.

There are many other approaches to content, and even those who insist that nothing useful can be said about content at this level of generality. Perhaps the main competitor to an approach based on accuracy conditions tends to put inferential connections at the heart of their picture of representation.[4] Inferential approaches must explain the scientific practice of evaluating representations in terms of their accuracy. Although there does not seem to be any barrier to doing this, I have found it more convenient to start with the accuracy conditions. On my approach, inferential claims about a given representation follow immediately from its accuracy conditions: a valid inference is accuracy-preserving.

Even granting this approach, there are still elements of this picture that could be questioned. Notice that there is an important difference between talking about a concrete system made up of objects and linked together by concrete relations involving quantities and properties and a set-theoretic structure $S = <D, R_1, R_2, \ldots >$. Definitions of structural relations typically invoke set-theoretic structures, and there

---

4. See Contessa (2007), Suárez (2010b), and Bueno and Colyvan (2011).

Content and Confirmation   29

might seem to be a troubling gap between such structures, which are presumably as abstract as any other set, and concrete systems like a group of five people. Also, the structural relations themselves are usually defined in terms of the existence of functions, and given that functions are usually thought to be sets, there again seems to be a disconnect between the contents generated by the structuralist account and anything concrete.

This gap has prompted van Fraassen to offer an elaborate account of representation based on essentially indexical propositions so that we can answer the question, "How can an abstract entity, such as a mathematical structure, represent something that is not abstract, something in nature?" (van Fraassen, 2008, p. 240). I prefer to take a less ambitious option. Suppose we have a concrete system along with a specification of the relevant physical properties. This specification fixes an associated structure. Following Suárez, we can say that the system instantiates that structure, relative to that specification, and allow that structural relations are preserved by this instantiation relation (Suárez, 2010b, p. 96). This allows us to say that a structural relation obtains between a concrete system and an abstract structure.[5]

2.3 ENRICHED CONTENTS

So far we have restricted our discussion to simple kinds of structural relations that are defined in terms of the basic elements of the concrete and mathematical systems. If we observe these restrictions, then we are limited to what I will call the *basic contents* of representations. Certain simple situations may call only for an understanding of the basic content, but as soon as we shift to more sophisticated cases, these contents are easily seen to be problematic. The problem is that the assumptions deployed in assembling most scientific representations involve claims that are not easily reconciled with basic contents. The tension comes when we wish to represent features of the system that go beyond those that correspond to the basic elements of the mathematical structure. When this happens I say that some *derived elements* are in play. These derived elements in the mathematics will be used to represent physical entities beyond those can be related directly to the mathematical entities that appear in the domain of the mathematical structure. Sometimes all of the entities of interest will be derived in this sense. This shift is especially common in cases of idealization. Here we employ assumptions that we believe to be false in the specification of the representation. If the contents of the resulting representations are to be reconciled with these sorts of beliefs, we need to shift away from the basic contents.

To focus our discussion, consider two famous partial differential equations. These are the wave equation and the heat equation. In their one-dimensional versions, they place conditions on a function $u(x, t)$. The wave equation requires

---

5. Shapiro introduces a distinction between systems and structures in Shapiro (1997) that serves the same sort of purpose.

$$u_{tt} = c^2 u_{xx} \qquad (2.1)$$

whereas the heat equation says that

$$\alpha^2 u_{xx} = u_t \qquad (2.2)$$

We can, of course, consider these equations as purely mathematical conditions on the otherwise unspecified function u. But issues of representational content come into play as soon as we ask what a target system must be like if it obeys either equation. The first thing to notice is that this question is badly formed. We use the wave equation and the heat equation to represent many different systems and the "waves" or "heat" in question can range all the way from the vibration of strings through to the correct price for stock options.[6] Even if we restrict our focus to one kind of target system, it is not easy to know what conditions the heat equation imposes on it.

Consider, for example, the use of the heat equation to represent the temperature changes in an iron bar as it is heated in the middle. If we impose suitable initial and boundary conditions, we can find the temperature $u(x, t)$ as time passes and the heat spreads out throughout the bar.[7] The parameter α is crucial here because it reflects how quickly this diffusion occurs. So, on a first pass, it looks like we should say that this representation is accurate when we have an isomorphism between the temperature at each point at each time and the set of ordered pairs (x, t) picked out by the solution to our equation, where the first coordinate denotes position and the second coordinate denotes time. This would be a strange requirement, though, as we do not actually expect such a degree of accuracy from our representation. Also, in this case at least, the temperature magnitudes that we are trying to track with our mathematics are not even defined on a single spatial point. It is instead a property of larger spatial regions in which the molecules of the iron bar aggregate and interact.

One option at this stage would be to stick with the basic contents that require isomorphisms or homomorphisms on the basic elements of the mathematical structure and conclude that all such representations are just inaccurate. This strategy cannot be ruled out completely, but it faces a serious challenge in specifying how we are to evaluate our representations if nearly all of them are inaccurate. Some, for example, rest on pragmatic tests of efficiency, or they focus instead on what inferences these inaccurate representations support. In this book I opt for a different approach, which insists that even when these basic contents lead to inaccurate representations, scientific practice can point us to other contents that allow for accurate representations even when idealizations are in play.

The first step here is to countenance what I call *enriched content*. In the heat equation, we have to work with small regions in addition to the points (x, t) picked out by our function. We should take these regions in the (x, t) plane to represent

---

6. An application of the heat equation in financial economics is discussed in chapter 7.

7. See figure 5.2 in chapter 5.

genuine features of the temperature changes in the iron bar. This representational option is open to us even if the derivation and solution of the heat equation seem to make reference to real-valued quantities and positions. We simply add to the representation that we intend it to capture temperature changes at a more coarse-grained level using regions of a certain size that are centered on the points picked out by our function. The threshold here can be set using a variety of factors. These include our prior theory of temperature, the steps in the derivation of the heat equation itself, or our contextually determined purposes in adopting this representation to represent this particular iron bar. My approach is to incorporate all of these various inputs into the specification of the enriched content. The enriched content has a much better chance of being accurate as it will typically be specified in terms of the aspects of the mathematical structure, which can be more realistically interpreted in terms of genuine features of the target system.

So far I have focused on the elements of the (x, t) plane and argued that some representations involve interpreting more than just these basic elements. Similar points need to be made for other aspects of the mathematics deployed in our representation. For example, in the heat equation the constant α does not represent a fundamental feature of the iron bar, but is instead derived from more basic aspects of the bar. The specification of the enriched content of the representation may involve these further details; if we get these parts wrong, the representation is then inaccurate. Or it may turn out that scientists are unsure what the α actually stands for and the claim that it represents some specific derived constant or even that it is a constant may drop out from the enriched content of the representation entirely.

We arrive at the enriched content of a representation, then, by allowing the content to be specified in terms of a structural relation with features of the mathematical structure beyond the entities in the domain of the mathematical structure. The resulting structural relations need to be more complicated than just simple isomorphisms and homomorphisms. In particular, we allow the specification of the structural relation to include mathematical terminology. For example, we may posit an isomorphism between the temperatures at times and u in the mathematical structure subject to a spatial error term $\epsilon = 1$ mm. Similarly, we may reinterpret α as standing for a derived feature of the bar or more instrumentally as an uninterpreted means to make claims about the temperature changes.

2.4 SCHEMATIC AND GENUINE CONTENTS

The sorts of adjustments necessary to move from the basic content to the enriched content are relatively minor. So far all we have done is include various derived elements and derived quantities and added in certain bounds on the precision of our structural relations. Something different is needed to make sense of another way in which the heat equation is used to represent target systems. One common move is to represent the iron bar as infinitely long. That is, the mathematical structure will have x extending to infinity in both the positive and negative directions. This aspect of the mathematical structure is used even though the bar being represented is of course of finite length. Here it does not really help to say that the bar is approximately infinitely long or that its length is within some ϵ of infinity. This sort of idealizing assumption needs a more careful interpretation if we are to see how it can contribute to an accurate scientific representation.

The strategy that I adopt for these sorts of cases is to insist that such assumptions decouple the mathematics from its apparent physical interpretation. When we set the length of the bar to infinity, we detach the relevant part of the mathematical system from its prior association with a physical quantity in the physical system. As a result, these structural similarities between the mathematical system and the physical system become irrelevant to the correctness of the overall scientific representation. It is not part of the content of the scientific representation that the bar is infinitely long or, perhaps, that it is any length at all. I call the content of the representation that results from such assumptions the *schematic content*. It is schematic in two senses. First, we have moved from an enriched content to a more impoverished content because we have severed some of the mathematics from its prior physical interpretation, for example, x is not always taken to represent a position on the bar. Second, there is now an unspecified parameter in the content of the representation that reflects what features of the system we wind up representing. In our iron bar case, it is not an option to represent bars as having no length, but as it stands our representation contains no information on what lengths we will take the representation to include. These schematic contents remain useful and can be investigated in their own right in a more purely mathematical fashion. Still, the content will remain schematic until the unspecified parameters are given some values. Only then will we arrive at what I take to be the genuine contents of such representations.

The fixing of the parameters of the schematic content can be a complex and case-specific affair. A given schematic representation of iron bars, for example, may be fleshed out differently for different target systems. These differences may result from the differences in the systems themselves, but it also may happen that the purpose of investigating the system will affect these choices. Here the scientific context makes yet another crucial contribution. The goals of a given representation of the bar may require that we have information about what is going on 5 cm from the center of the bar. This would lead a scientist to adjust the parameter so that the lengths represented were greater than 5 cm. This contribution serves merely to fix the content. It is not part of the content that the goals require this sort of representation.

I claim that this sort of multistage process is needed to do justice to the subtleties of our best mathematical scientific representations. To review, we started with a notion of basic content that involved simple structural relations between basic concrete elements and their corresponding mathematical entities. We then introduced a notion of enriched content that allowed an appeal to a wider range of structural relations, including those whose specifications required mathematical terminology, along with mathematically derived elements. This led to a notion of a schematic content that results from an enriched content through the decoupling of some parts of the mathematics from its physical interpretation. This deinterpretation engendered an unspecified parameter that prevented the schematic representation from being used to make definite claims about a given physical system. Finally, the genuine content resulted from the schematic content through the contextually guided specification of these parameters. Here what guided the specification could include the goals and purposes of a scientist, but these goals are not included in the resulting content. That is, it is no part of the accuracy conditions of the representation that the representation serve the goals and purposes of the scientist.

2.5 INFERENCE

A crucial test for my account of the content of mathematical scientific representations is that it contribute to a useful theory of scientific inference. My approach to this theory is to insist that we first understand a scientific representation by grasping its content and then work out which inferences it permits. But there are special challenges related to scientific representations that make a story about their inferential relationships especially challenging. To focus our discussion, it should be clear that I wish to understand primarily how these representations can be confirmed. I start with the prima facie plausible assumption that representations are confirmed by entailing, perhaps in conjunction with auxiliary assumptions, some predictions that are then experimentally verified. So, we need to focus on this inferential link between our representations and these predictions to understand confirmation.

The main benefit of approaching content in the way I have is that it fits nicely with our standard account of the inferential relationships of ordinary propositions via their truth-conditions. The reason that "All humans are mortal" and "Socrates is a human" entails "Socrates is mortal" is that if the premises are true, then the conclusion must be true. That is, in every possible world that the premises are true, the conclusion is also true. In introductory logic courses we learn how to paraphrase sentences of English into a formal logical language where the distinction between logical and nonlogical terms is explicit. As a result, we see which terms are subject to variable interpretation and which must be held fixed. This allows a systematic study of logical entailment and vindicates our prior conviction that our argument about Socrates is logically valid.

Still, it is always kept in mind that our simple formal logical languages are not adequate to handle all cases. I want to suggest here that simply extending our introductory logic approach to our mathematical scientific representations will distort their genuine inferential relations and obscure the difficulties in confirming them. The problem arises because mathematical claims are true in all possible worlds. When we consider the contribution of mathematical claims to content, then, we need something more subtle than possible worlds and truth-conditions talk would suggest. Consider the simple inference from "The number of fish is greater than the number of cats" and "The number of cats is greater than the number of dogs" to the conclusion "The number of fish is greater than the number of dogs." It is true in all possible worlds, or metaphysically necessary, that if $a > b$ and $b > c$, then $a > c$. So there is no possible world in which the premises are both true and the conclusion is false. Still, a standard paraphrase of the argument into a formal logical language would give it the form "$Nx(x \text{ is a fish}) > Nx(x \text{ is a cat})$, $Nx(x \text{ is a cat}) > Nx(x \text{ is a}$

dog), therefore Nx(x is a fish) > Nx(x is a dog)" or "Gab, Gbc, therefore Gac". This paraphrase signals that there are no logical terms in the premises or the conclusion, and so none of the terms have a fixed interpretation. Taking this approach, then, leads us to conclude that the argument is not logically valid.

These two different verdicts present us with a choice. We can insist that inference respect metaphysical necessities and side with the result that the inference is valid. Or we can widen our notions of necessity and possibility further by invoking a sense of logical possibility according to which the inference turns out to be invalid. I opt for the latter option, although I do not want to be too dogmatic and insist that this reveals the true nature of inference. My reason for taking this route is that this is the best way to make clear how mathematical beliefs can make a genuine contribution to the confirmation of scientific representations. To see why, note that the view that inference respect metaphysical necessities makes a strong link between the content of a given representation and what is metaphysically necessary. For example, any claim will entail that $2 + 2 = 4$ or any other mathematical truth. But this stops us from telling any informative story about how learning mathematical truths can lead to genuinely new inferences from scientific claims. All that we wind up doing, on this metaphysical necessity approach, is making explicit what was already implicit in the content of the original representation.[8]

On the approach I adopt, some metaphysical necessities are not part of the content of our scientific representations. As a result, learning some mathematics, which is then combined with our representations, can produce valid inferences that are otherwise invalid. To see why this is important, recall the heat equation representation. Suppose we have fully specified all the parameters in what I was calling the schematic content, and so our representation sets genuine conditions on the system. To determine whether or not the representation is accurate, that is, whether the system meets these conditions, we must first find out what the conditions actually are and then go on to see if they are realized in the concrete system. Solving a system of partial differential equations is far from trivial. Still, if the problem is well posed, so there is a unique solution, it is metaphysically necessary that the function takes on some specific values.

Our problem is to understand this inference. I would insist that we can understand the content of the representation and yet not be able to work out what the solution is. This suggests that the solution is not part of the content of the representation. This result comes through clearly once we notice that the link between the system of equations and its solution is generally a number of mathematical truths. These truths are, in my view, metaphysically necessary, but that is not relevant to the conception of inference I am presently articulating. All that is relevant is the nonlogical character of the mathematical terminology. This blocks the entailment relation between the system of equations and its solution. To get a genuine entailment, we need to add additional premises to the argument

---

8. An anonymous referee has suggested that learning mathematical truths may serve only to make us aware of valid inferences that were already available from a given representation. I believe that this approach is too limited, but cannot argue this point here.

corresponding to the features of the mathematical entities that are sufficient to pin down the interpretation of the mathematical terms. When this is done, we get a valid inference.

We wind up, then, with a fine-grained notion of content according to which the scientific representations that involve mathematics have much less content than might have initially appeared. Adopting such a representation while remaining ignorant of the additional mathematical facts will not lead to any substantial restrictions on how the system has to be. Only when we conjoin the representation with more mathematical claims can we articulate the representation to the point where we can derive predictions from it. The basic point has been long appreciated in connection with the need for auxiliary assumptions about instruments. What I have done here is extend this point to mathematical assumptions as well.

2.6 CORE CONCEPTIONS

In my attempt to remove the role of metaphysical necessities from logical inference, it may seem like I have gone too far. There should be a middle ground between saying that all metaphysical necessities must be taken for granted when considering entailment relations and that none of them can be used. Conceding this point will not undermine my general picture as long as it clear that many of the metaphysically necessary truths involving mathematical entities that are used in scientific reasoning are not part of the background against which we define entailment. A range of options is in fact possible. So far we have only invoked two cases, which turn out to be at opposite ends of this sort of spectrum. With the counting case, it is somewhat hard to imagine someone who understands the premises, and so succeeds in thinking about numbers and the greater than relation and yet does not realize, on some tacit level, the relevant fact about the greater than relation, namely, that it is transitive. We can label this point by saying that this fact about the greater than relation is part of the core conception or implicit conception of greater than.[9] To think about this relation or possess this concept, one must have this belief. If this is right, then anyone who is in a position to understand the premises of the number argument can also reasonably add the relevant premise and make the argument valid in our strict sense. The same cannot be said for the heat equation and its solution. Here the techniques required to move from the system of equations to its solution go far beyond what could be reasonably required to understand the equations themselves.

This said, it can remain unclear in particular cases what gets into what I call the core conception. It is tempting to think that some preferred axiomatization of the relevant mathematics might match up precisely with what we are after. It seems to me, though, that typical axiomatizations of a mathematical domain go far beyond

---

9. See Peacocke (1998c) for one view of what this comes to. In chapter 14 I criticize Peacocke and argue that the relevant core conception is quite minimal.

what we would want to include in the core conception. For the numbers, we have of course the Peano axioms given, perhaps, in a first-order language with an induction schema. But for such axiomatized theories, even though there are some true claims about the numbers that are missing, it does not follow that all the truths about numbers that are in the theory are properly part of the core conception. For example, given that Fermat's last theorem is true, it is reasonable to expect that it is a theorem of Peano arithmetic. But this is not the sort of metaphysical necessity that should be taken for granted as part of the notion of entailment. A scientific inference that depended on the truth of Fermat's last theorem should not be judged valid unless that dependence is made explicit.

The problem recurs no matter how impoverished the core conception winds up being for the simple reason that we can grasp several core conceptions at the same time and not be in any reasonable sense in a position to grasp the ramifications of this combination. What is needed, then, is some further check on the complexity of the contents that we ascribe to our scientific representations in addition to what we have offered so far. This point receives some independent support from Humphreys, who has argued for a more "humane" notion of content that takes account of the limitations that we must work under (Humphreys 2004). To adapt a case of Humphreys's, an array of data with a million entries fixes the mean value of the data. But it is not feasible for a person to hold all the entries in their mind to calculate the mean value without the aid of some kind of calculation procedure, either on paper or using a calculator. In some sense, then, while the array of data entails its mean value, this is not a useful notion of entailment for understanding how scientific representation works, in particular, what is gained when we learn a calculation algorithm.

Unfortunately, Humphreys does not present an account of content that would meet these demands, but instead operates informally with restricted notions of content and logical entailment. This is usually how I proceed as well, except to offer a brief suggestion here for how this sort of entailment could be made more precise. My suggestion is to consider the length of a formal derivation from the premises to the conclusion in a formal proof system like a natural deduction system. For the calculation of the mean value, this derivation would take the form of implementing an algorithm that yields the desired value as output. For example, we could calculate the sum of the first two entries, then add the third, and so on, and complete the calculation by dividing by one million. Other algorithms might be more efficient, but the rough order of the length of these derivations provides a natural representation of just how far the conclusion is from the premises. Finally, to impose a more humane notion of content and entailment, we can stipulate that only a certain level of complexity in the derivation can be tolerated. Beyond this bound, we are considering conclusions that are independent of the premises.

2.7 INTRINSIC AND EXTRINSIC

Our discussion of content and inference allows us to offer a more precise characterization of the distinction between intrinsic and extrinsic mathematics that was introduced in chapter 1. Recall that the intrinsic mathematics of a scientific representation is the mathematics that is used in directly specifying how a system has to be for the representation to be correct. If our representation uses some equations to describe the dynamics of some system, for example, the mathematics used in giving those equations is intrinsic. However, other contributions by mathematics involve other, extrinsic mathematics. For example, the mathematics used to derive and apply that system of equations might be quite different than the mathematics of the equations themselves. Based on our discussion of content, it should be clear that the intrinsic mathematics is what appears in the mathematical structure involved in the content. Any other parts of mathematics beyond what explicitly appears in this structure is then classified as extrinsic.

Despite this attempt at a sharp division, it is often very difficult to decide exactly what gets counted as a part of the intrinsic mathematics. Suppose our representation makes use of functions on the real numbers, but that these functions only take values on some interval of reals. Then real numbers above a certain bound need not appear in the mathematical structure that contributes to a specification of the content. But it would seem extreme to relegate these real numbers to the extrinsic mathematics. Similarly, we may have a system of first-order ordinary differential equations. That is, we have d/dt, but not d^2/dt^2, d^3/dt^3, and so on. But these further derivatives seem so closely tied to the first-order derivative that it would be strange to call them extrinsic to the representation.

Some progress can be made on this problem if we deploy the notion of a core conception that was already used to specify the inferential connections between representations. Arguably, all the real numbers and all the orders of differentiation are included in the core conception of real numbers and differential equations. So, even though these mathematical entities are not, strictly speaking, part of the mathematical structures in question, they nevertheless are linked to these structures via the core conceptions of the mathematics. When this happens, I treat these parts of mathematics as also intrinsic.

This more generous notion of "intrinsic" has its limits, though. To consider just one central example, there is a close link between the real numbers and the complex numbers. Exactly what this link is depends on how we think about these numbers. One common way of approaching a complex number $a + bi$ is as a point in a two-dimensional plane, where a indicates its x position and b fixes its y position. The real numbers then appear in the plane as the x axis, and it is tempting to conclude that the real number a is identical with the complex number $a + 0i$. Thought of this way, the relationship between the real numbers and the complex numbers is just like the relationship between an interval of real numbers and the rest of the real numbers. This suggests that even a deployment of complex analysis in a derivation from a representation that just used real numbers would remain intrinsic. I want to resist this conclusion, though, because it would essentially trivialize the intrinsic/extrinsic distinction. Here and in similar cases I would insist that there is no link between the core conception of real numbers and these features of the complex numbers. This is so even though there is a metaphysically necessary connection between the real and complex numbers, perhaps even amounting to partial identity. Again, our intuitive test will be that a person can fully understand representations involving the real numbers and not be in a position to discern the link to the complex numbers. Making this connection involves a new contribution from mathematics and will be treated as extrinsic.

We see two places, then, in which mathematics enters into scientific representations, and these will correspond to two different problems with confirmation. The simplest kind of case is when extrinsic mathematics is used to derive a prediction from a mathematical scientific representation. Here our core example will be the solution of a system of differential equations, as with our heat case. Other applications of extrinsic mathematics are also worth keeping in mind. For example, the derivation of one representation from another may or may not involve a transfer of some degree of confirmation. Whether this transfer occurs will be affected, in part, by the mathematical character of the steps of the derivation and how they mesh with our additional commitments about the system.

A more difficult question concerns the link between confirmation and the intrinsic mathematics of a representation. In the last chapter I suggested that the intrinsic mathematics can contribute by limiting the content of the representation, for example, to an equilibrium state or a particular scale. We need to consider how this might work and what sorts of principles of confirmation are consistent with this result.

2.8 CONFIRMATION THEORY

Confirmation theory is the systematic attempt to explain how scientific representations are confirmed in scientific practice. I take it as given that the notion of confirmation here is what is sufficient to give an agent a reason to believe the representation. These reasons are defeasible in the light of further scientific developments, and they may also be overridden by competing considerations at a given time. It may turn out, for example, that two representations are both confirmed to some degree by a given experiment, yet an agent who becomes aware of this situation may not reasonably adopt both representations because they are inconsistent in some respects. In such cases, I take it that confirmation theory should also give some account of what considerations are appropriate in breaking ties.

We can distinguish several different approaches to confirmation. An initial division can be made based on whether the theory employs the probability calculus to represent degrees of confirmation or relative measures of confirmation. The most popular deployment of the probability calculus in a theory of confirmation is of course Bayesianism.[10] In its subjective variant, Bayesianism associates degrees of belief with competing scientific hypotheses and explains under what circumstances these degrees of belief should go up or down in light of learning new information about the world. These degrees of belief, in turn, are grounded in what sorts of odds an agent would accept on a certain kind of bet. It is then argued that rational agents should try to assign their degrees of belief using the rules of the probability

---

10. A very helpful overview of Bayesianism is Earman (1992).

calculus, and that they update or conditionalize their degrees of belief over time using Bayes's theorem. One distinctive feature of subjective Bayesians is that they have a permissive conception of rationality that allows different agents to start with different prior probability assignments. This seems to many to undermine the link between degrees of belief and what is reasonable. The subjectivity is removed by various attempts to make a stronger link between degrees of belief and objective features of the world. Objective Bayesians insist that prior probabilities must be fixed by genuine probabilities in the world, for example, the objective chance that a given belief will turn out to be true. Although objective Bayesians seem more in line with our ordinary conception of rationality as objective, it remains a challenge for them to justify their prior probability assignments.

One problematic feature of the rules of the probability calculus is that they require that logically equivalent propositions be assigned the same probability. When probabilities are interpreted as degrees of belief, this assumption can easily be seen to be unrealistic. As typically understood, all logical and mathematical truths are logically equivalent, but agents do not know which sentences written in logical and mathematical terms express true propositions. That is, agents are not logically and mathematically omniscient. As a result, they do not know which propositions are logically equivalent, so it is impossible for them to follow this rule of the probability calculus. This point seems especially difficult for the objective Bayesian as she insists that degrees of belief track objective chances. In any viable sense, the objective chance of all mathematical truths is 1 and the objective chance of all mathematical falsehoods is 0, but not even the best current mathematicians are in a position to make the right assignments for all but a few cases. A subjective Bayesian may seem to have more options because she can permit an agent to violate some rules of the probability calculus based on their subjective limitations. Still, subjective Bayesians typically have nothing to say about degrees of belief for mathematical propositions, and it is unsatisfying to allow massive violations of the rules of the probability calculus in such a central part of science.

The consequences of the assumption of logical and mathematical omniscience are easy to see if we note that a boost in confirmation on learning some new information E is represented as a change from $Pr(H)$ to $Pr(H|E)$ via conditionalization. However, $Pr(H|E) = Pr(H\&E)/Pr(E)$. If E is any proposition with degree of belief of 1, then $Pr(E) = 1$. If E is consistent with H, then this means that $Pr(H\&E) = Pr(H)$. As a result, $Pr(H) = Pr(H|E)$. So, if every mathematical and logical truth is assigned a degree of belief of 1, then they make no contribution to the confirmation of scientific representations. A rational agent would already have factored these truths into his of her probability assignments and so is precluded from deploying these beliefs in the practice of confirmation.

More sophisticated Bayesians are aware of this sort of problem and tend to offer two responses. The first is to restrict their focus to contingent propositions. Though this may seem reasonable, it excludes from the Bayesian framework a whole host of scientific cases where mathematical beliefs play a central role. I would not insist that every case of scientific confirmation must have a prominent role for the mathematics, but it remains clear that mathematics often makes a crucial contribution. More ambitious Bayesians can offer a second response. This is to attempt to revise the Bayesian framework to accommodate failures of logical and mathematical omniscience. Independently of my motivations to do this, there has already been an extensive discussion of these issues in connection with what is known as the problem of old evidence. The problem arises when we think of evidence that is believed with degree of belief 1 prior to the formulation of a new scientific hypothesis. Based on the argument from the previous paragraph, if the evidence is consistent with the hypothesis, then it cannot lead to any boost in the degree of belief in the hypothesis over the prior probability that the agent assigns to it. This seems contrary to actual scientific practice. Quite often, the ability of a hypothesis to accommodate evidence that was already accepted provides a decisive reason to accept the hypothesis.

Although formally similar, I would suggest that the correct solution to the problem of old evidence should be quite different from our difficulties with logical and mathematical omniscience. When the evidence is contingent, it is possible to consider an agent A' quite like the agent A who believes the evidence with degree of belief 1. We can then consider how A''s degrees of belief in the relevant hypothesis would change on learning the evidence. This can give us some insight into whether A has a reason to believe the hypothesis based on the evidence. This shows that it is possible to factor out the contribution of the evidence to the rational degree of belief in the hypothesis even when it is old evidence.[11] This promising strategy is not possible for the case of mathematical truths. Consider again our primary case of the extrinsic deployment of some body of mathematics M in solving a system of differential equations D and obtaining solution S. It is metaphysically necessary that $D \to S$. As a result, $Pr(S|D) = Pr(S|D\&M)$. But here the fact that the agent believes M is not relevant to the probability assignments. There is thus no solution to the problem based on some attempt to factor out the agent's belief in M by considering similar agents. Something different is required to handle the problem.

The way out that I suggest falls out naturally from the proposals from earlier in this chapter concerning the content of these sorts of representations. The basic point there was that we need to restrict the content so that it does not include the contributions of metaphysical necessities and so that it is limited to reflect our cognitive limitations when working with representations. If we adopt this perspective when we consider logically equivalent propositions, then it follows that many mathematical truths are not logically equivalent and there is no reason for an agent to ascribe all such truths a degree of belief of 1. That is, even though it is metaphysically necessary that S is the solution of D, an agent may not realize this. Furthermore, based on the extrinsic link between S and D, it is not the case the D logically entails S. So, $D \to S$ is not logically equivalent to a tautology $A \vee \neg A$, and an agent may assign a low degree of belief to the former while assigning the latter a degree of belief of 1. On considering M and bringing it to bear on D, an agent can come to realize that $D\&M$ logically entails S. In such a case, she should raise her belief in $D\&M \to S$ to 1 if there are no worries about the derivation based on nonrigorous mathematics

---

11. Howson and Urbach (1989, p. 274): "the support of h by e is gauged according to the effect a knowledge of e *would* now have on one's degree of belief in h, on the (counter-factual) supposition that one does not yet know e."

or the complexity of the derivation. We can reconstruct such cases of the extrinsic contribution of mathematics to science by sticking with this restricted notion of content and logical entailment.

To see how this might work in a simple case, let D be the fully specified representation of an iron bar system. S is in fact the unique solution to D, but an agent is not in a position to realize this unless he deploys the extrinsic mathematics M to solve D. Let E_i be a series of observations that accord with D, for example, the results of the measurements of the temperature at different parts of the bar at different times. Prior to solving the system of equations and obtaining D, the various E_i fail to stand in any systematic confirmation relation to D. That is, even if the agent has obtained the E_i, she does not recognize them as evidence for or against D. Typically, then, $Pr(D) = Pr(D|E_i)$. We can further assume that $Pr(D)$ is relatively low. This happens even when D results from a well-motivated theory because the derivation of D from this theory involves a host of questionable assumptions that may or may not work for a given iron bar system. Suppose then that the agent comes to recognize that S is the solution based on the use of M. That is, $Pr(D\&M \to S)$ is adjusted upward. Assuming the agent can also recognize that $S \to E_i$, it follows that he is now in a position to see the relevance of the E_i for D. I would represent this as $Pr(D|E_i) < Pr(D|E_i\&M)$. It is the greater degree of belief that the agent should use to conditionalize on the E_i after they have deployed M to find S.

Here I have not offered much of a story about how an agent comes to believe M or what is involved in deploying M to obtain S from D. I will not say anything more about this until chapters 10 and 14, where we discuss the possibility of accounting for the nonempirical justification of mathematical beliefs. For now it should just be clear that it is very hard to explain how mathematics can usefully contribute to scientific confirmation unless we assume some sort of nonempirical justification. Otherwise, M comes to play the role of another auxiliary hypothesis, and it becomes difficult to assign a boost or drop in confirmation to D as opposed to $D\&M$ when the solution S is determined to be empirically correct or incorrect. We do not need M to be believed with degree of belief 1, but only relatively highly when compared with D.

2.9 PRIOR PROBABILITIES

What we have seen, then, is that a modified form of Bayesianism can accommodate simple cases of the extrinsic contribution of mathematics by taking on our restricted notion of content. This leads to violations of omniscience, but these violations are well motivated. It is too much to hope that this sort of maneuver can be used for the other main category of how mathematics contributes to confirmation. When the contribution is intrinsic, it is less clear how we are to proceed. The basic problem is that the content of the representation involves the mathematics in such a central way that it is not possible for an agent to understand the representation unless they believe the mathematics. This tight connection precludes the strategy that we deployed for extrinsic cases based on a limited notion of entailment. Consider, for example, two different representations R and R' of the same system that involve different intrinsic mathematics, M and M'. In certain cases, it seems like M' is making a crucial contribution to R' that is lacking in the case of M and R. In the equilibrium case summarized in the last chapter, I argued that moving to an equilibrium representation can greatly increase our ability to confirm the representation. So, there is something about M' and R' as compared to M and R that relates to confirmation.

As with the problem of old evidence, there is an analogue to this sort of problem in the standard Bayesian framework. This is the problem of fixing the prior probability for a given belief. What we seem to want to say is that, prior to the collection of evidence, R' has a higher prior probability than R. But neither traditional Bayesian strategy of handling priors is of much help here. A subjective Bayesian would require no systematic link between kinds of mathematical representations and relatively high prior probability. For different reasons, the way that objective Bayesians fix priors is also of little help. If we focus on the objective chance of R and R', there seems to be no reason to think that always or most of the time the objective chance of the equilibrium representation would be higher than the objective chance of the dynamic representation.

The same worry seems to hold for approaches to confirmation that drop the probability calculus. One example of this is Norton's material theory of induction. Norton argues that nondeductive inferences like enumerative induction are not grounded in subject-independent rules of inductive logic, but instead in local, material facts that obtain in a given sort of system. For example, the reason a survey of samples allows me to conclude that a given kind of metal melts at a certain temperature, but that a formally identical inference concerning the melting point of wax would fail, is that there is a fact that obtains concerning metals and melting that is absent in the wax case. As Norton summarizes his view,

> All inductions ultimately derive their licenses from facts pertinent to the matter of the induction. I shall call these licensing facts the material postulate of the induction. They justify the induction, whether the inducing scientist is aware of them or not, just as the scientist may effect a valid deduction without explicitly knowing that it implements the disjunctive syllogism. (Norton 2003, pp. 650–651)

The risk is that the material approach will sever the link between the inference being justified and the agent having a reason to believe the conclusion based on a belief in the premises. A number of options remain for Norton, though, in handling the cases that he is most interested in, like the metals case. For example, he may think of justification and reasons in a more externalist way and say that perceptual or causal access to the relevant material postulates is sufficient to provide the agent with a reason.

Still, a focus on facts that obtain in the world is not going to help in clarifying what the mathematics might be contributing in the sorts of intrinsic cases that are my current focus. Consider the task of trying to confirm a dynamic representation as opposed to an equilibrium representation. If I am thinking about their content in the right way, then both involve a complicated structural relation between some class of physical systems and some mathematical structures. If both representations are correct, then whatever material postulates are relevant to Norton's account will obtain for both representations. As the agent is interacting causally and perceptually with both systems, whatever additional story that Norton may wish to add concerning justification will again apply to both cases. So it looks like the material theory has nothing to say about the relevant differences between these two sorts of representations. This suggests that the problems we have been noting for approaches to confirmation based on the probability calculus can be generalized to other promising approaches to confirmation. It seems that the role of metaphysically necessary mathematical truths is hard to take account of no matter how we are thinking of confirmation, especially with these intrinsic cases.

My proposal is to settle for a relative ranking based on the relative range of constraints that a representation places on a system for that representation to be accurate. In certain cases, we can have pairs of representations that are related in the following way. R includes R' as a special case or at least as a case that results from R through the specification of certain parameters or other simplifications of this sort. As a result, there is a sense in which R' says less than R. This diminished content shows itself in two ways. First, the number of systems that R' purports to capture is less than what R purports to capture. This follows as a natural consequence of the simplifications that produce R' from R. Second, even when the two representations have the same target system, R' says less about it than R. That is, R' imposes conditions on the system that are easier to meet than R. When both of these tests are passed, then I say that R' has *less content* than R.

With this notion of the amount of content in mind, we can articulate two principles that I would argue any confirmation theory must include. The principles are

(P1) Other things being equal, if R' has less content than R, then R' starts with a higher relative confirmation than R.
(P2i) Other things being equal, if R' has less content than R and a piece of evidence E supports both R' and R, then R' will receive a larger boost in confirmation than R.
(P2ii) Other things being equal, if R' has less content than R and a piece of evidence E undermines both R' and R, then R' will receive a larger drop in confirmation than R.

It is worth seeing how these principles help us reconstruct our central case of intrinsic mathematics contributing to the confirmation of a scientific representation. If R is a dynamic representation of some kind of system and R' is an equilibrium representation that results from R via certain simplifications, then it is plausible to think that R' has less content in my sense than R. First, almost all of the systems of the given kind are not equilibrium systems, so the targets of R' are a proper subset of R. Second, when we consider what R says about equilibrium systems, insofar as R is a dynamical representation, it will include details of how the system stays in that equilibrium for an extended period of time. Other things being equal, R' will not include that information because it will just represent what the system is like while it stays in the equilibrium state. So, our principles should be in force.

(P1) is the most debatable of our principles and has some similarity to principles of simplicity that some might use to fix priors in an objective Bayesian framework. I do not see how one could make it plausible as a quantitative principle, but in the ordinal way that I have presented it here, it seems plausible. (P2i) and (P2ii) come into play after we have collected evidence relevant to both representations for a system that they both represent. Because R' has an impoverished content as compared to R, R' is more susceptible to being confirmed or disconfirmed by a single relevant piece of evidence. For example, if an equilibrium representation R' predicts that the system will be stable in a given equilibrium state, and it turns out that the system is unstable in that state, then it is more or less completely disconfirmed and should be discarded. By contrast, a failure of R to capture this feature may be explained away more easily based on the large number of simplifications necessary to cash out stability from a dynamic perspective. Even if R gets this stability feature wrong, it can still get the dynamics largely correct, and so can remain a viable representation.

In the more positive case, where (P2i) is in force, we have evidence that an equilibrium representation R' gets some feature, for example, stability, correct. This gives the representation a larger boost in confirmation than what is received by a corresponding dynamical representation R. R, although it has gotten things right, involves a wide range of other commitments about the system. Just because we have seen that it gets this part correct, we are not entitled to raise its confirmation as much as we did for R'. Additional tests concerning the dynamics itself are needed.

I am not optimistic that every approach to confirmation theory can find a way of motivating the principles I have given here in their own terms. Notice also that an objective Bayesian may try to justify these principles, but even if this is successful, it is not clear that the objective Bayesian can take on the innovations that I used earlier to handle extrinsic contributions from mathematics. This is because their degree of belief assignments are supposed to be constrained by objective chances, so it seems that the objective Bayesian has reason to resist some changes in probability assignments that subjective Bayesians could embrace.

In the next five chapters my aim is not to develop a comprehensive confirmation theory, but instead to develop an understanding of cases. At this point of our discussion we have been content with some simple examples and general principles that allow us to understand what mathematics might have to do with scientific confirmation. Starting in chapter 3, I move more into the details of particular cases. This increase in detail results in a greater appreciation of the many different ways the classification scheme that I have developed can be fleshed out. It will also make clearer how confirmation works in actual scientific practice, and thus illuminate and further support the general principles I have just presented.

3

Causes

3.1 ACCOUNTS OF CAUSATION

In chapter 1 I made the prima facie case for three claims. First, there is an important distinction between causal and acausal representations. Second, mathematics plays a crucial role in formulating many acausal representations. Third, the contribution of the mathematics here is to the confirmation of the acausal representations. After our discussion of content and confirmation in chapter 2 we are now in a better position to understand what these three claims amount to. I support them through a consideration of some examples. Before we get to the examples, though, it is necessary to survey some of the main conceptions of causation. Once these are on the table, we can better see what acausal representations would lack and what positive benefits they might have.

It is something of an oversimplification, but still roughly correct, to say that since Hume many empirically minded philosophers have been suspicious of the role of causes in science. A core worry is that if scientists are involved in finding the causes of phenomena, then scientific debates will become intractably bogged down in metaphysical stand-offs. One manifestation of this concern was the way in which logical empiricists and their descendants stopped short of invoking causal relations in their otherwise ambitious attempt to clarify scientific concepts like explanation, reduction, and confirmation (Hitchcock 2008, p. 317). The hope was that a broadly formal or logical framework was sufficient to clarify these core concepts and that the obscure "metaphysical" notion of cause could be passed over in silence. Mathematics, of course, was present in this logical framework, but was thought not to pose any problems because it was deemed to be analytic. For the logical empiricists, this meant roughly that the content and epistemic status of mathematics could be taken care of using logical resources alone.

In retrospect, the failure of the logical empiricist program seems largely tied to their hostility to causal notions. Especially in the areas of explanation and confirmation, but also in discussions of laws, theory change, and unobservable entities, causal concepts have played a central role in the extant viable positions. I believe that the resurgence in interest in causation has contributed to the difficulty in understanding the role of mathematics in science precisely because contemporary philosophers of science have typically continued to identify the mathematical with the formal or the logical. As a result, we see a simple opposition between the causal and the logical, with little room left for genuine contributions from mathematics.

At its limit, there is the view that all scientific representation should be causal, and so should aim to be completely free of mathematics. For now, I want to agree with almost all philosophers of science that causal representations are extremely important to science and that we have little sense of how to produce such representations without an appeal to some mathematics. In this chapter, we investigate how mathematics also enters into acausal representations and what their value might be.

As a final preliminary remark, let me distinguish the main focus of this chapter, namely, causation, from a separate issue of explanation. It may be that most or even all explanations involve causal representations. Still, a discussion of causation can be pursued independently of debates about explanation. Later, in chapters 9, 10, and 11, explanation will be discussed in its own right. There I argue that there are noncausal explanations, but I do not want to presuppose that in my discussion of the relative merits of causal and acausal representations. The merits I am interested in here are simply correctness or descriptive accuracy.

The working assumption of this chapter is that causal representations can be distinguished from acausal representations based on their content. That is, there is something about the conditions that a causal representation places on its target systems that differs from the associated conditions of acausal representations. Among philosophers there is little agreement on what this content really amounts to. To start, we can distinguish between reductive and nonreductive accounts of causation. A reductive proposal aims to reduce the content of causal representations to some kind of acausal representations. In this respect, they are the heirs to the logical empiricist project of downplaying the importance of causation, although they may invoke concepts that logical empiricists were uncomfortable with, for example, counterfactuals. Reductive proposals face many challenges, including the basic difficulty of making a clear distinction between causal and acausal representations. For example, the test for whether a relationship is causal may involve global features of the world or our representation of it, and so the causal character of a representation would not turn on the features of any particular representation.[1] A clearer and more tractable distinction arises from nonreductive accounts of causation. These accounts explain the notion of cause in terms of other concepts that might seem equally problematic, especially from the perspective of a sparse metaphysics. Nonreductive approaches, in turn, have been typically divided into process theories and difference-making theories. Process theories view causation as a process of a certain special kind. A popular process theory is the Salmon/Dowe view that causation is a process that transmits conserved quantities like mass or energy (Dowe 2007). A more recent and more expansive kind of process theory links causation with mechanisms, where "mechanisms are entities and activities organized such that they are productive of regular changes from start or set-up to finish or termination conditions" (Machamer, Darden, and Craver 2000, p. 3).[2] For both, causal relations

---

1. I have in mind the view that causes are fixed by laws, and laws are fixed using the best overall system of description of the physical world. See Lewis (1999a).

2. See Machamer et al. (2000, p. 7) for the contrast with Salmon.

involve linking two entities or events with a special kind of happening over time. They are nonreductive accounts because they do not hold out much hope of specifying the relevant kind of process in completely noncausal terms.

The second main kind of nonreductive account of causation is the difference-making proposal. Here the idea is that causal relations result in kinds of outcomes that are systematically different than what obtains in noncausal cases. Such results may obtain in virtue of the sorts of processes emphasized by the process theories, but these processes are not necessary for causation to exist. A recent and influential difference-making account of causation is Woodward's manipulability theory. The key part of Woodward's proposal is that causal claims have as their content what would result from certain sorts of manipulations of the cause, namely, that they would bring about a variation in the effect: "No causal difference without a difference in manipulability relations, and no difference in manipulability relations without a causal difference" (Woodward 2003, p. 61).[3] For example, smoking causes lung cancer and stained teeth, but stained teeth do not cause lung cancer. Woodward unpacks this distinction by noting that varying the smoking state of a person would result in variations of lung cancer and stained teeth, but varying the stained teeth state, while holding other potential factors fixed, would not result in any variations in the occurrence of lung cancer.

We have three accounts of causation on the table, then: conserved quantity, mechanism, and manipulability. Each assigns a different content to a claim like "C causes E," but all three see causation as a genuine feature of the world that we can get right or wrong in our scientific representations. Although we can confirm or disconfirm such representations by appeal to noncausal representations such as experimental data, there is little hope of reducing the content of causal representations to some collection of noncausal features of the world. Causation, then, is a constituent of the world that science must get right if our theories are to be complete.

Based on their flexibility, and whether they think of causation as tied to processes, our three accounts may disagree on whether certain cases count as causes. More often, though, the proposals will agree that we have a case of causation and merely offer competing diagnoses of why causation appears.[4] For example, the mechanism account can surely agree that smoking causes lung cancer, and stained teeth and that stained teeth do not cause lung cancer. But for this proposal the basis of these claims lies in the existence or nonexistence of mechanisms linking smoking, lung cancer, and stained teeth. Presumably they would grant that the results of manipulations would constitute evidence for the existence of these mechanisms. Similarly, Woodward would accept that establishing the existence of mechanisms would be good evidence for what would result from manipulation even if causal claims are not claims about mechanisms.

This allows us to capture a few shared characteristics of causal relations. When these characteristics all appear as part of the content of a given representation, then it will be a causal representation. When these characteristics are all absent, then we

---

3. See Woodward (2003, M, p. 59) for a more careful formulation.

4. Important exceptions are action-at-a-distance and omission. See Woodward (2003, p. 148).

have what I have been calling an acausal representation. A central feature of causal representations on any of these accounts is, of course, change. Coinciding with this is a temporal dimension to the representation. This is explicit in process theories. Difference-making theories must also include change and time because they consider what would change or make a difference under various, perhaps nonactual, circumstances. A further shared feature of these causal representations is the link to counterfactuals of a certain kind. This feature is at the core of the manipulability account. But process theories like conserved quantities and mechanism must also include counterfactual content in their causal representations. This is because part of the notion of a causal process is something that could be disrupted and if it were disrupted, the effect may not result. This is why results from manipulation are good evidence for the existence of these causal processes.

3.2 A CAUSAL REPRESENTATION

In this section I present what I take to be an uncontroversial instance of a causal representation. We will see how our three accounts of causation can handle this case and how this corresponds to the presence of the central features of causal representations summarized in the last section. Although this case will involve mathematics, it is not clear how our theories of causation come down on the issue of the role for mathematics in causal representations. After discussing the contribution that mathematics makes in some causal representations, I turn to some acausal representations and the contribution that mathematics makes there.

Suppose we have a system of N cars traveling down a single-lane road. The lead car's front bumper's position will be represented by the real-valued function of time $x_1(t)$, the second car's position by $x_2(t)$, and so on through $x_N(t)$. With L representing the length of each car, we impose the constraint that $x_i(t) + L < x_{i-1}(t)$, that is, no car can be closer to the front of the car ahead of it than the length of the car. Violation of this constraint would imply a collision. Our representation also invokes a constant reaction time τ for each driver, which is intuitively the time between when the driver sees a change on the road ahead and begins her accelerating/braking maneuver. Two qualitative assumptions can be used to motivate a representation of the braking force of each car after the lead car:

a. The greater the difference in velocity between the two cars, the greater the braking force will be.
b. The greater the distance between the two cars, the lesser the braking force will be.

This yields, for $1 < i \leq N$,

$$\ddot{x}_i(t+\tau) = \lambda \frac{\dot{x}_i(t) - \dot{x}_{i-1}(t)}{|x_i(t) - x_{i-1}(t)|} \tag{3.1}$$

where $\lambda = A/m$, m the mass of each car, and A some constant. Appealing to our constraint allows us to simplify the equation further, and integrating gives us the velocity for each car after the lead car in terms of its relative position at an earlier time: for $1 < i \leq N$,

$$\dot{x}_i(t+\tau) = \lambda \ln |x_i(t) - x_{i-1}(t)| + \alpha_i \tag{3.2}$$

where α_i is a constant. Through the specification of λ, τ, the α_i and some further choices, we could develop this representation into a complete specification of the trajectory of each car over some time interval. In particular, the trajectory of the lead car is crucial to the evolution of the system as it is totally unconstrained by our equations.

Equation 3.2 and these additional components constitute a causal representation of a given traffic system. Using the terminology from previous chapters, we can distinguish its intrinsic and extrinsic mathematics. In particular, 3.1 and the mathematics that gets us from 3.1 to 3.2 is extrinsic to 3.2. An account of how the traffic system has to be for 3.2 to be correct need not invoke the extrinsic mathematics of the representation. To see the value of this distinction, note that 3.2 could wind up being accurate even if 3.1 was not. Similarly, the main mediating link between 3.1 and 3.2 is

$$\frac{\dot{x}_i(t) - \dot{x}_{i-1}(t)}{|x_i(t) - x_{i-1}(t)|} = \frac{d}{dt} \ln |x_i(t) - x_{i-1}(t)| \tag{3.3}$$

when $x_i(t) < x_{i-1}(t)$. Equation 3.3 does not enter into the content of the representation. Even when such mediating equations are false, it need not undermine the accuracy of the resulting representation. By contrast, the mathematics involved in specifying the mathematical structure that is part of the content of 3.2 is intrinsic.

Equation 3.2 is a special kind of causal representation that I call a *dynamic* representation. These representations include information about how a given system will evolve over time as a result of causal interactions between their constituents. In our example, the cars are represented as unstructured constituents of the system that obey equations based on their relative positions and velocities. Nobody using the representation to represent a traffic system believes that the cars are actually like this, but as far as the correctness of the content of this representation goes, any further structure to the cars, like their drivers and their further causal features, is irrelevant. So, on my conception of a dynamic representation, the entities standing in causal relations need not be the ultimate entities of microphysics or any other fundamental physical theory. This accords better with scientific discussions of causation and our three main competing philosophical accounts of causation.

As presented so far, 3.2 includes one of the key characteristics of a causal representation as it describes change over time, but it may seem to lack any link to counterfactuals. We might imagine a preliminary version of 3.2 that included only correlations between the cars' positions and velocities, but was not meant to represent the causes of these observed patterns. This precausal representation would become causal once additional contents were added that tied these correlations to the right kind of counterfactuals. Here, process and difference-making accounts part ways. A process theory like conserved quantity or mechanism requires a content like "There is a process or series of processes such that 3.2," whereas a manipulability theory mandates a content like "The results of the right kind of intervention on the traffic system are captured by 3.2." Either way, a series of mathematical equations is not sufficient for a causal representation. At the same time, the sorts of equations typically found in a dynamic representation are tailor-made to the specification of the sorts of counterfactuals required by our accounts of causation. For a process theory, the equation either includes information about a process or else is viewed as encoding the results of some otherwise unrepresented process. On a difference-making theory, the parameters of the equations indicate what would result if these parameters were to be manipulated in the right way.

This last point helps us see what the contribution of mathematics to causal representations might be given the view that causal relations are genuine features of the world. The right kind of mathematical structure for these representations is one that tracks the dynamic development of the system, not only in its actual state but also in a preferred class of counterfactual situations. This might happen at a maximal level of detail, so that the process or results of manipulation are encoded in the content of the representation to a perfect extent. Or, more usually, there is much about the process or results of manipulation that remains unknown, and here the representation stays neutral on these various possibilities. Hence, mathematics can be used not only to track what is known, but also to limit the scope of the representation to the known causal features of the system.

There is a vast amount of philosophical discussion of how causal representations can be confirmed using the sort of experimental and observation data at our disposal. The general consensus seems to be that data cannot univocally support a causal representation unless certain background assumptions of a more or less causal character are provisionally made.[5] These background assumptions provide the bridge from mere correlation or description to conclusions about causation. Here again, mathematics plays a central role, now in making the extrinsic links between the raw data, some model of the data, and the causal representation under investigation. Statistical tests of significance, for example, involve nontrivial assumptions about the distribution of the potential measurements and how the actual measurements fit into this distribution. This allows a well-motivated comparison between the model of the data and the predictions of the causal representation.[6] In our traffic case, we could imagine setting up a series of traffic systems with different drivers and different prescribed behaviors to the lead car. This raw data would then be processed into a data model that would take account of random fluctuations and other interference with the development of the traffic system. Finally, this data model could be compared to the predictions of the dynamic traffic representation across various sets of parameters. If our representation did well through a series of such tests, then we would have a well-confirmed causal representation of traffic systems.

---

5. In Woodward (2003) these assumptions take the form of the adoption of a causal graph. In chapter 6 we explore the need for assumptions along the lines of Kuhnian paradigms.

6. The importance of such tests has been extensively discussed in Mayo (1996).

It should not come as much of a surprise that 3.2 does poorly when faced with this sort of experimental testing. This shows that 3.2 is not an accurate causal representation of actual traffic systems. What is missing are a variety of causal factors and variations relevant to actual traffic systems such as the changes across drivers, the multiple lanes of traffic, and the shape of the road. More successful dynamic representations of traffic are of course possible through more complicated mathematical representations. If we investigated these further, we would see the same intrinsic and extrinsic roles for mathematics. First, the intrinsic mathematics of the representation would aim to track the actual and relevant counterfactual interactions between the cars. Second, mathematically processed models of traffic data would be used to compare the predictions of these representations with what is going on in the relevant systems. A large collection of accurate causal representations have been successfully confirmed through this sort of activity, and although mathematics is not the only factor in this achievement, it is surely a crucial contributing factor.

3.3 SOME ACAUSAL REPRESENTATIONS

Too much philosophy of science assumes that an adequate understanding of the successful confirmation of causal representations is sufficient to solve our philosophical questions about the success of science. The implicit thought seems to be that the only useful acausal representations are the sorts of precausal representations already noted that record correlations and that form the inputs into the process of testing causal representations. I want to argue now, though, that this misses a wide variety of scientific representations and that mathematics plays an equally (if not more important) role in the formulation and confirmation of these representations. The representations I have in mind lack both of the key characteristics of causal representations that we already saw: change in time and supporting a certain kind of counterfactual variation. Some representations that lack these characteristics are merely the precausal representations that report correlations. However, as we will see, others are a distinct kind of acausal representation that make important contributions to the success of science.

To start, consider the simple example of the representation of the bridges of Königsberg using a non-Eulerian graph. A *graph* is an ordered pair, where in the first position is a set of objects called *vertices* and in the second position is a set of (unordered) pairs of vertices called *edges*. A *path* of a graph is a series of edges where one of the vertices in the nth edge overlaps with one of the vertices in the $n + 1$th edge. *Connected* graphs have a path between any two distinct vertices. The *valence* of a vertex is how many edges it is part of. Finally, a graph is *Eulerian* just in case it is connected and has a path from an initial vertex v that includes each edge exactly once and that ends with v. Euler showed that a connected graph G is Eulerian iff every vertex of G has even valence.[7] To see these mathematical concepts in action,

---

7. For further explanation see Carlson (2001). I discuss this case in Pincock (2007) and the example is also noted independently in Wilholt (2004). For additional similar examples, see Franklin (1989), although I do not endorse his interpretation.

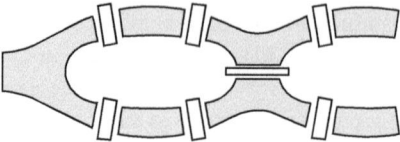

Figure 3.1: The Bridges of Königsberg

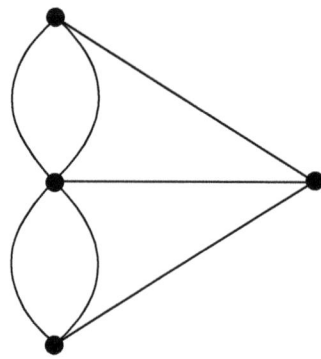

Figure 3.2: A Non-Eulerian Graph

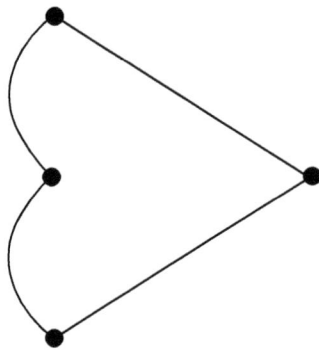

Figure 3.3: An Eulerian Graph

consider the sketch of the arrangement of the bridges of Königsberg (figure 3.1). If we treat the islands and banks as objects, and use the bridges to form edges, the physical system forms a graph (figure 3.2). Another bridge system might exist in a nearly identical town down the river where the residents had refrained from building three bridges so that their bridge system formed the graph given in figure 3.3. The graph in figure 3.2 is non-Eulerian, and the graph in figure 3.3 is Eulerian.

We can deploy the account of the content of a mathematical scientific representation from chapter 2 to unpack what a representation of the bridges using a graph might come to. Given our theorem, it is fair to summarize the content of figure 3.2 as the claim that the bridges of Königsberg form a non-Eulerian graph. More carefully, there is an isomorphism between the regions of land and islands and the vertices of the graph in figure 3.2 that maps the physically connected regions to edges in the graph. An inspection of the graph reveals that the valence of each vertex is an odd number, so it follows that the graph is non-Eulerian. That is, there is no path that starts at a vertex v and makes a circuit through each edge exactly once and that ends back at v. This imposes the following condition on the bridge system: there is no way of traversing each bridge exactly once that brings the traveler back to her starting region. This is the feature of the bridge system that had puzzled some of the residents of Königsberg and that Euler sought to capture with his graph-theoretic representation. The impossibility of this sort of circuit is a simple consequence of the claim that the graph in figure 3.2 is an accurate representation of the bridges.[8]

It seems clear that our representation of the bridges is not a causal representation for the simple reason that it does not represent change over time. We could certainly add another component to the representation that represents the trajectory of a hypothetical traveler across the bridges, but this would amount to a shift to a different representation. As it stands, the graph representation captures just the noncausal structure of the bridges. An accurate representation of this sort of structure is often adequate to represent features of scientific interest. In this case, it was sufficient to capture the fact that a certain kind of circuit is not possible. The mathematics here is not tracking genuine causal relations, but is only reflecting a certain kind of formal structure whose features in the physical system have some scientific significance. One important kind of acausal representation, then, represents such structures without representing any changes over time.

This is relatively easy to see for process theories. It would be implausible to insist that our graph representation is making tacit appeal to some conserved quantities or to some mechanisms that are responsible for the feature of the bridge system we are interested in representing. Again, we could shift to a different kind of representation in terms of the movement of individuals across the bridges, and here we would recover some kind of causal process. Still, the representation of all these processes and their features is a different representation than what our graph represents. Sticking with the more structural representation is sufficient to capture what we are interested in even in the absence of causal processes.

It might be more controversial to insist that a difference-making approach to causation would also deem our representation acausal. This is because the manipulation account emphasizes relations of counterfactual dependence and not the temporal change that is clearly absent in our graph case. An advocate of a broad manipulation account might read some claims of counterfactual dependence into our representation and so insist that it is causal after all. For example, some of the bridges could be a cause of the impossibility of the circuit because removing those bridges and shifting to a system with the graph in figure 3.3 would make the circuit possible. I agree that it is true that if some of these bridges were removed, then a circuit would be possible. Still, I would insist that the original representation that I described does not have this additional fact as part of its content precisely because it does not involve any other graphs besides the one pictured in figure 3.2 in its content. To bring in other graphs

---

8. Again, I set aside issues of explanation for now.

is to change the representation, just as adding a component to track the trajectory of a person would involve a shift to a different representation.

Even though this series of counterfactual claims is not part of the representation I am describing, there is another series that I would grant is part of the content. This is that the graph structure of the bridge system is unaffected by the physical constitution of the bridges. Supposing the bridges are made of stone, we can see clearly that changing the bridges to gold, while preserving the isomorphism to the relevant graph, would keep the impossibility of the circuit intact. So we see that the graph representation not only captures the feature of interest but also has as part of its content that various aspects of the system are also irrelevant to this feature. This is a kind of negative contribution that some manipulation theorists might be tempted to call "causal." I would reply that the counterfactuals here are quite special and should not be mixed in with the sorts of counterfactuals that are central to causal representations. Among other things, it makes little sense to consider how these very bridges would change if they were made of gold as it is arguably essential to the bridges that they be made of stone.[9] So there is no possible manipulation of these bridges that would change their constitution and preserve the non-Eulerian feature. I conclude that we gain important information about the bridges from the graph representation, but that this information is not easily classified as causal.

This bridges example is somewhat artificial, and it might seem like the benefits of this sort of acausal representation would be restricted to these sorts of cases. The next example is meant to show that this is not the case. Many so-called steady-state representations are acausal in precisely the same sense as the bridges case: they do not represent causes but do represent important features of systems, and some of their patterns of counterfactual dependence. For our purposes, a steady-state representation is one where the main features of interest of the system do not change over time.[10]

A preferred kind of steady state for our traffic systems arises directly from the observation that a solution of 3.2 with a series of evenly spaced cars would fix a constant velocity for each car. If we posit a series of steady states of this kind, we can generate a new series of representations of traffic systems that provide access to important features of these systems. More carefully, we suppose that

$$\forall i, j \;\; \dot{x}_i = \dot{x}_j$$
$$\forall i > 1 \;\; x_i(t) - x_{i-1}(t) = d$$

When the system is in this kind of state, we have a well-defined density function ρ:

$$\rho(x_0, t) = \frac{\text{number of cars in } [x_0 - \epsilon, x_0 + \epsilon] \text{ at } t}{2\epsilon} \qquad (3.4)$$

---

9. See Woodward (2003, p. 220) for the admission that the manipulability account has limits. It is not clear to me what Woodward intends to say about the case I discuss here.

10. A special case of a steady-state representation is an equilibrium representation where the main features are not changing because there is no change in any of the subcomponents of the system. I often speak more loosely of equilibrium states even when this condition is not met.

where $L \ll 2\epsilon \ll$ the length of the road. Here ρ is more or less independent of the choice of ϵ and can be treated as (Illner, et. al. 2005, p. 131)

$$\rho = \frac{1}{d+L} \tag{3.5}$$

We assume that $v(x, t)$, the velocity of cars at point x at time t, is a function just of the density $\rho(x, t)$. This allows us to distinguish two crucial states for the system. First, there is the point of minimal velocity, which will occur at the maximum density $\rho_{max} = \frac{1}{L}$ when the distance d between the cars goes to 0. At this point, $v(\rho_{max}) = 0$ and the cars stand still. Second, there is a maximal velocity for which we label the density ρ_{crit}. We are interested in how the velocity varies as a function of the density as we move from $v_{max} = v(\rho_{crit})$ to $v(\rho_{max})$. Intuitively, as the density is increased, the velocity will decrease.

Based on these assumptions and 3.2, we can find v as a function of ρ when $\rho > \rho_{crit}$:

$$v(\rho) = v_{max} \ln\left[\frac{\rho_{max}}{\rho}\right]\left[\ln\frac{\rho_{max}}{\rho_{crit}}\right]^{-1} \tag{3.6}$$

Notice how the otherwise undetermined λ and α_i of 3.2 have been removed. To plot 3.6, we need only determine the density at which the maximum velocity is reached.

A second magnitude of interest is the flux at a given point, that is, the number of cars that cross a given point per unit of time:

$$j(\rho) = \rho v(\rho) \tag{3.7}$$

Using 3.6, we can find the density ρ_0 with maximum flux by solving $j'(\rho) = 0$. This yields

$$\rho_0 = \frac{\rho_{max}}{e} \tag{3.8}$$

$e = 2.71828\ldots$ here is the base of the natural logarithm. See figure 3.4 for a graph of the density and flux.

How are we to make sense of the content of this kind of representation using the framework developed in chapter 2? It might seem like all we are doing is extracting a part of the content of 3.2. Although there is something right about this initial thought, it ignores the additional mathematics and novel assumptions necessary to get from 3.2 to 3.6. In the jargon of chapter 2, this corresponds to a shift from the basic content to the enriched content. The definitions of velocity, density, and flux at a point bring new content to the representation beyond what we had in our original representation. Furthermore, there is also an implicit shift to a schematic content when we dropped the temporal variables for these quantities. It is simply not an option to represent a traffic system as if it were timeless, just as we could not represent an iron bar system as if it had no length. Instead, this abstraction away from time allows us to represent features of the traffic systems that would be more difficult (if not impossible) with a temporal representation. We see, among other things, how an increase in density is related to velocity. But we do not represent this relationship using a causal representation. In particular, there is no attempt to represent how a traffic system might reach a steady state or what might keep it there if it did. Our information is instead about how systems with different densities would behave if they were able to reach a steady state.

Our shift to 3.6 involves a genuinely new representation of features of traffic systems that is not part of the content of the original representation. Here we see something analogous to what we saw with the Königsberg bridges. The new feature of the traffic system concerns the velocity, density, and flux relations among a special kind of system. Further investigations of 3.6 could involve what is known as "stability analysis." Here we check to see what happens to a system in a steady state if it is subjected to small perturbations or deviations from that state. For example, given a system with a definite velocity, density, and flux, we might quickly vary the velocity of the lead car. The resulting investigation would reveal whether the system would return to this steady-state or some other, or whether it might result in some kind of crash. Here we would go back to 3.2 and use its features, for example, the value of τ, to answer our questions. Equations 3.2 and 3.6 can be used together, then, to represent features of traffic systems that neither representation can accomplish independently.

Now, there is a clear sense in which the shift from 3.2 to 3.6 has involved a shift from a causal representation to an acausal representation. We see this most obviously in the lack of change through time that we have in 3.6. This automatically rules out the sort of representation of a causal process responsible for the steady state that the process theorists require. The steady state representation, of course, does not deny that there are underlying causal processes responsible for keeping a traffic system in this kind of steady state. But just as with the bridges, these processes are in no way connected to the content of the representation. As with the bridges, it might seem that the representation is nevertheless causal if we adopt a manipulability account of causation. What, in the end, does figure 3.4 represent if not the

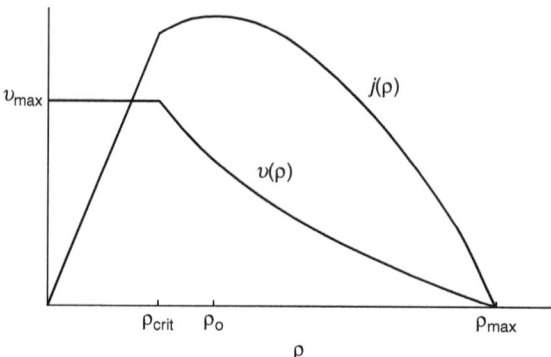

Figure 3.4: Velocity and Flux

Source: Illner et al. (2005, p. 155)

results of manipulating the density of cars on the road in terms of a difference in velocity? My response here is the same as with the bridges. If we are not to offer a trivial account of causal representation, then only some manipulations are relevant to causal content. Woodward captures this with his notion of an intervention. If we think of an actual traffic system in a given steady state, there is no clear sense in which a possible intervention is an increase in the density of the cars on the road. So, although the acausal representation gives us important information about the traffic systems, including some counterfactuals, these counterfactuals are not the sort that are rightly tied to causal representations. We get information about how a wide range of systems will behave, but no information about how a particular system would behave under interventions. This sort of acausal representation would thus be very useful in the context of designing a road or helping improve a traffic problem in an existing road. But it is too much to read into the representation any information about the causes of a given traffic system's quantities. For this, we must return to a causal representation like 3.2 or a more accurate successor.

Along with the important differences in content, there is a relevant difference in the confirmation of the two traffic representations. Recall my observation that 3.2 is not an accurate representation of its target systems, and that a more complicated system of equations is needed to handle the complexities of actual traffic. Nevertheless, it remains possible that 3.6 is not only more accurate but that we can be in a position to establish this with available experimental data. In fact, here we have precisely the sort of relation between representations that I discussed at the end of chapter 2: 3.6 has less content than 3.2. To begin with, the range of target systems of 3.6 is a proper subset of the target systems of 3.2. Second, 3.6 imposes fewer conditions on its targets than 3.2. This is because the quantities of 3.6 are aggregates or averages of the quantities of 3.2. We can see that many distributions of cars would give rise to the same velocity, density, and flux values predicted by the steady state representation, whereas the causal representation contains more details and so is consistent with fewer distributions of its quantities. This helps explain the possibility of using a steady state representation even when we are not confident in the underlying causal representation of the system. We can directly test the predictions of the steady state representation and see that it is accurate for its intended target systems. This sort of successful test would give us little evidence for the correctness of the causal representation precisely because the causal representation makes many more claims about the systems and their counterfactual variations.

I would also argue that this acausal representation is easier to disconfirm when it is inaccurate. We could call this an indirect contribution to confirmation because disconfirmation will alert a scientist of the need to move on to another representation. My claim about indirect confirmation is a different and perhaps more controversial point than the claim about direct confirmation that I have just defended.[11] The key assumption needed to reach this conclusion is that this acausal representation requires fewer auxiliary assumptions than its associated causal representation. If this is right, then it will be harder to insulate the acausal representation from refutation

---

11. I am grateful to Neil Tennant for emphasizing this to me.

when it leads to a failed prediction. By contrast, the failure of the causal representation to yield a correct prediction can be more easily blamed on the auxiliary assumptions. In the case of 3.6 I think it is clear that the main auxiliary assumption needed to generate a prediction is that a given traffic system is actually in a steady state. On the other hand, a whole range of assumptions relating to the specific cars is needed to relate 3.2 to a traffic system. For example, a failed prediction might be blamed on the values assigned to the parameters or on the otherwise unrepresented behavior of the first car. The diminished role of auxiliary assumptions for representations like 3.6 helps scientists find out that they have gotten things wrong. This indirectly helps scientists make progress because they are able to move on to a new and hopefully more accurate representation rather than wasting time with a representation that is fatally flawed.

To summarize our discussion of the traffic steady-state case, an acausal representation may be important for science for two reasons. First, it affords a representation of features of interest of a system that may not be represented by any causal representations. Second, it is typically easier to confirm or disconfirm because it has less content (in my sense) than its causal associates. Because it says less about its targets, a successful test will provide significant support. By contrast, the associated causal representation will remain harder to confirm because of its rich causal content and detailed representation of a wider range of target systems.

There is a special case of this relationship between causal and acausal representations that is especially common outside of physics. This is when no causal representation has even been formulated because the causal workings of the system are unknown or too complex to be represented by anything that we can actually work with. In population biology, for example, scientists may simply begin with a representation of aggregates or averages of the sort that we saw in our acausal traffic example. A much discussed example is the Lotka-Volterra equations for predator-prey systems. As Weisberg has explained, biologists engage in a certain form of investigation of these representations, known as "robustness analysis," which enable them to be confident in the accuracy of certain aspects of these representations even without any kind of accurate causal representation. This, in turn, allows the accurate representation of features of biological systems that would be otherwise unavailable. A quick summary of this sort of robustness analysis will again illustrate the importance of acausal representations for the success of science.[12]

The goal of these representations is to capture how a population of predator and prey vary as a result of the processes of predation and reproduction. Intuitively, the more prey there are, the more there will be for the predator to eat, and so the more predators there will be through reproduction. However, if there are too many predators, this will cut down on the population of prey, and so the population of predators will drop as well. These assumptions are captured in the following two equations:

$$\frac{dV}{dt} = rV - (aV)P \qquad (3.9)$$

---

12. See Weisberg (2008) for more discussion.

$$\frac{dP}{dt} = b(aV)P - mP \qquad (3.10)$$

Here V is the number of predators and P is the number of prey. r is the rate of reproduction for the predators, and m is the rate of death for the prey. Crucially, a is a parameter representing how many prey are eaten by the existing predators. b measures how readily the consumed prey are converted into new predators. Combining 3.9 and 3.10 gives us the Lotka-Volterra representation of a predator-prey system.

Although there is a sense in which this is a causal representation, it represents the causes of the phenomena in question in such an impoverished way that no biologist should take it to be an accurate representation of the causes operating in any biological system. Part of the problem is the way the parameters are given as simple constants that do not reflect the variations among the populations or the complex effects of factors like the environment and other species. More fundamentally, though, as with the cars, the genuine causal actors in these systems are the individual animals and not anything like the number of predators or prey in the system. For this reason, even if there is an interpretation of the representation so that it meets the conditions on a causal representation, any such interpretation would involve a host of inaccuracies. Nevertheless, an investigation of biological systems using this representation remains possible. An example emphasized by Weisberg is the Volterra principle: "the population of prey will increase relative to the number of predators upon application of a pesticide" (Weisberg 2006, p. 735). This is a well-confirmed principle that applies to a wide array of biological systems. How are scientists able to motivate and eventually confirm such a principle using an analysis of a clearly inaccurate causal representation?

Weisberg's answer is that the claim is included in a wide range of representations that result from varying the parameters of 3.9 and 3.10. Although there is no reason to think any particular representation in this family is accurate, there is reason to believe that a claim entailed by all the members of this family is correct. This is what is meant by a robust result: it holds for a wide range of representations as we vary the values of the parameters of the representation. Robust results allow us to capture features of a system even when we fail to understand enough about the system to accurately represent its causal workings. With this terminology in mind, we can go back and see some similarities with both the bridges and the traffic case. The non-Eulerian feature of the bridges is robust under changes in the material constitution of the bridges. Similarly, the velocity-density relation is robust under changes in the particular features of the drivers, for example, their specific reaction time. Our Lotka-Volterra example shows that shifting to a steady-state representation is just one way to achieve this kind of result. With 3.9 and 3.10, there is no steady state, but instead a series of oscillations where the numbers of prey and predator shift up and down in a regular pattern. By altering the representation in line with the supposed effects of a pesticide, we can see how these oscillations will shift. The shift corresponds to Volterra's principle, and the details of the analysis help us to see how much we can vary the parameters of (3.9) and (3.10) and preserve this result. This allows us to predict what will happen in a given biological system. When this prediction is confirmed, so is Volterra's principle. Nobody would take the further step of using these sorts of observations to confirm a particular choice of parameters for 3.9 and 3.10.

Is Volterra's principle a causal principle? It certainly seems so given either a process approach or a manipulation approach. If this is right, we can use robustness analysis to confirm causal claims about systems even when we lack acceptable causal representations of those systems. In this case, we fail to have an accurate dynamic representation of the animals, and so we fail to represent the detailed causal evolution of these systems. Nevertheless, we can generate well-confirmed causal claims that are highly specific and useful. A similar point has been made by Parker in her discussion of models of climate change (Parker 2006). The reason we are confident that an increase in human-produced CO_2 emissions is the cause of the overall increase in temperature is because we have a robust result across many different representations of the Earth's climate. Our confidence in this result is greater than our confidence in any particular climate representation because the CO_2 result is stable under dramatically different parameters and assumptions about how the climate works. Perhaps none of these representations accurately represents the causal processes at work in the atmosphere. Still, we can remain confident that we understand the cause of the observed rise in temperature.

We have seen with our detailed examples of the bridges and traffic, and our outlines of the biology and climatology cases, how important acausal representations can be in the support of scientific claims about systems of interest. To conclude our discussion, we need to summarize the role for mathematics in the success of this kind of acausal representation and reassess the limits of causal representations in light of this.

3.4 THE VALUE OF ACAUSAL REPRESENTATIONS

The contribution of mathematics in successful causal representations is to track a genuine feature of the world. When the mathematics is filled out judiciously, what results is a representation whose parts correspond to genuine constituents of the system under investigation. Furthermore, the mathematics encodes a series of possible trajectories for the system in terms of these constituents and represents the right sort of counterfactual relationship between the trajectories. Such dynamic representations are a goal of many of the sciences, and we can readily understand why they are desirable. Not only do they provide a clear understanding of the workings of a system, but they indicate how a given system can be changed so that it might produce a preferable result.

Not all successful scientific representations work this way. On the one hand there are the precausal representations that are needed to arrive at and confirm accurate causal representations. On the other hand, there are the acausal representations like those involved in steady-state representations and robustness analysis. Is there anything to be said about the contribution of mathematics to this kind of representation? Looking past the differences between them, which of course should not be underestimated, it does seem clear what is special about these representations: that they can be confirmed. That is, we can come to know that the representations are accurately capturing genuine features of the systems in question. This confirmation is possible even when a corresponding causal representation cannot be confirmed. I want to argue now that this link between acausal representations and confirmation is established by the mathematics. That is, the contribution that the mathematics makes to these acausal representations is a boost in potential confirmation.

Our discussion has emphasized the peculiar content associated with causal representations. This added content makes it clear both why causal representations are desirable and why they are so hard to obtain. Accurate causal representations encode a tremendous amount of information about the actual and possible behaviors of a system. At the same time, they leave out many other features of a system, especially what we might call the acausal associates of causal relations. Causation, be it cashed out in terms of conserved quantities, mechanisms, or manipulability, requires a focus on special kinds of features. In particular, we need to zero in on genuine constituents and a certain kind of counterfactual dependence between them. Additional important features of the system and its parts are tied to the acausal associates of the causal constituents. This may be because they are formal, structural features as with the bridges of Königsberg. Or the features may be properties of the system that only arise in a special kind of state, as with the steady-state traffic case. Finally, in our last example, we saw a causal feature of predator-prey systems summarized by the Volterra principle. This feature is due to the patterns of nongenuine constituents, that is, the total numbers of predators and prey, that is not represented by any causal representation we can work with. The first thing the mathematics brings in these cases, then, over and above what a successful causal representation has provided, is a representation of a novel and important feature of the system under investigation.

Representing new features would be scientifically idle, though, if it were not possible to eventually confirm the accuracy of some of these acausal representations. I think it is possible to argue that the sort of representations that result from moving from causal to acausal features are generally easier to confirm than typical causal representations. This is not meant to imply that they are easy to confirm in some absolute sense, but only that by comparison to their causal cousins they are more amenable to successful testing. Two kinds of boost in confirmation result. First, for a given acausal representation, it is typically more feasible to see that it is accurate, when it is accurate. I call this a *direct* boost in confirmation. When this happens, we can add a new accurate representation to our store of scientific knowledge. Second, when the acausal representation is inaccurate, it is typically easier to see this than when we are working with a causal representation. When we have an inaccurate acausal representation and are able to find this out, then we know that we need to move on and consider other representations based on different assumptions about the system. This is an *indirect* contribution to confirmation because it helps us avoid believing in inaccurate representations and may suggest new representations that we should try to confirm. Mathematics helps in the process of both direct and indirect confirmation for our acausal representations.

The argument for direct confirmation relies on the two principles about relative confirmation from chapter 2 and the general claim that typical pairs of causal and acausal representations stand in my relation of less content. Recall that representation *A* has less content than representation *B* when *A* has as its target systems a proper subset of the target systems *B* and *A* imposes less on its target systems than *B*. In the traffic case, the acausal steady-state representation 3.6 has less content than the causal representation 3.2. Is there any reason to think that this relationship will hold more generally? I think the answer is quite clearly "yes" based on the special character of causal contents. Nonreductive accounts of causation like our conserved quantity, mechanism, and manipulation theories all ascribe a rich set of conditions to causal claims. These conditions are included in the content of any causal representation of a target system. By definition, acausal representations abstract away from this content and stay neutral on the causal processes at work in its target systems. This means that acausal representations will typically impose less demanding conditions on its target systems than an associated causal representation. The qualification "typically" here is needed because the new features represented in the acausal representation might be very complex and hard to meet, so that they more than compensate for the removal of causal content. Still, I think our examples indicate how the new features of the acausal representation tend to be few in number and fairly simple. They do not rise to the level of complexity of the causal networks involved in causal representations.

If we accept this point about less content and the two associated principles from chapter 2, then it follows that an accurate prediction generated from an acausal representation will give it a bigger boost in confirmation than a similar prediction generated from a causal representation. That is, when a given acausal representation is accurate in some respect, and we exploit this part of the representation in generating an accurate prediction, and are subsequently able to experimentally check this prediction, there is a stronger boost in our confirmation of the representation and we are able to confidently add it to the store of our scientific beliefs. This is the simplest case of direct confirmation. Other cases of direct confirmation may involve other representations in generating the prediction or in linking the successful prediction back to the acausal representation. These links surely would dilute some of our confidence in the acausal representation for the usual reasons tied to confirmational holism. At the same time, there is every reason to think that acausal representations will outperform their causal cousins in the presence of these holistic partners.

A contribution to indirect confirmation arises when the acausal representation is inaccurate, and a failed prediction reveals this. In this case, the lack of complexity of the content of the acausal representation allows us to diagnose the inaccuracy much more easily than in the case of an associated causal representation. This is because a failure with a causal representation can be explained away as a result of complicated causal interactions or the dubious assumptions that were needed to connect the causal representation with an experimental outcome. By contrast, the failure of a prediction from an acausal representation is harder to explain away and will lead more directly to a rejection of the representation. The contribution to confirmation here is indirect, though significant. By allowing us to zero in more easily on which of our representations are inaccurate, we can more quickly move on to testing more viable representations. If we restricted ourselves to causal representations, this sort of progress would be much harder to achieve.

In the face of a failed prediction, it is tempting to look to the mathematics as a source of suggestions about how to revise the mathematics and so formulate or discover a potentially more successful representation. Here, talk of the "unreasonable effectiveness" of mathematics in scientific discovery becomes tempting. As I argue in chapter 8, I am not convinced that there is any systematic reason to try to discover new scientific representations using some kind of nonempirical reflection on the mathematics by itself. So, in suggesting an indirect contribution to confirmation here, I do not mean to suggest a mathematical contribution to the discovery of new scientific representations. All I am focusing on here is how mathematically formulated acausal representations can contribute to and enhance the ordinary scientific process of testing, confirmation, and refutation. Something more substantial will result in the next chapter, where we consider the benefits of having a family of variously interpreted mathematical representations. But again, the discoveries that arise from this family will not be motivated by mathematical considerations in any mysterious way.

3.5 BATTERMAN AND WILSON

In different ways, the work of Robert Batterman and Mark Wilson have also made connections between mathematics, a shift to acausal representations, and the success of science. A brief discussion of their respective projects rounds out our discussion of acausal representations and also signals some respects in which my proposal is different from theirs.

Batterman has often emphasized the scientific benefits of representing systems at a level that stops short of including all the complex details of each system (Batterman 2002b). If we assume that it is at this most fundamental level that we have the constituents responsible for the causal workings of each of the systems under consideration, then we can see how Batterman's focus would lead him quickly away from causal representations and toward what I have been calling acausal representations. He has preferred to put the benefits of this shift in terms of explanation, emphasizing especially that we can only explain a phenomenon that recurs across systems by representing what these systems have in common. Leaving the point about explanation for chapter 11, we can happily concede that there are many common features across systems that will be missed if we attempt to represent each system in all its fundamental details. Still, it is not immediately clear which shifts away from fundamental details involving microphysical constituents will lead to acausal representations. Our three preferred accounts of causation each allow for genuine causal relationships between entities that are themselves complex, as with a car in a traffic system. We can distinguish two shifts, then, from a fundamental microphysical causal representation. First, a scientist might adopt a nonfundamental but still essentially causal representation based on treating complex constituents of the system as causal relata. This perspective is often sufficient to represent what is common to a variety of systems of a given kind because, even though each system has its own microphysical causal details, each has the same causal features at the new level. Second, the shift might be a genuinely acausal representation in our sense along the lines of our steady-state case or the robustness analysis associated with Volterra's principle. Still, this shift need not go along with a shift away from the microphysical level of representing the system as we may have a steady-state representation of microphysical processes just as well as any other level of the system.

Typically, though, both shifts are made together. In a simple example of Batterman's, we consider the maximum or critical load P_c that a given strut can bear before buckling (Batterman 2002b, p. 12). The equation that represents this feature of a strut involves only a constant E associated with the material that makes up the strut, the length L of the strut, and a factor I corresponding to the cross-sectional area of the strut:

$$P_c = \pi^2 \frac{EI}{L^2} \qquad (3.11)$$

This is a genuinely acausal representation because it merely sets a threshold above which buckling will occur and does not represent the causal relations that would bring about the buckling. At the same time, the representation is not of the microphysical constitution of the materials making up the strut. Although the representation is informed by some understanding of the causal structure of struts, the actual causal network of the microphysical constitution of a given strut is not something we can easily determine experimentally. It is easier, then, to confirm that this acausal representation of the strut is correct than it is to formulate and support a causal representation of the features of the strut involved in load bearing.

Although there is a clear agreement between this point of Batterman's and my own discussion here, Batterman often emphasizes the special way in which the acausal representation is derived and places great weight on "asymptotic" representations that result from taking a given quantity to a limit like 0 or infinity: "the best way to think of the role of asymptotic methods in mathematical modeling is that they are a means for extracting *stable phenomenologies* from unknown and, perhaps, unknowable detailed theories" (Batterman 2002a, p. 35). Here we have the most important difference between Batterman's approach and my own. Based on the way I have explained the content of a representation, the steps that appear in the derivation of the representation involve extrinsic mathematics that need not figure in any way in the content of the resulting representation. So it turns out that the steps of the derivation are often irrelevant to the content of the acausal representation. It is true that I have conceded in chapter 2 that, from the perspective of interpretation, we can often only tell what the intended content of a representation is by seeing how a scientist derived it. Still, the assumptions involved in the derivation need not be part of the content of the representation, and we have a special reason to adopt this perspective when the derivations result from asymptotic limits.[13] By contrast, Batterman seems to trace the ability of the acausal representations to represent important features of systems precisely to the reliance on these limits, especially when those limits involve singularities: "in taking such limits we are often led to focus on mathematical singularities that can emerge in those limits" (Batterman 2010, p. 20). Although he

---

13. I return to this point in chapter 5 when I discuss scaling and idealizations.

claims in this paper only that such limits are just one way to isolate stable features, there is no broader examination of what other ways are open or what they might have in common. I think Batterman has focused too narrowly on some special cases here and so missed the more general category of acausal representations discussed in this chapter.

Although Mark Wilson's views receive extended treatment in chapter 13, it is worth noting now his repeated insistence on the need for techniques of what he calls "variable reduction" (Wilson 2006, p. 190). Such techniques arise from the daunting complexity of representations that result from an attempt to treat a system using a fully dynamic representation of the sort that I discussed with 3.2. For more complex systems, such as physical materials like iron bars, we cannot even represent the system in this way even if we have some kind of understanding of what the bar must be like at the molecular level. For Wilson, this sort of understanding must be supplemented by a different kind of representation of the complex system that we can actually formulate. We can actually implement experimental testing and other forms of empirical success such as engineering design for these nonfundamental representations while a focus on other kinds of fundamental representations is scientifically useless. This leads Wilson to develop an inventory of a number of ways these workable representations can be articulated. They typically involve aggregation or other kinds of parameters that we are aware involve an abstraction away from the causal workings of the system. The gain achieved by these is, as I have emphasized, a representation of new features of the system that would be otherwise missed.

Where Wilson goes beyond the points made so far is his emphasis on the costs imposed on the resulting representation due to the "descriptive policies" (Wilson 2006, p. 195) that the mathematics often brings with it. Among other things, the need for highly circumscribed and often highly artificial boundaries makes it difficult to extrapolate the success in representing one kind of complex system to other kinds of systems. This supports Wilson's lack of enthusiasm for traditional scientific realism because even when we confirm the accuracy of a given acausal representation, we may not have a good understanding of the relevant feature of the system that is responsible for our success or how it might change in different circumstances. The resulting conception of scientific representation and theory "facades" receive some discussion in chapter 13. For now it is enough to note that Wilson has also emphasized the need for acausal representations in science and their role in whatever success science has achieved to date.

4

Varying Interpretations

4.1 ABSTRACTION AS VARIATION

In chapter 1 two different notions of "abstract" were distinguished. There was first the notion of an abstract acausal representation that fails to include causal content in the conditions it places on its target systems. This was the focus of chapter 3. Now we turn to the second kind of abstract representation. These are the abstract varying representations that result from varying the interpretation of a given mathematical scientific representation. When different variations are considered, more or less extensive families of representations of more or less related target systems arise. An investigation of these families will show how the mathematics common to members of the family have important benefits for the success of science. When this process of abstraction and reinterpretation goes smoothly, we wind up with an accurate representation of each of the target systems as well as some of their commonalities. As we will see, the mathematical links between the members of the family aids in the process of confirming the accuracy of the members of the family. Even when the generation of accurate families fails, there can still be indirect confirmational benefits created by these mathematical links. This happens when failed predictions for one member of the family suggest what needs to be changed to restore accuracy. Here we find a point of contact between mathematical scientific representations and the discovery of new, accurate representations.

The example noted in chapter 1 of a family of abstract varying representations is the linear harmonic oscillator. We say a system is a linear harmonic oscillator when it has a distinguished equilibrium point and a linear restoring force. This means that we can summarize the forces active in such a system by

$$F = -kx \qquad (4.1)$$

where k is a constant and x represents the distance from equilibrium. Two clear examples of systems that can be accurately represented in this way are metal springs and pendulums. Here we have not only the trivial variation of considering many systems of a given physical type, like a collection of springs made of the same metal, but also a variation across springs of different kinds of metal or even other kinds of materials. The particular value of k will vary from spring to spring, and it is fair to see a certain kind of variation in interpretation here from metal to metal. A more significant variation in interpretation is needed to handle pendulums, for here x is taken to represent the *angular* displacement from equilibrium, and not a length as with the springs. The linear harmonic oscillator is a fairly simple representation, so there are not many components to vary. Still, it happens that a wide range of interpretations of k and x lead to a large family of representations whose accuracy has been experimentally tested within appropriate domains.

The first step in extending a fixed interpretation into a family of variously interpreted representations is to deinterpret a given component of the mathematics. This process need not deprive the representation of its causal content, and so the members of the resulting family may retain their causal character. This is the most natural way to view the harmonic oscillator example where we end up with a causal representation of the springs and the pendulums. As much as possible I want to set aside issues of causation in this chapter and focus on the families of variously interpreted representations in their own right. When a deinterpreted representation winds up being acausal in the sense of chapter 3, then I would expect to see all the benefits of that sort of representation. But it is important not to confuse the contributions of mathematics from acausal representations with what we get from varying interpretations. The benefits from varying interpretations turn on the mathematical relationships between representations and how these fit with the changing interpretation of the representations. By contrast, the contributions of acausal representation are due to the relatively simple contents that these representations have in their own right, at least when compared to their causal counterparts. Here the mathematical relationship is often much less clear as the mathematical assumptions needed to derive the acausal representation from the causal representation may be poorly understood. When the varying interpretation strategy is implemented, however, what results is typically a straightforward overlap of the mathematics. This does not mean, of course, that it is easy to tell how significant this overlap is for the proper interpretation of the mathematics.

The contrast between the two sorts of contributions comes out clearly if we consider the practice of robustness analysis as we discussed it in connection with Weisberg's example of the Lotka-Volterra equations and the Volterra principle. Robustness analysis considers a family of representations with a fixed interpretation where what changes across the family is simply the values of some parameters whose true value is unknown, or whose interpretive status is otherwise in doubt. This allows us to be more confident in the common claims entailed by all the members of the family. In our varying interpretation cases, we have a family of representations where each member of the family represents a different kind of system. The aim of investigating this family is not to extract some claim that they have in common. Among other things, there will typically not be any common claim about any of the target systems across the family. Instead, the common mathematical structure that is variously interpreted may reveal common features that these systems share that would be otherwise missed. Exactly how this contributes to the success of science will come through in the examples discussed and will be summarized at the end of the chapter.

A final benefit of our discussion of families of variously interpreted representations is that it highlights the contributions of the well-articulated theories presented in applied mathematics textbooks. Here theories are developed and results are proven with an eye toward the needs of scientists and engineers, and the emphasis is clearly on working out a "toolkit" of mathematical structures that are variously interpreted and applied throughout the sciences. We can consider, then, a completely deinterpreted mathematical structure and develop a theory of its features independently of any concerns about its actual application. This proves to be a powerful resource for the scientist who encounters a new problem. The success of drawing on a limited number of wholly mathematical models to understand completely different kinds of systems may seem initially mysterious, but we will see that it can be given a benign interpretation. To highlight the power of this abstract approach, I outline the basics of a theory of differential equations and explain what this theory contributes to the success of our families of variously interpreted representations. The chapter concludes with a discussion of the area of mathematics known as applied mathematics.

4.2 IRROTATIONAL FLUIDS AND ELECTROSTATICS

A much-noted case of varying interpretation of the same mathematics is the use of scalar potentials to represent vector fields. We think of a vector field for a given magnitude, like velocity or force, as an assignment of a vector to each point in space at a time. In fluid mechanics, for example, we treat the fluid as continuous and assign a velocity vector to each fluid element at each time. In the study of systems operating under gravitational or electromagnetic forces, we might begin by assigning vectors to each point in space. These vectors are meant to represent what accelerations a unit mass or charge would experience were it to be placed at that point. Take this initial vector field representation at face value; it is a mathematical theorem that whenever such a vector field F satisfies the equation $\nabla \times F = 0$, then there is a scalar field ϕ such that at each point $F = \nabla \phi$. ϕ here is typically called a potential (or scalar potential). It allows us to represent a given physical system in terms of the mathematically simpler assignment of scalars to each point at each time rather than the vector field representation. Part of the gain in simplicity results from moving from vectors to scalars, that is, individual numbers. But a deeper reason for the gains from working with ϕ rather than F is that there is some flexibility in fixing the particular values for ϕ. To see why, notice that for any constant C, $\nabla \phi = \nabla(\phi + C)$. We can transform ϕ by adding a constant, then, and not affect our claims about the associated vector fields. This "gauge freedom" allows one problem to be treated in one gauge and another problem to be treated in another gauge.

Our focus, however, is not just on why scientists aim to recast vector fields in terms of scalar potentials, but on how the variety of resulting interpretations contributes to the success of science. Although there is a temptation to see the shift as a mere mathematical trick, we will see that it highlights a great deal of scientific interest about the systems under discussion. The central issue for us, then, is what these mathematical similarities reveal about the physical systems. Two cases will occupy our attention: irrotational fluids and electrostatics. In both cases, we not only find a scalar potential ϕ to represent the vector field of interest but also show that ϕ satisfies what is known as Laplace's equation: $\nabla^2 \phi = 0$. As a result, the investigation of ϕ for both interpretations can be handled using a well-articulated mathematical theory of the functions that satisfy Laplace's equation. They are known as the harmonic functions.

The basic equations for fluid mechanics are the Navier-Stokes equations. These equations relate the velocity vector field to the pressure p at a point with reference to the density of the fluid ρ and additional forces such as gravity. A first simplification is to treat a fluid as *incompressible* and *homogenous*. That is, the density ρ is given as a constant that does not vary throughout the fluid. A second simplification assumes that the flow is *steady*, that is, none of the quantities is time-dependent. This is not realistic for the beginning of a fluid flow, but once the initial transient effects die down, many flows approximate this steady state. Finally, our third simplification is to ignore the viscosity μ of the fluid. This means that the fluid elements do not resist internal circulation. This greatly simplifies the Navier-Stokes equations because it allows us to drop certain terms with second-order partial derivatives. In chapter 5 we return to this particular simplification and its limitations as it highlights important issues about scale and a method in applied mathematics known as scaling.

Under our three assumptions, the Navier-Stokes equations reduce to the Euler equations for an ideal fluid. We restrict our discussion to two-dimensional fluid flow, that is, flows where the fluid can be divided into slices that do not interact. This is partly for reasons of presentation, but it is a surprising fact that some of the mathematical techniques that I describe cannot be directly extended to the three-dimensional case. In component form for two dimensions x and y, the Euler equations are

$$u \frac{\partial u}{\partial x} + v \frac{\partial u}{\partial y} = -\frac{1}{\rho} \frac{\partial p}{\partial x} \tag{4.2}$$

$$u \frac{\partial v}{\partial x} + v \frac{\partial v}{\partial y} = -\frac{1}{\rho} \frac{\partial p}{\partial y} \tag{4.3}$$

Here u is the x-component of the velocity and v is the y-component of the velocity. When the fluid is incompressible, we also obtain the following continuity equation:

$$\frac{\partial u}{\partial x} + \frac{\partial v}{\partial y} = 0 \tag{4.4}$$

Finally, the absence of viscosity entails that the fluid is *irrotational*, that is, that for velocity vector $\vec{v}$, $\nabla \times \vec{v} = 0$. Again, in components, this entails that

$$\frac{\partial v}{\partial x} - \frac{\partial u}{\partial y} = 0 \tag{4.5}$$

Equations 4.2–4.5 give us all we need to derive an instance of Laplace's equation and deploy the associated theory of harmonic functions.

The first step is to introduce the potential function ϕ that satisfies

$$u = \frac{\partial \phi}{\partial x}, \quad v = \frac{\partial \phi}{\partial y} \tag{4.6}$$

This is licensed by 4.5. Similar reasoning entails the existence of a streamline function ψ where

$$u = \frac{\partial \psi}{\partial y}, \quad v = -\frac{\partial \psi}{\partial x} \tag{4.7}$$

Combining these two sets of equations tells us that ϕ and ψ meet the Cauchy-Riemann conditions

$$\frac{\partial \phi}{\partial x} = \frac{\partial \psi}{\partial y}, \quad \frac{\partial \phi}{\partial y} = -\frac{\partial \psi}{\partial x} \tag{4.8}$$

and so also satisfy the Laplace equation. That is, both ϕ and ψ are harmonic functions: $\nabla^2 \phi = 0$, $\nabla^2 \psi = 0$.

Taking the partial derivative of the potential function ϕ in any direction gives the component of the velocity in that direction. It follows that lines of constant potential are lines along which the velocity component vanishes. Conversely, lines along which the streamline function ψ is constant correspond to the vanishing of the velocity component perpendicular to the streamline. Plotting such streamlines with the equipotential lines thus produces a grid or *flow net* of squares. Fluid elements travel along streamlines, and the spacing of the streamlines in such a grid provides information about velocity variations as the fluid flows, for example, around an obstacle. The crucial mathematical benefit of treating a given problem using ϕ and ψ rather than directly in terms of u and v is that 4.2 and 4.3 contain product terms like $u\frac{\partial u}{\partial x}$, which render the equations nonlinear and difficult to solve analytically. By contrast, the Laplace equation is linear and can often be solved uniquely once the right kind of boundary conditions are presented. Even this solution is not always trivial, and a standard textbook topic is the "inverse" method of presenting simple cases of ϕ and ψ that solve these equations and then combining these solutions to represent more interesting kinds of fluid flows. The challenge is then to interpret the results so that natural boundary conditions can be seen to be associated with the solution. This process works because the equations are linear, and so any superposition of solutions will again be a solution to the equations.[1]

To see the power of this inverse method, consider the result of superposing two simple cases: the cases of uniform flow and a fluid source. In studying the ϕ and ψ for these cases and their combination, we bring in one last mathematical theory: complex analysis or the theory of functions with a complex variable. Recall that a complex number z can be thought of as $a + ib$, where a and b are real numbers and

---

1. There are additional complications associated with the determination of pressure. I skip these here for the purposes of exposition. See Kundu and Cohen (2008, p. 169) for further discussion.

$i^2 = -1$. Similarly, a complex function w with variable $z = x + iy$ can be thought of as $f + ig$, where f and g are real-valued functions of both x and y. At the heart of complex analysis is the study of a special class of functions known as the *analytic functions*. They can be identified in several different ways, but for our purposes the nicest feature of a function that is analytic in a given region is that it satisfies 4.8 in that region.[2] This means that we can take the ϕ and ψ for a given flow and represent it using a function known as the *complex potential*:

$$w = \phi + i\psi \quad (4.9)$$

Conversely, any analytic function for a region corresponds to a instance of ϕ and ψ, and so can be analyzed as a case of irrotational fluid flow.[3]

Our first case, then, is the case of uniform flow, from left to right in the xy-plane. Here the velocity vectors all have the same magnitude U and point in the direction parallel to the x-axis. The complex potential has the form

$$w = Uz = Ux + iUy \quad (4.10)$$

That is, $\phi = Ux$ and $\psi = Uy$.[4] As our second case, consider the complex potential

$$w = \frac{m}{2\pi} \ln z \quad (4.11)$$

Any point in the complex plane $a + ib$ can be represented in polar coordinates where r gives the distance of the point from the origin and θ gives the angle from the x-axis. The change of coordinates is carried out by $r = \sqrt{a^2 + b^2}$ and $\theta = \tan^{-1}(b/a)$. Any complex variable z can thus be expressed as $re^{i\theta}$.[5] So for the case of 4.11 we can express ϕ and ψ in polar coordinates as[6]

$$\phi = \frac{m}{2\pi} \ln r, \quad \psi = \frac{m}{2\pi} \theta \quad (4.12)$$

Taking the gradient along the radial direction r and along the angular direction θ yields

$$u_r = \frac{m}{2\pi r}, \quad u_\theta = 0 \quad (4.13)$$

---

2. The function has a singular point just where it fails to satisfy 4.8.

3. This does not mean that the mathematical representation will have a natural interpretation in terms of a physically possible fluid flow.

4. To see why, notice that $u = \frac{\partial \phi}{\partial x} = U$ and $v = \frac{\partial \phi}{\partial y} = 0$ at any point.

5. The surprising role of e here is discussed in some detail in chapter 13.

6. For ψ the crucial equation is that $\ln(e^q) = q$.

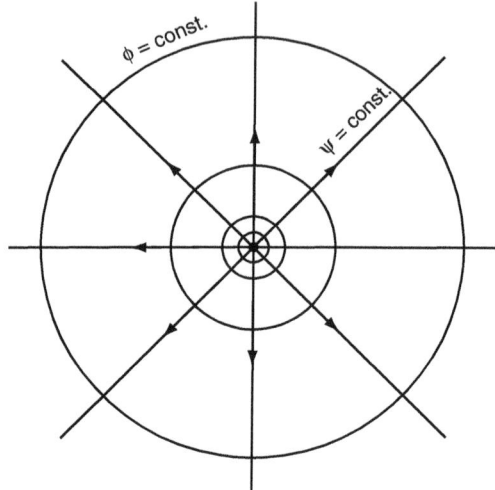

Figure 4.1: Source

Source: Kundu and Cohen (2008, p. 173)

That is, if m is positive, the fluid flows in straight lines outward from the origin at a velocity that increases as we approach the origin. See figure 4.1. Here, then, is our first instance of the inverse method. A complex potential is investigated and determined to correspond to an important case of fluid flow known as a *source*. If m is negative, then the fluid flows in the opposite direction into the origin, and we have a *sink*.

As noted earlier, the solutions to Laplace's equation can be combined to yield new solutions, as the equation is linear. We can add the two complex potentials for the uniform flow and the source to yield

$$w = Uz + \frac{m}{2\pi} \ln z \qquad (4.14)$$

Focusing on ψ, the streamline function, we see that it corresponds to the imaginary part of w, and so is[7]

$$\psi = Ur \sin \theta + \frac{m}{2\pi} \theta \qquad (4.15)$$

The streamlines, or lines of constant ψ, are given in figure 4.2. Outside the line $\psi = m/2$ we can see lines corresponding to a flow around a symmetric object that extends to the right indefinitely. The streamlines form boundaries that a fluid element cannot cross, so as they come closer above and below the object, the velocity of the fluid elements must be increasing. This corresponds to a decrease in pressure both above and below the object. Proceeding by superposition of previous solutions, then, we

---

7. For Uz, the key identity is $z = re^{i\theta} = r(\cos\theta + i\sin\theta)$.

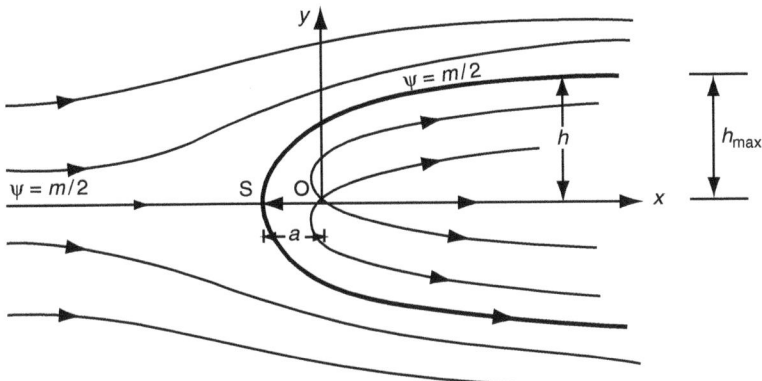

Figure 4.2: Flow Around an Object
Source: Kundu and Cohen (2008, p. 177)

can obtain representations of new kinds of fluid flow and directly understand their features.

Precisely the same mathematical framework can be deployed in electrostatics. Here the physical magnitudes are completely different, but we can use a scalar potential that satisfies Laplace's equation to erect a parallel mathematical representation. The crucial difference is that now the scalar potential is used to represent a force field, not a velocity field. As with the fluid mechanics case, we can think of electrostatics as a special case of the more general equations for classical electrodynamics, that is, Maxwell's equations. In particular, we assume a region with no charged particles in it, but where charged particles are held in place at the boundary. This means that the following two equations hold for the electric field E inside this region:

$$\nabla \cdot E = 0 \quad (4.16)$$

$$\nabla \times E = 0 \quad (4.17)$$

E is a vector field that is meant to represent the force that a positively charged particle would undergo if placed at a given point in space. As with the fluid case, one of our equations (4.17) guarantees that there is a scalar potential ϕ such that $E = -\nabla\phi$. Here a negative sign is applied to the result of taking the gradient, unlike in the fluid case. This is a small difference associated with the goal of recovering the force that a positive charge would undergo. It is important to track such differences, though, so as not to mistakenly interpret the similarities between the fluid case and the electrostatic case. It highlights that not only have we shifted from a velocity field to a force field, we are also now dealing with a case where positive and negative charges respond differently to the field. There is, of course, no analogous feature in the fluid case.

The key mathematical similarity is the satisfaction of Laplace's equation $\nabla^2\phi = 0$. Here this result follows from simply substituting $E = -\nabla\phi$ into 4.16. Based on the result that the scalar potential for our electrostatic case is a harmonic function, we can again deploy the notion of a complex potential $w = \phi + i\psi$. That is, each coherent electrostatic case will correspond to a complex potential, and often we can move in the other direction from a given complex potential to a coherent electrostatic system. The streamline function ψ deployed in the fluid mechanics case of course now has a different interpretation. Lines of constant ψ now correspond to what are called *flux lines*. These are the lines along which the force is directed for a positive charge. As before, two solutions for ϕ can be added together to yield a new solution.

Our simplest fluid case of uniform flow transfers directly to a simple electrostatic case. Here $w = Uz$, so $\phi = Ux$. This means that the E field has a constant value of $-U$ at each point, pointing in the x-direction. That is, a positive charge would experience a constant repulsive force directing it in the negative x-direction. Similarly, direct interpretations can be given for the source example discussed for fluids. As we saw, when the complex potential is given by 4.11, then the gradient in the radial direction is $m/2\pi r$. For the electrostatic case, this means that the E field points outward from the origin (when $m < 0$), and decreases in strength as we get further from the origin. In both cases it is not immediately clear what arrangement of charges would have to be in place for such an E field to arise. Additional constraints here come from our theory of the origin of the E field and its relation to charges. Similarly, we have ignored questions about how a particular fluid flow could be generated, or when it might be inconsistent with the properties of the fluid in question. The same point extends to two other cases where the existence of harmonic functions satisfying Laplace's equation make the apparatus of scalar and complex potentials applicable: heat conduction and gravitation.[8] In all four cases it is not as if the mathematical similarities solve all our problems so that a good understand of irrotational fluid flow immediately gives us insight into heat conduction or electrostatics. In fact, as we discuss briefly in chapter 7, we can blame the mistaken caloric theory of heat, at least in part, on the mathematical similarities between some kinds of fluid flow and the effects of heat.

This point of caution should not diminish the striking similarities revealed by our discussion of harmonic functions and Laplace's equation. In each application, the existence of harmonic functions in a domain opens the door to the rich results of potential theory. An ability to vary the interpretation of this or that mathematics, then, can greatly increase our understanding of a family of physical systems. After presenting one more, slightly more interesting case of varying interpretation, we turn to the benefits of this sort of situation for the success of science.

4.3 SHOCK WAVES

Recall the discussion in section 3.3 of the steady-state traffic representation. Here we assume that all the cars are equally spaced and are each traveling at the same velocity. A further assumption is that the velocity v is solely a function of the density $\rho(x, t)$.

---

8. See Brown and Churchill (1996, ch. 10).

Varying Interpretations

This allowed us to define a flux function $j(x, t) = v(\rho)\rho(x, t)$. Now we extend that representation by assuming that the material in question is conserved and making various continuity assumptions. This allows us to derive the conservation law

$$\rho_t + j_x = 0 \qquad (4.18)$$

Here subscripts indicate partial differentiation. But j is solely a function of ρ, so we obtain

$$\rho_t + j'(\rho)\rho_x = 0 \qquad (4.19)$$

The prime indicates differentiation with respect to ρ. Assuming that the initial density ρ_0 is given, we have what is called an initial value problem.

A quick note on the difference between 4.19 and our earlier example of Laplace's equation is in order. Both are partial differential equations, but our traffic equation explicitly invokes time, whereas for Laplace's equation, the differentiation involved only the two spatial dimensions. This corresponds to the way problems are typically posed for the two equations. For Laplace's equation, it is most natural to consider a problem where the boundary conditions are given and then work out what steady conditions meet those boundary conditions. Even when we adopted the inverse method, all we did was start with the steady state of the system and then worked out or conjectured which boundary conditions would be consistent with that state. Now, with the explicit time dependence of 4.19, it is more natural to start from some specified initial conditions and try to determine how the system will evolve over time.

One technique for approaching an initial value problem for an equation like 4.19 is known as the method of characteristics. Essentially, we start by assuming that a unique solution to the problem exists and isolate it using ρ_0. Applying the method delivers curves, known as the characteristic base curves, along which ρ is constant. The arrangement of these curves tells us whether there really is a unique solution. When this assumption breaks down, the base curves can still provide useful information on the evolution of the system. For example, they can be used to isolate where discontinuities in ρ known as shock waves develop. We consider a simple example of such an analysis so the insight it provides into such systems becomes clearer.

We start by making $j(\rho) = 4\rho(2 - \rho)$ and imagine the initial density ρ_0 to be 1 for all $x \leq 1$, $1/2$ for $1 < x \leq 3$, and $3/2$ for $x > 3$. In the traffic case, this corresponds to the maximal density being 2, so to the left of $x = 1$ we have the cars spaced with one car length between them, and in the adjacent domain they are less densely spaced. We would expect a rightward flow of cars then, increasing the density. To say something more precise, we find the characteristic base curves along which the density remains constant as time increases. Their slope turns out to be

$$0 \text{ if } x \leq 1$$
$$4 \text{ if } 1 < x \leq 3$$
$$-4 \text{ if } x > 3$$

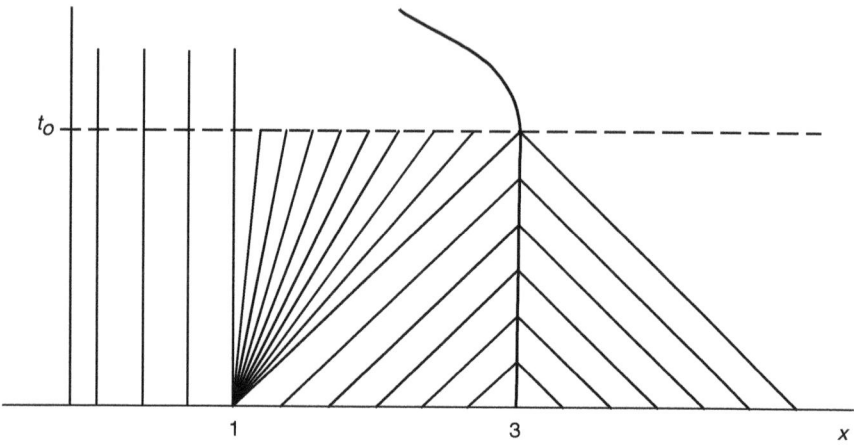

Figure 4.3: Characteristic Base Curves (Illner et. al. 2005, p. 187)

The details of this derivation are in Appendix A. If we plot a representative sample of these base curves emanating from the x-axis, as in figure 4.3, we notice two things. First, there is a gap at $x = 1$ where our method has given us no information about how the density will change as time increases. Second, there is a line coming from $x = 3$ where the base curves from two initial domains intersect, seemingly indicating that there will be two different densities at the same location. The former defect can be remedied in a physically intuitive way by smoothing out the transition in the density at $x = 1$ from 1 to $1/2$. This produces a rarefaction wave where a fan of lines of gradually increasing density are inserted into our original diagram. The second discontinuity is more serious, however, as it shows the existence of a shock wave where our representation of the changes in density over time breaks down. The representation can still be useful as we can trace how the density will develop on both sides of the shock and how the shock will propagate over time, for example, how it will be affected by the development of the rarefaction wave.[9]

From our traffic perspective, we can consider what a driver would experience as she crossed the rarefaction wave and then the shock. At the beginning of the rarefaction wave the density decreases, that is, the cars suddenly have more space between them. This leads the driver to accelerate. This density levels out, so the driver then assumes a steady speed. However, as she enters the left side of the shock, there is a dramatic increase in density. This requires sudden deceleration, presumably to a zero velocity associated with the maximal density. As she emerges from the shock regime on the right side, she finds a density below the maximal density, allowing her to accelerate. Her velocity after the shock is less than her original velocity as the density has increased. We see, then, that whatever its artificial features, our simple

---

9. An explicit solution for this system is given at Illner, et. al. (2005, p. 189).

Varying Interpretations

example allows us to reproduce something that drivers are all too familiar with: a traffic jam produced by the volume of cars on the road.[10]

Here the core mathematical notion is the discontinuous change in density that we have a labeled a shock wave. Our informal argument explained how the proper interpretation of this piece of mathematics in the traffic case is the existence of a traffic jam. Now we turn to the deployment of the same mathematical notion in a dramatically different context, namely, fluid flow. To make the comparison with the traffic case as straightforward as possible, we imagine a fluid that is restricted to a thin pipe that is oriented in the x-direction. To represent the formation of shocks we must make some changes from the fluid case (4.2–4.5). The most important shift is to introduce the density ρ as a variable. We continue to think of the fluid as inviscid and also ignore questions of temperature variation. But now we think of the pressure p as a function of the density ρ. This allows us to see the flow in the pipe as governed by the equations:

$$\frac{\partial u}{\partial t} + u\frac{\partial u}{\partial x} = -\frac{1}{\rho}\frac{\partial p}{\partial x} \qquad (4.20)$$

$$\frac{\partial \rho}{\partial t} + u\frac{\partial \rho}{\partial x} + \rho\frac{\partial u}{\partial x} = 0 \qquad (4.21)$$

As the flow is not steady, some new ∂t terms appear. Additional complexity arises from the relationship between pressure p and density ρ.[11]

The engine driving the development of a shock wave in such a fluid is the interaction between the pressure, density, and speed at which a disturbance from an initial state propagates through the fluid. It turns out that we can capture this speed c using the ratio of pressure to density: $c^2 = \frac{\gamma p}{\rho}$.[12] γ is a constant that varies from fluid to fluid. If we think of a fluid like air at some initial pressure p_0 and density ρ_0, then $c_0 = \sqrt{\frac{\gamma p_0}{\rho_0}}$ is the speed at which a disturbed pressure and density wave would travel down the pipe. It is thus labeled the *speed of sound* for the air in that state because sound waves are instances of this kind of disturbance. Imagine, then, an initial density distribution that is analogous to our traffic case: 2 for all $x \leq 1$, 1 for $x > 1$. The air in the more dense region will expand to the right with some speed c_1. But this compressed air will have a higher local speed than the air just to the right of $x = 1$. As the air accumulates in the region around $x = 1$, more air comes in than goes out. This further compresses the air, making the local speed even higher. What develops, then, is a discontinuous front where the density drops dramatically.

---

10. See Glaskin (2008) for the claim that shock waves have been experimentally observed in traffic systems, based on Sugiyama et al. (2008). The models employed there are different from the simple representation presented here.

11. For more discussion see Lin and Segel (1988, §15.3).

12. We return to this case in chapter 7, as the confirmation of this equation was taken by many to confirm the caloric theory of heat.

A shock wave appears in the compressible fluid case because the velocity of the disturbance is affected by the compression of the air.

Lin and Segel note that it remains "a remarkable fact" (Lin and Segel 1988, p. 549) that this discontinuity can be handled by adopting the policy of refraining from using this representation to understand what is going on within the shock wave while still taking seriously what is represented in the other regions. Conservation relations between the two sides of the shock tell us, for example, what is happening on either side of the shock, and in certain circumstances we can plot where the shock will travel over time. As Wilson has put it when discussing this sort of case, also with reference to the analogy with traffic jams, "if we can examine a situation from several sides and discern that some catastrophe is certain to occur, we needn't describe the complete details of that calamity in order to predict when it will occur and what its likely aftermath might be" (Wilson 2006, p. 188). The procedure of selectively interpreting such representations is not some strange abberation associated with some peculiar mathematics. To the contrary, it often provides our only means of representing phenomena of widespread scientific significance.

4.4 THE VALUE OF VARYING INTERPRETATIONS

Based on our two extended examples, we can classify cases of varying interpretation of a piece of mathematics based on the answers to two questions. First, to what degree has the physical interpretation of the same bit of mathematics been changed? At one end of the spectrum are cases where we remain squarely within the same kind of physical system, but simply shift to another system of that kind. For example, we move from a given copper spring to another copper spring. At the other end of this spectrum are cases where a completely different physical magnitude is associated with the mathematics. This is what occurred in both of our examples. The scalar potential was used to represent velocities for fluids and forces for the electrostatic systems. Similarly, the density for the traffic case was a density of cars while for the compressible fluid it was a density of the fluid. The second question concerns the degree to which the intrinsic mathematics of the members of the family of representations is genuinely shared. With the linear harmonic oscillator, there is a nearly total overlap between the mathematics intrinsic to the representation of the spring and the pendulum. But with the irrotational fluids and electrostatic case, the common mathematics is much more limited. We have isolated the existence of a scalar potential and argued that for the cases under discussion this potential satisfies Laplace's equation. But the idealizations and simplifications needed to find this overlap were fairly severe. As these simplifications are relaxed, and we shift to more realistic representations of fluids or charged particles, there is no guarantee that the mathematics will remain shared. A similar case of selective matching of mathematics can be observed in the shock wave examples. Here the basic equations of the representation are quite different, although they turn out to have a high-level property in common that is responsible for their production of shock waves. Similarly, the appearance of the discontinuous change in density unites the two representations, but our deeper understanding of what occurs in the shock and around it is different.

For example, heat effects would become important in representing the fluid in the shock, whereas this is not relevant to understanding the experience of traveling through a shock in a traffic jam.

Still, I want to argue that these differences do not undermine some scientific benefits that arise for a scientist who is mathematically sophisticated enough to notice cases where the same piece of mathematics is being variously interpreted. On a first pass, we can use the notion from chapter 1 of a flow of confirmation within such a family of representations. This flow is mediated by the mathematical similarities between the representations, so here we have a new kind of epistemic contribution from mathematics over and above what we saw in chapter 3. To see how this can work in a best-case scenario, imagine a scientist who knows the mathematics that we have reviewed for Laplace's equation, has extensive experimental confirmation of its application for the case of irrotational fluids, and has tentatively adopted the hypothesis that Laplace's equation is an accurate representation of the scalar potential for the electric field for electrostatic systems. We have, then, a two-member family of variously interpreted mathematical scientific representations, where one member has accrued a high degree of confirmation through ordinary scientific testing.

The key claim I want to defend for such a case is that a small amount of experimental testing of the electrostatic representation should give it a larger boost in confirmation than if that representation was not mathematically linked to the successful irrotational fluid representation. This is because the independent confirmation of the way the mathematics is deployed for the fluids gives the scientist a template against which to judge the success of the electrostatic representation. Initially when using the notion of a complex potential and harmonic functions in the fluid case, the scientist would reasonably have some doubts about the appropriateness of using mathematics this way. She may worry that the mathematics will lead to absurd predictions or predictions that fail to track any genuine features of the fluids under investigation. This is not so much a concern about the mathematics itself, which I am taking for granted is confirmed independently by purely mathematical standards. The scientist who knows the mathematics still has a right to wonder if it is being deployed coherently for this kind of physical system. As we saw, the mathematics involves a host of nonempirical aspects of the fluid and imports additional mathematical structure, which seems to have no natural physical interpretation. In particular, the entire apparatus of complex analysis seems ill-suited to the fluid because its two-dimensional character is not easily assimilated to the intricate structure of the two-dimensional complex plane. I claim that these doubts should be allayed by the success of the irrotational fluid representation at prediction and description of actual fluid flows. This success gives the scientist a reason to expect that a limited success with deploying the same mathematical apparatus in the electrostatic case will continue. Analogous doubts about the appropriateness of the mathematics for the electrostatic case should be more quickly put to rest based on the success with treating irrotational fluids.

Suppose that everything proceeds as I have described and as a result the scientist ends up with two highly confirmed mathematical scientific representations that are linked by some shared mathematical features. Are there any additional epistemic benefits to having this sort of family for the ongoing success of the scientific enterprise? Humphreys's discussion of analog computers shows one important argument for answering this question positively. As he puts it, "the essence of an analog computer is to set up a physical system whose behavior satisfies the same equations as those which need to be solved. By adjusting the input to this system, and measuring the output, the latter will give the solution to the desired equation(s)" (Humphreys 2004, p. 126). Further emphasizing the fact that "a small number of equation types covers a vast number of applications" (Humphreys 2004, p. 126), Humphreys explains how electrical circuits can be constructed and investigated to determine the solution to problems involving oscillators that are more complicated than the simple linear harmonic oscillator. This is really just a special case of the more general epistemic benefits that arise from a family of a variously interpreted mathematical representations. If we are confident in the accuracy of both representations, we can use experiments with one kind of system to justify claims about the other kind of system. When the elements of one system are much easier to work with, the benefits of this kind of transfer are considerable.

Indeed, Humphreys's discussion of analog computers is just an instance of his more general investigation of the scientific significance of computer simulations. Although analog computers are relatively difficult to construct and evaluate, when we make the shift to a digital computer with a suitable suite of software, the cost of developing a system that will track the mathematics of a given physical system is minimal. We can understand the epistemic benefits of computer simulation in terms of a two-stage transfer of confirmation. First, there must be some independent experimental confirmation of the representation of the physical system. Based on this confidence, we can then construct a computer simulation that shares the relevant mathematical features with our representation of the physical system. So, the first step is to transfer our confidence in the coherence of the physical representation over to the computer simulation. It is not always trivial to ensure that this has been done correctly, especially when computational or programming limitations force adjustments. Second, we can run the simulation for a new instance of the physical system under investigation. Here, based on our confidence in the mathematical similarity and the reliability of the computer itself, we can generate a predication that we believe to be a genuine consequence of the simulation. Again, breakdowns can occur due to hardware failure or software bugs. But if we have taken reasonable steps to ensure that these did not occur, we can transfer the conclusion from the simulation back to the physical system.

4.5 VARYING INTERPRETATIONS AND DISCOVERY

Our discussion so far has focused squarely on the confirmation of already existing representations. As a result, I have not said anything about how the existence of mathematical similarities contributes to the formulation or discovery of new scientific representations. This is the main focus of chapter 8. For now, it suffices to say that the epistemic benefits of having families of variously interpreted representations can help us understand why they are desirable. That is, a scientist might hope to formulate or discover a mathematically linked series of representations of different physical systems because if she found such a family, then this would contribute significantly to the success of those representations. This point is reinforced once we grant the difficulty of actually working with mathematical scientific representations. Extracting solutions to systems of equations is not a trivial task, so it is more or less a waste of time to formulate such systems if the scientist knows of no way to solve them or at least approximate a solution. This implies that it is a sensible heuristic on scientific discovery for a scientist, other things being equal, to aim to deploy a mathematical theory that is well understood and has a track record of success.

This point does not entail that there is any good reason, prior to experimentation, to expect physical systems made up of different things to be accurately represented by similar mathematical structures. Instead, both cases we have discussed indicate how rare and exceptional even a partial overlap is. To return to the shock wave example, there are any number of disanalogies between the traffic system and the compressible fluid system that have been ignored in an attempt to highlight the salient mathematical similarity. It is not too hard to imagine the genesis of an inaccurate representation of a traffic system based on an attempt to transfer some of the success from the fluid case to a new domain. Suppose, for example, that a scientist knows all the relevant mathematics and has obtained some empirical success treating compressible fluids along the lines discussed. In particular, she has learned how to handle the genesis and development of shock waves in her representations and in her fluids, and as a result has made a number of successful predictions. It is tempting, then, to take the entire mathematical representation of shock waves in fluids and transfer it over to the traffic case based on the following two similarities. First, traffic systems (like compressible fluids) have varying density. Second, both fluids and traffic systems are constituted out of individual objects that it seems can be fruitfully ignored in an idealization as a continuous medium. The scientist might try to use her understanding of the role of variation in the speed of sound to somehow represent how traffic shock waves develop. Based on our discussion, we can see that this is a mistake because the underlying reason for the evolution of the traffic system has nothing to do with this. Instead, it is because we assumed that velocity was a certain kind of function of density that we were able to derive 4.19. The velocity here was the velocity of the cars, not anything like the velocity of a disturbance in pressure and density. We see, then, how an error could be committed by this kind of erroneous transfer of mathematics. It is a delicate matter to decide which mathematics to transfer over to a new case and how its interpretation should be adjusted based on the new subject matter.

Such a turn of events can still lead to a case of indirect confirmation, though, when the flawed representation is experimentally tested and found to be in error. Recall from the last chapter that indirect confirmation occurs when an inaccurate representation is discovered to be inaccurate. Although there is nothing about families of variously interpreted representations per se that makes them any easier to test, I want to argue that when a representation in the family is found to be inaccurate, the scientist has more information about how to proceed than if the representation was not in the family. Essentially, the family of mathematically related representations gives the scientist a framework within which to probe for the source of the inaccuracies. Two possibilities are especially likely. First, each member of the family may be inaccurate because there is some genuine underlying mathematical similarity to all the systems, but the current family of representations has missed this in some respects. A scientist can determine whether this is the case by testing other members of the family for the features that were determined to be inaccurate for one member. If all the members are found to be inaccurate in the same way, then the scientist knows where to adjust her family of representations. A new term may be needed for the equations, for example, or perhaps a magnitude that was treated as a constant should be treated as a variable. The second kind of scenario is what I had in mind when describing the overextension of the fluid representation to the traffic case. This is where one member of the representation is accurate, but a second representation is inaccurate in some respects, even though it does ultimately agree with the original representation in other respects. In this case, the failure of the second representation can help pinpoint precisely how the second kind of system differs from the first kind of system. As a result, a failure of a prediction can alert the scientist to some new and perhaps unappreciated features of the second system, here the traffic system. She might then investigate what is responsible for the difference between the two systems. This is a limited tool for discovering new accurate representations based on the partial failure of a mathematical analogy. As Lin and Segel put it when discussing the failure of a model to recover a prediction, "disagreement means that there is more to the phenomenon than meets the eye" (Lin and Segel 1988, p. 19). When this disagreement occurs in the context of a family of successful variously interpreted representations, the scientist can exploit her understanding of the mathematics to help formulate new proposed representations that have a better chance of being accurate.

In summary, then, the epistemic benefits of varying interpretations of some mathematical scientific representations fall into both the direct and indirect categories. When the members of the family turn out to be accurate, the mathematical similarities make it easier to confirm the accuracy of the representations. Even when some parts of the reinterpretation wind up being inaccurate, the scientist can use the mathematical links to more easily diagnose the source of the failure and even to help in the quest for the formulation of a more accurate representation of the system in question.

4.6 THE TOOLKIT OF APPLIED MATHEMATICS

The main theme of chapter 3 is the need to appreciate the importance of acausal representations for the success of science and the corresponding contribution of mathematics to the direct and indirect confirmation of such representations. A central target of that discussion is philosophers who overemphasize the importance of causal representations or who make the mistake of thinking that causal representations are the only way for science to successfully represent a physical system. Developing the main points of chapter 3, then, did not require much mathematical detail or appreciation of the subtleties of the mathematics involved. By contrast, the point of this chapter has been to highlight the benefits of a thorough understanding of the mathematical relationships between representations of different systems. This has required some engagement with the details of how mathematics is actually applied, at least for some celebrated success cases. If we continued along these lines, then our discussion would come to resemble a textbook in applied mathematics. Here the emphasis is on studying types of mathematical representations in their own right, with a consequent stress on mathematical results and their implications. Individual examples are introduced, but the point remains that the student should develop a toolkit of well-understood mathematical structures with which she can approach a novel scientific challenge.

As an example of this, we can briefly indicate how our two main examples could be extended and integrated into a more systematic branch of applied mathematics, namely, the theory of partial differential equations. Laplace's equation $\nabla^2 \phi = 0$ and its close cousin, Poisson's equation $\nabla^2 \phi = k$, for k some constant, can be studied in their own right, and systematic results can be obtained for when such equations have a unique solution for a given kind of boundary condition. But these equations, in turn, can be assimilated to a much more general category of *elliptic* partial differential equations that share many important mathematical features. Crucially, solutions to such equations never develop singularities within their domain. This tells us right away that the sorts of equations like 4.19 and what we used for compressible fluids must fall into a different type because these equations permit the formation of shock waves, which are just one important kind of singularity or discontinuity. It turns out that such equations, like the widespread wave equation $u_{tt} = c^2 u_{xx}$, are classified as *hyperbolic*. The theory of elliptic and hyperbolic equations starts from a fairly abstract characterization of the difference between these kinds of equations and then goes on to explain how these differences lead to different kinds of solutions. Equally important, the techniques that can be used to obtain these solutions are quite different, as are the conditions under which unique solutions to these equations exist.[13]

We can now provide a preliminary answer concerning the benefits that this body of mathematical knowledge has for the scientist. It allows a scientist to work with families of variously interpreted mathematical scientific representations. Taken to its most abstract level, these families are united by nothing more than their abstract mathematical features, for example, they satisfy Laplace's equation. But an understanding of the mathematical theory of such features gives the scientist a body of resources that can be deployed in the investigation of scientific representations. A vast array of conditional knowledge is thus made available that would be absent without a knowledge of the mathematics. This knowledge is so important that it is often accumulated by a specialist within the scientific community, namely, the applied mathematician. Understanding the success of science, then, requires us to grant this specialist a crucial role in the scientific process, alongside other specialists like the theoretician, the experimentalist, and the engineer.

To understand more what applied mathematicians contribute to the scientific process, we briefly consider two introductory surveys of applied mathematics. These collections are most useful for our purposes because presumably here students can find out what they are getting themselves into if they continue to pursue

---

13. For an overview, see Pinchover and Rubinstein (2005).

applied mathematics as a specialization. Volume 1 of the SIAM series "Classics in Applied Mathematics", for example, is Lin and Segel's *Mathematics Applied to Deterministic Problems in the Natural Sciences*. The book "is concerned with the construction, analysis, and interpretation of mathematical models that shed light on significant problems in the natural sciences" (Lin and Segel 1988, p. vii). Part A of the book discusses "Deterministic Systems and Ordinary Differential Equations," "Random Processes and Partial Differential Equations," and "Superposition, Heat Flow, and Fourier Analysis." Part B continues with several chapters under the heading of "Some Fundamental Procedures Illustrated on Ordinary Differential Equations." These procedures include dimensional analysis, perturbation theory, and stability analysis. Finally, there is part C, called "Introduction to Theories of Continuous Fields," with discussions of the wave equation, potential theory, and various kinds of fluid flow (inviscid, viscous, compressible). A similar approach is taken in Fowler's (1997) *Mathematical Models in the Applied Sciences*, volume 17 in the Cambridge "Texts in Applied Mathematics" series. Here the first four chapters set out methods such as nondimensionalization, asymptotics, and perturbation theory. These chapters are followed by four on "classical models," eight on "continuum models," and four on "advanced models." Each model discussed is of a specific physical system, ranging from "heat transfer" and "electromagnetism" through "spruce budworm infestations" and "frost heave in freezing soils."

The organization of both volumes suggests that applied mathematics is thought of primarily as the development of a limited number of mathematical theories and techniques with an eye on formulating and analyzing acceptable mathematical models of specific sorts of physical systems. This proposal finds some confirmation in the programmatic remarks that begin each volume. Lin and Segel say:

> The purpose of applied mathematics is to elucidate scientific concepts and describe scientific phenomena through the use of mathematics, and to stimulate the development of new mathematics through such studies. The process of using mathematics for increasing scientific understanding can be conveniently divided into the following three steps: (i) The formulation of the scientific problem in mathematical terms. (ii) The solution of the mathematical problems thus created. (iii) The interpretation of the solution and its empirical verification in scientific terms. (Lin and Segel 1988, p. 5)

Later, they add "(iv) The generation of scientifically relevant new mathematics through creation, generalization, abstraction, and axiomatic formulation" (Lin and Segel 1988, p. 7). In a similar way, Fowler divides the practice of applied mathematics into six stages: problem identification, model formulation, reduction, analysis, computation, and model validation. For good measure, he adds,

> Applied mathematicians have a procedure, almost a philosophy, that they apply when building models. First, there is the phenomenon of interest that one wants to describe or, more importantly, explain. Observations of the phenomenon lead, sometimes after a great deal of effort, to a hypothetical mechanism that can explain the phenomenon. The purpose of a model is then to formulate

a description of the mechanism in quantitative terms, and the analysis of the resulting model leads to results that can be tested against the observations. Ideally, the model also leads to predictions which, if verified, lend authenticity to the model. (Fowler 1997, p. 3)[14]

Both Lin and Segel's and Fowler's programmatic remarks fit with the details of their later discussions. In each case, it is not sufficient just to learn the mathematics. An applied mathematician, we are told, must learn how to judge which areas of mathematics are likely to be useful in formulating a given scientific problem as well as how to interpret the results of their analysis in physical terms. The remaining books in these series seem to fit this framework as they are either further elaborations of the mathematical theories introduced in the survey textbooks or else detailed investigations of particular kinds of mathematical models of physical systems.

If we take these textbooks to be accurate indications of the difference between pure and applied mathematics, then we can draw the following sociological conclusions. Although both pure and applied mathematicians are concerned with the development of mathematical theories, applied mathematicians have different priorities and tests for the adequacy of their work. An applied mathematician aims to have her mathematics available to provide mathematical formulations for scientifically motivated problems. Furthermore, even if her mathematics succeeds in the formulation task, it will only be judged adequate if she can also solve or otherwise analyze the resulting mathematical problem. An even more challenging requirement is that she must remain available to assist the scientist in interpreting her solutions to mathematical problems. That is, it must be possible to assess the links between the simplified or analyzed mathematical model and whatever experimental data can be collected.

In contrast, the task of the pure mathematician seems different. The tests for adequate pure mathematics are more internal to pure mathematical practice. The pure mathematician either takes on a problem that has been previously identified by mathematicians as mathematically important, or formulates new conjectures and problems of intrinsic mathematical interest. What is expected seems more fixed and amenable to sustained treatment. This is not meant to suggest that achieving these goals in pure mathematics is any easier than reaching the different goals of the applied mathematician. Still, the applied mathematician seems constantly left to the whims of scientists and their theoretical deficiencies, and so the goals here are often less well defined.

In a book like this, it is not possible or prudent to try to review all the various topics covered by specialists in applied mathematics or even develop one topic in the level of detail appropriate for a thorough understanding of the relevant mathematical techniques. Instead, I have elected to touch on a number of different examples so that the different kinds of contributions to the success of science will come through in a

---

14. The notion of mechanism here remains unclear. As our discussion in chapter 3 emphasizes, not all adequate mathematical models represent causal mechanisms. (This passage is partly given in Chihara 2004, p. 263.)

way that is comprehensible to nonspecialists like philosophers or others interested in understanding the success of science. Very roughly, chapter 3 noted some of these topics as they related to acausal representations, such as stability analysis. Now, in a bit more detail in this chapter, we have seen some other areas of applied mathematics in the study of kinds of differential equations and their appropriateness for different physical systems. In chapter 5 I turn to what I take to be the most important area of applied mathematics for an understanding of why mathematics contributes so much to science: scale. As we will see, scale brings together both kinds of contributions that have been surveyed in chapters 3 and 4 but in a new and important way.

5

Scale Matters

5.1 SCALE AND SCIENTIFIC REPRESENTATION

When we model a physical system we often believe that certain features of the system are more important than others. One way to signal this importance is to speak of the relative *scale* of these features, as compared to other features that we deem unimportant. The scale of a feature can be understood in terms of its relationship to whatever parameters are used to approach the system. Central examples for our discussion here are distances and times. Other parameters that are sometimes of importance are energy or more complicated results of combining parameters like area, velocity, or, as we will see, dimensionless numbers like the Reynolds number. When we assign a scale in terms of these parameters to a feature of a system, we are saying that this feature has a relative magnitude that is comparable to the chosen parameter. This often allows a vast simplification in the complexity of our representation without sacrificing accuracy. For example, as I argue in this chapter, many idealizations can be understood in terms of this scaling procedure. When done poorly, scaling appears as nothing more than a mathematical trick whose success remains a mystery. When done with the proper motivation, scaling can reveal genuine features of a system that would otherwise be obscure or epistemically inaccessible to a scientific modeler.

The main claim of this chapter is that mathematics often contributes to the scale or scales of a successful scientific representation. The ways this happens are so diverse that it would take more than one chapter to give a comprehensive survey and classification. Rather than attempt this, I take as my starting point the two kinds of abstract contributions discussed in the last two chapters. One kind of contribution from mathematics to scale will be reminiscent of the abstract acausal representations of chapter 3. This occurs when we use mathematics to limit the content of our representation to a single scale. I group these cases under the heading of "scale separation." After describing some cases of this sort, I argue that epistemic benefits obtain for these representations that are analogous to what we saw for abstract acausal representations. A second kind of contribution from mathematics to scale will accord with the abstract varying representations of chapter 4. These cases can be fruitfully treated under the heading of "scale similarity." Here what will result is a family of mathematically unified representations where what varies are the scales at which we represent a system of a given kind. Again, epistemic benefits become clear as we review some important cases.

Important additional aspects of the role of scale in scientific representation result from trying to understand Lin and Segel's remark that "perturbation theory is often used implicitly when we formulate simplified physical models" (Lin and Segel 1988, p. 48). Perturbation theory is a mathematical theory which we will see in sections 5.4–5.6 is closely linked to scale. In section 5.4 I argue that it can be used to motivate an account of how some idealizations work and what these successful idealizations tell us about the world. As the story here is grounded in certain mathematical techniques and results, we have an important contribution from mathematics to the success of science. Bringing in the basic ideas of perturbation theory also lets us understand what is going on when we successfully combine representations of a single target system that have different scales. These "multiscale" representations have a much greater domain of application than single-scale representations, but they also pose certain interpretive difficulties that we turn to at the end of this chapter. After treating one case in some detail in section 5.6, I consider two other recent discussions of such representations in section 5.7.

Although the material in this chapter is more mathematically demanding than what we have discussed previously, I hope that the insight provided into the importance of scale compensates for the mathematical difficulties.

5.2 SCALE SEPARATION

The best-case scenario for the representation of a complex system obtains when we can specify a single scale at which all features relevant to the phenomena of interest occur. This is what we can call *scale separation*. It is discussed, for example, by Hillerbrand in "Scale Separation as a Condition for Quantitative Modeling" (Hillerbrand n.d.). The basic idea is that when we can be assured that the relevant features of a system all appear at a certain scale, or, as it is sometimes put, in a certain *regime*, then we can represent the system as operating exclusively at that scale. Hillerbrand mentions several examples of this sort of success, and one famous case is Poincaré's investigation into n-body Newtonian systems like the solar system. Despite the intractably complex interactions between the planets, the "motion of the Earth can be well described by its motion around the Sun only because the forces exerted on the Earth by other bodies, even by the large planets like Saturn and Jupiter, are smaller by orders of magnitude than the force exerted by the Sun" (Hillerbrand n.d., p. 5). We effect a scale separation here by discounting all forces except the Earth–sun relationship. Delicate mathematical arguments are necessary to show that this move is feasible, and Poincaré's genius lay not only in giving these arguments but also in showing how greater accuracy could be achieved using perturbation theory. Here, as we will see, we find a way to improve the accuracy of a representation by including the effects of other terms to a certain degree. Although any attempt to achieve perfect accuracy by these means would return us to the original intractable problem, we can see that only a few other terms are needed to achieve any reasonable demand for accuracy. We return to perturbation theory later in this chapter and see that it can be used even when scale separation fails.

Hillberbrand's main point is that the complexity of a system is not a barrier to representing it mathematically because scale separation allows us to cope with and minimize the complexity. Although I would agree that scale separation is sufficient here, it is not clear that it is necessary, at least if we have a fairly restricted conception of what scale separation presupposes. As I have presented it, scale separation requires that the features of the system of interest really are restricted to a certain scale. In the solar system case, there is a threshold below which the forces are neglected, and it so happens that all forces except the Earth–sun interaction fall into this neglected category. There are other cases where mathematical modeling has been very successful, but scale separation in this sense fails. Because they are analogous to the solar system example, though, I call these all cases of scale separation in an extended sense.

An important example is large-scale meteorological modeling. In this case, the content of the model is restricted to the large-scale changes in the ocean–atmosphere system. It is simply not the case, though, that these features of the climate are unaffected by smaller-scale features of the planet, for example, human activity and patterns of vegetation. Despite this failure of strict scale separation, there are several ways to obtain accurate models of the large-scale features of interest. We consider two options. The first is to go acausal to some extent and rest content with a representation that fails to capture all the causal relations of interest. This sacrifices some causal content and so makes it easier to accurately represent the phenomenon of interest. A second possibility is to represent the interscale interactions using a highly simplified parameter. This parameter might be chosen based on the results of another kind of model or may be fixed using available data.

As an example of a successful acausal representation we can consider the El Niño Southern Oscillation (ENSO). Writing in 1996, Allan, Lindesay, and Parker noted that "although there is no overall theory that can explain all aspects of the ENSO phenomenon, research efforts over many decades have resolved some of its most important physical and dynamical aspects" (Allan, Lindesay, and Parker 1996, p. 3).[1] We can think of ENSO as characterized by a dominant state of normal conditions, punctuated by an El Niño event, followed by a return to normal conditions (see figure 5.1).[2] In the atmosphere, normal conditions over the Pacific Ocean near the Equator are characterized by a low-altitude westward (east to west) air flow, with the air rising in the western equatorial Pacific and returning eastward (west to east) at high altitude. This coincides with a warm mass of water pooling in the western equatorial Pacific, which contributes moisture and precipitation in that region. At the surface of the ocean, the result is a sea surface temperature (SST) gradient increase from east to west. Below the surface, the region of greatest temperature gradient, known as the thermocline, varies from being quite close to the surface in the east to being relatively deep in the west.[3]

---

1. I follow their helpful summary of current ENSO research and have also benefited from their overview of the history of ENSO studies.

2. This ignores the so-called La Niña event, which typically follows an El Niño event. The importance of this phase in the ENSO cycle was not generally recognized until the 1980s and does not play a part in the model of ENSO that I discuss.

3. According to Allan, Lindesay, and Parker (1996, p. 25), the depth ranges from 50 to 150 m.

Normal Conditions

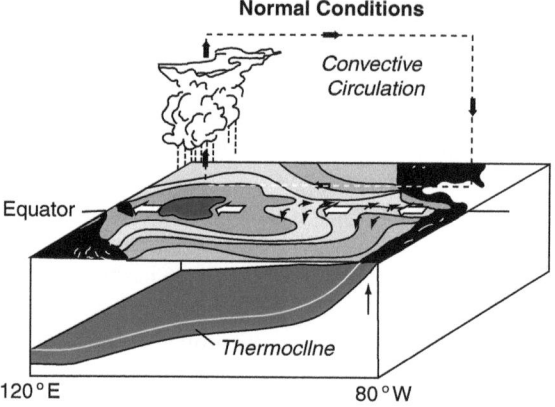

El Niño Conditions

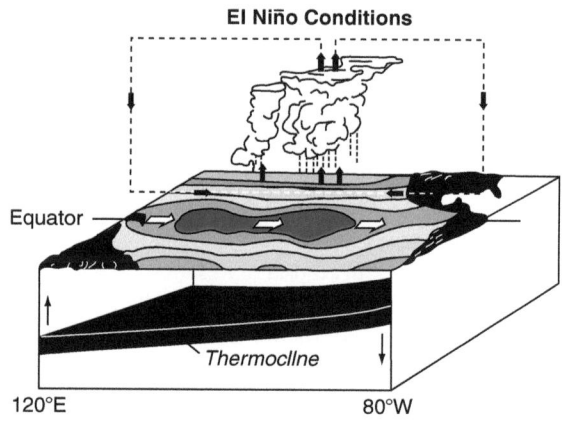

Figure 5.1: Normal versus ENSO Conditions
Source: Courtesy of NOAA/PMEL/TAO

An El Niño event differs from these normal conditions in a variety of respects. First, in the atmosphere, the Pacific-wide circulation pattern is suppressed, leading to air rising further east. Precipitation patterns in the western equatorial Pacific are thus disrupted, and increased precipitation occurs over the eastern equatorial Pacific. The SST gradient is also removed, with relatively warm water now being spread across the entire equatorial Pacific. Beneath the surface, what was before a slanted thermocline is considerably flattened, leading to warm currents off the coast of South America. The El Niño event ends when normal conditions are restored.

This description of ENSO leaves the dynamics of the process unspecified and the exact nature of the causes of the start of an El Niño event remain a subject of debate. However, even in the absence of a good representation of what causes the event, scientists have developed good representations of how the event unfolds once it begins. One leading proposal, called the "best explanation of the physical mechanism responsible for interannual ENSO behaviour to date" (Allan, Lindesay, and Parker 1996, p. 28), is the "delayed action oscillator" model. On this model, an extended run of normal conditions produces a somewhat unstable state, as warm water continues to pool in the western equatorial Pacific. A relatively small eastward air current, say due to a cyclone, can then momentarily disrupt the Pacific-wide circulation pattern. Once this occurs, the pooled warm water rushes toward the eastern Pacific, evening out the SST and flattening the thermocline. When the mass of warm water reaches the eastern Pacific, it is partly reflected back. After these water waves are exhausted, the westward air currents return to dominance, and the pooling of warm water in the west resumes.

A few points about this picture are worth emphasizing. To start, the dynamic picture that results from the interactions between the ocean and the atmosphere was not considered until the 1960s, beginning with the work of Jacob Bjerknes. Prior to this, earlier investigators like Walker isolated only the atmospheric fluctuations and labeled it the "Southern Oscillation."[4] But even after Bjerknes's proposal was accepted, it remained very hard to understand the dynamics of any particular ENSO event. What scientists seem to have pursued, then, is a two-track approach. On one track there are attempts to represent the causal processes that take the ocean–atmosphere system through the stages of the ENSO event once it has begun. Success here is an impressive achievement because the cyclic character of ENSO is initially quite surprising. It is not clear, for example, why the ENSO events are so regular and why similar sorts of events in the Atlantic are different in character.

A second track concerns attempts to understand what triggers the ENSO event in the first place. This is, of course, of great interest for forecasting purposes. It appears that there is still no causal representation that will capture when an ENSO event will begin based on ocean–atmospheric conditions that we can collect. Instead, various detectors are used to tell forecasters when an ENSO event has begun and to try to gauge its character. In this respect, ENSO modelers are in a similar position to forecasters interested in modeling much smaller meteorological phenomena like tornadoes and hurricanes. The causal models of both phenomena are quite advanced and have led to a clear understanding of how tornadoes and hurricanes evolve once they form and the conditions necessary for their formation. Still, it is not possible to predict the actual formation of a tornado or a hurricane using a causal representation and the data that we can actually collect. Limited causal understanding, then, coexists with representational accuracy of the parts of the phenomena that we do understand.

A similar pattern can be observed in climate change modeling. Here scientists clearly recognize that strict scale separation fails, but they cope with the failure somewhat differently than what we saw in the ENSO case. As Parker has shown, climate modelers deploy models that include highly simplified parameters to account for effects across scales that the dynamics of their representations typically ignore. An example that she notes is clouds:

> Although scientists would like their climate models to incorporate some representation of the effects of clouds, individual clouds occur on scales that

4. More discussion of this history is given in Pincock (2009).

the models do not explicitly resolve. In addition, there is genuine uncertainty about how clouds interact with larger-scale dynamical processes in the climate system and hence about how the effects of clouds can be best parameterized. As a result, several different parameterizations of clouds have been developed, reflecting different approaches to representing clouds within the bounds of present scientific uncertainties. (Parker 2006, p. 352)

This accounts for the existence of several different representations of the climate in our best climate change science. Different groups of scientists pursue different parameterizations of these sorts of interscale phenomena. The hope is clearly that these different choices will not affect the content of the representation for those aspects of the climate that we care about capturing accurately.

Here, again, we see a sacrifice in causal content and the corresponding belief that it does not compromise the accuracy of the representation in other respects.[5] Setting climate change skepticism aside, there are important methodological questions about how climate change scientists interpret the results of their families of incompatible models of climate change. The models tend to agree, for example, on the long-term results of doubling the CO_2 in the atmosphere, but disagree about the shorter-term and small spatial scale impacts of these changes. Scientists take this agreement to be sufficient to support their conclusions about long-term results but are more cautious about more specific predictions. As it is often more specific predictions that policy makers ask for, there is some tension within the community of scientists about how specific to get. One manifestation of this is the debate about whether climate change caused by human activity will increase the intensity of hurricanes.[6] Although some climate scientists feel comfortable extending their large-scale models down to the scale relevant to hurricanes, others resist this extrapolation.

The main argument from chapter 3 was that we can understand the link between acausal representations and confirmation using a notion of one representation having less content than another representation. In cases where we restrict our focus to a single scale, it is fairly plausible that a single-scale representation will have less content than a more comprehensive representation. When scale separation is viable, it follows that all the arguments from chapter 3 carry over to these cases as well. Mathematics plays a crucial role in restricting the content of a representation to a single scale, so we have a further kind of epistemic contribution from mathematics.

Much of the time, though, strict scale separation fails, and other techniques need to be deployed to develop accurate representations. As I have summarized these techniques here, the basic idea is to suppress a certain amount of causal content even when we recognize its importance for the phenomena of interest. For the ENSO case, scientists of course believe that there are causes of ENSO events, even if they cannot accurately represent them in specific cases. For the climate modeling case, cloud cover does play an important causal role in fixing the temperature of the Earth, but the representations deploy only highly simplified parameterizations. This means

---

5. This case was also briefly noted in section 3.3 in connection with robustness analysis.

6. For discussion and references, see Pincock (2011).

that we have a mixture of causal and acausal content. As a result, it is harder to draw any general conclusions linking the mathematics here to direct and indirect confirmation. Still, we can say something about how the mathematics related specifically to the acausal components is working. It is allowing us to represent the phenomena of interest in a way that we take, after suitable investigation and testing, to be accurate in certain respects. Based on the complexity of the systems in question, we have no more direct, causal way of representing these systems. So the mathematics allows us to do what would otherwise be impossible. In line with my previous discussions, I take it to be quite natural to take this contribution to be broadly epistemic. Improved access to the phenomena might allow certain parameterizations to be further refined or perhaps even eliminated. For now, though, we need to stick with our acausal mathematical surrogates. Many of these cases, then, achieve accuracy with a mixed causal/acausal representation when a causal representation is not feasible.

5.3 SCALE SIMILARITY

A second kind of successful scaling can be grouped under the heading of scale similarity. A core instance is the representation of self-similar phenomena. Batterman has made a strong argument for the importance of self-similarity for such standard philosophical topics as explanation and reduction (Batterman 2002b). Sometimes an investigation of a kind of system will reveal that the same phenomenon appears repeatedly at different scales, for example, lengths or times. When this occurs, it is sometimes possible to represent all the relevant instances of the phenomenon with a suitably adjusted single representation. In such cases, we say that the phenomenon is self-similar. As a simple example, Batterman notes the case of the one-dimensional heat diffusion equation with a constant κ for the thermal diffusivity. We are interested in tracking the temperature θ from some initial distribution over the x spatial coordinate. For a wide class of initial distributions, the same curve captures the later temperatures. The crucial step, though, is to shift from the plot of θ vs. x to a suitably scaled set of coordinates, namely, $\theta \frac{\sqrt{\kappa t}}{M_0}$ vs. $x \frac{1}{\sqrt{\kappa t}}$. Compare figure 5.2 to figure 5.3. Here M_0 is the area of the initial distribution, not the distribution itself. So we see

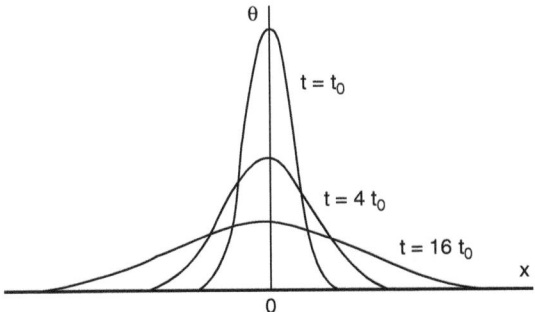

Figure 5.2: Temperature versus Distance

Source: Batterman (2002b, p. 48)

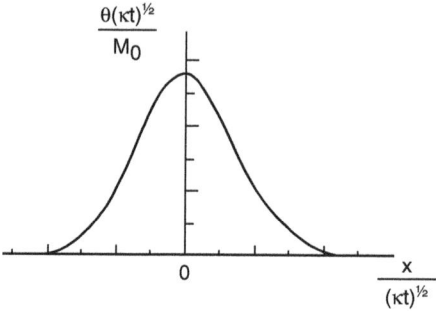

Figure 5.3: Self-Similar Coordinates
Source: Batterman (2002b, p. 49)

that for any initial distribution with the same area, the same relationship obtains. Among other things, this reveals what a number of different systems across many different scales have in common.

There is a related notion of dynamic similarity between physical systems that can be understood in terms of more complicated relationships between the scales of features across these systems. The core kind of case here is the desire to build a small concrete model of a larger system that is hard to understand directly or may be impractical to build as a prototype. Kundu and Cohen note the clear epistemic benefits of building a model that is dynamically similar to a given physical system: "in a class of dynamically similar flows we can predict flow properties if we have experimental data on one of them ... The concept of similarity is ... indispensable for designing models in which tests can be conducted for predicting flow properties of full-scale objects such as aircraft, submarines and dams" (Kundu and Cohen 2008, p. 279). Susan Sterrett has recently discussed a striking case of this with a model of the Mississippi River basin, and Michael Weisberg is currently working on the intriguing case of a model of the San Francisco Bay.[7] More ordinary models include aircraft wings and the hulls of ships. It is a nontrivial task to find out when two systems are dynamically similar. Several methods can be used, but the most intuitive begins with a well-motivated system of differential equations and considers the parameters that are given for the system.

Suppose, for example, that we wish to see what drag a ship of a certain shape would experience if it was built to be a certain size and moved through water at a given velocity. The drag that the ship experiences is the force of resistance that is applied to it by the water in the direction opposite to its motion. Minimizing drag, then, is an urgent design issue when constructing ships that would travel at high velocity with minimal expenditure of fuel. Based on our understanding of fluids like water we can see drag as a product of two factors. First, there is the frictional drag induced by the viscosity of the water. Second, there is the wave drag associated with the water waves caused by the motion of the ship. For cases where both sorts of drag are relevant an analysis of the Navier-Stokes equations reveals that there are

---

7. Sterrett (n.d.), Weisberg (n.d.).

two dimensionless parameters that must have the same value for the two cases to be dynamically similar. This sort of analysis is further examined in sections 5.4 and 5.5. The parameters that result are the Reynolds number and the Froude number:

$$Re \equiv \frac{Ul}{\nu}, \quad Fr \equiv \frac{U}{\sqrt{gl}} \tag{5.1}$$

Here U is the velocity of the ship, l is its length, ν is the viscosity of water, and g is the acceleration due to gravity. In addition to preserving the shape of the ship as we build the model, we must also make sure that the values of the two parameters are preserved. This ensures not only geometric similarity but also dynamic similarity.

Consider, then, the case of a ship that is 100 m long that we aim to have a velocity of 10 m/s (Kundu and Cohen 2008, p. 290). We build a model of the same geometric shape that is 1/25th the size, that is, 4 m long. How fast should we move the model ship through the water so that the model system is dynamically similar to the proposed ship? A naive proposal is $10/25 = 0.4$ m/s. However, this system would have a Froude number of $\frac{U}{\sqrt{gl}} = \frac{.4}{\sqrt{4g}} = \frac{0.8}{\sqrt{g}}$, whereas the ship system has a Froude number of $\frac{10}{\sqrt{100g}} = \frac{1}{\sqrt{g}}$. To preserve the Froude number, then, we require that our unknown model velocity U_m satisfy

$$\frac{U_m}{\sqrt{4g}} = \frac{1}{\sqrt{g}} \tag{5.2}$$

That is, $U_m = 2$ m/s. Measuring the drag on the model at this speed allows us to calculate the drag that would be experienced on the ship if it were built.

The full treatment of this case actually involves a more subtle treatment based on the fact that it is not practically possible to ensure that both the Froude number and the Reynolds number are the same across both systems. The reason for this is that the viscosity ν and the gravity g are constants, and we are required to decrease l. It may be possible to shift to a liquid with a different viscosity, but this is not always practical. But the definitions of Fr and Re then require different changes in U if both parameters are to be held constant. For example, repeating the argument for Re instead of Fr would require that the velocity of the model ship actually be 250 m/s instead of 2 m/s! Similar problems occur when new effects become important at smaller scales, as with the role of surface tension in water waves for models of rivers. Both problems are handled by sacrificing full dynamic similarity in favor of either partial dynamic similarity with respect to the most dominant effects or even a sacrifice of geometric similarity. In our drag case, focusing on Fr indicates the assumption that friction plays a less important role in the drag for the actual ship. The model is then built using just this parameter, and a later correction is used to interpret the data.

For both self-similar phenomena and dynamically similar systems, we can arrive at a single scientific representation that is adequate to represent the relevant features of the phenomena across suitable variations. This confers a unifying power that is an instance of the abstract varying representations surveyed in chapter 4. In this case, though, the epistemic contribution is more directly apparent than the case of mathematically similar representations with varying physical interpretations. This is because the suitably scaled mathematics is identical across the family of representations and the physical interpretation is the same except for the appropriate change in scale. These techniques let us immediately recognize how the members of the family are physically related. Two epistemic benefits are especially clear in the two cases noted. For the self-similar case, we come to understand that it is only the area of the initial distribution that is relevant to the temperatures at later times. For the ship model case, our dimensionless parameters tell us which smaller scale models are dynamically similar to which larger-scale systems. In both cases, the unifying power of the mathematics allows us to come to know features of the world that would otherwise be very hard (if not impossible) to recognize.

5.4 SCALE AND IDEALIZATION

So far we have seen how the scale of a representation can be important to confirmation based either on a narrow specification of the scale of interest or on the links between representations of some phenomenon across many scales. A further development of both kinds of contributions occurs if we investigate the link between scaling and idealization. Idealization is typically introduced as the assumption of something that we take to be false. I focus on the idealizations necessary to understand water wave dispersion to illustrate the more general links between idealization and scale. After describing this case in some detail, it will be possible to see it as an instance of regular perturbation theory.

Water wave dispersion is the often observed phenomenon in which an irregular pattern of waves gradually spreads out from the center of some disturbance of the water, like a rock being dropped in an otherwise calm ocean. What we observe in this case is that waves of longer wavelength move more quickly away from the center. As a result, the wave pattern becomes more regular. It turns out that an accurate representation of this phenomenon depends on two different idealizations. First, we must deploy the small amplitude idealization. This allows us to move from the nonlinear Navier-Stokes equations for incompressible, inviscid fluids to a set of linear equations. Second, we use the deep-water idealization. It allows us to represent the relationship between the phase velocity of a wave and its wavelength. In a popular fluid mechanics textbook we find that both steps are treated as *approximations*, that is, assumptions that will not introduce significant errors into the representation (Kundu and Cohen 2008). However, a better perspective, offered by Segel's *Mathematics Applied to Continuum Mechanics* (Segel 2007), lets us see that these assumptions actually correspond to applications of perturbation theory. The difference between the two presentations corresponds to the priorities of the textbooks. Where as Kundu and Cohen need to only assure their readers that these assumptions are appropriate in this particular case, Segel aims to explain in general when assumptions of this type are appropriate. The latter goal also contributes to an understanding of these idealizations, for they show how apparently false assumptions can be used to reveal important features of real systems. On the picture of idealization that I aim to defend, an idealization transforms a representation that only obscurely represents a feature of interest into one that represents that same feature with more prominence and clarity. This involves, among other things, decoupling some parts of the representation from their original interpretation. This basic idea was introduced in chapter 2, where we noted that the gap between a mathematical structure and a physical system can require a schematic content that drops any physical interpretation from parts of the mathematical structure. At that point, though, the procedure for when or how to do this was left fairly open-ended. Now we develop a case where the details indicate precisely how this more schematic content is arrived at and what its value is. Because the techniques here are highly mathematical, we get another kind of contribution from mathematics that is linked to scale and confirmation.

As we have seen, our best representation of fluids like the ocean treat it as a continuous-medium subject to various internal stresses and external forces. The key magnitudes that we try to capture are the velocity vectors at a point at a time and the pressure magnitude at a point at a time. In our simplified case, we treat the density of the fluid as a constant and assume that the only force at work is gravity, with a constant acceleration downward on a unit mass given by g. Therefore, we ignore the viscosity or friction due to the interaction of the fluid elements as well as the surface tension of the ocean at the surface boundary with the air. In these circumstances, there is a scalar field, or potential, ϕ defined on the fluid elements such that the velocity v at a point is given by $\nabla \phi$.[8] We treat the two-dimensional case, so this gives the x and z components of velocity as u and w, respectively:

$$u = \frac{\partial}{\partial x}\phi, \; w = \frac{\partial}{\partial z}\phi \tag{5.3}$$

We set $z = 0$ to be the surface of an undisturbed ocean system, H as the depth of the ocean, a as the maximum amplitude of the ocean wave, and λ as the wavelength of the wave, that is, the length between one crest and the next crest. The shape of the wave is given by the displacement $\eta(x, t)$ from the undisturbed case. See figure 5.4.

The first step in developing our representation is to move from the Navier-Stokes equations for such fluids (incompressible, inviscid) to the results of deploying the small-amplitude wave idealization. This is how Kundu and Cohen motivate the idealization:

> We shall assume that the amplitude a of oscillation of the free surface is small, in the sense that both a/λ and a/H are much smaller than one. The condition $a/\lambda \ll 1$ implies that the slope of the sea surface is small, and the condition $a/H \ll 1$ implies that the instantaneous depth does not differ significantly from the undisturbed depth. These conditions allow us to linearize the problem. (Kundu and Cohen 2008, p. 219)

---

8. Scalar potentials were discussed in section 4.2.

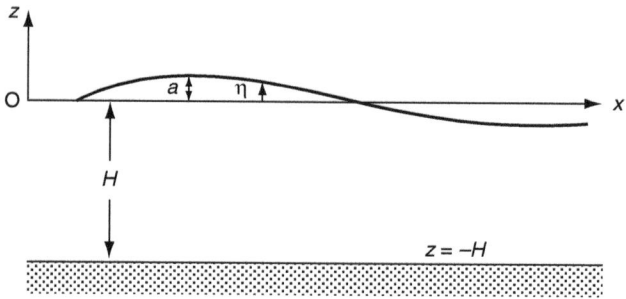

Figure 5.4: Surface Waves

Source: Kundu and Cohen (2008, p. 219)

The Navier-Stokes equations are nonlinear in the sense that they have terms like $u\frac{\partial u}{\partial x}$ that complicate their solution. In particular, we cannot use a principle of superposition to the effect that any two solutions can be combined by linear combinations to yield new solutions.[9] Considering only waves of small amplitude allows us transform the problem into the solution of the Laplace equation for ϕ:

$$\frac{\partial^2 \phi}{\partial x^2} + \frac{\partial^2 \phi}{\partial z^2} = 0 \tag{5.4}$$

subject to the three conditions

$$\frac{\partial \phi}{\partial z} = 0 \text{ at } z = -H \tag{5.5}$$

$$\frac{\partial \phi}{\partial z} = \frac{\partial \eta}{\partial t} \text{ at } z = 0 \tag{5.6}$$

$$\frac{\partial \phi}{\partial t} = -g\eta \text{ at } z = 0 \tag{5.7}$$

The first two conditions are kinematic because they rely on the features of the boundaries, here the bottom of the ocean and the ocean surface. The third condition is dynamic because it turns on the interaction between the pressure and velocity of a fluid element at the surface. Crucially, the latter two equations are evaluated at constant $z = 0$ and not at the varying surface of the water. Again, this simplification is only possible because of the small-amplitude idealization.[10]

We aim for a solution to η of the form

$$\eta = a\cos(kx - \omega t) \tag{5.8}$$

The key features of a wave are its amplitude a, the displacement at a point, the wavelength λ, and the *period T*, the time it takes for a given phase of the wave to

---

9. See, again, section 4.2 for some of the benefits of superposition.

10. In my reconstruction I ignore the subtleties needed to find the pressure throughout the fluid.

repeat itself at a given point or, equivalently, the time it takes for the wave to travel one wavelength. However, for convenience we also use k, the *wavenumber* equal to $2\pi/\lambda$ as well as ω, the *circular frequency* equal to $2\pi/T$. We can determine c, the *phase speed*, or speed of propagation of a part of the wave with a constant phase, like a wave crest, by $\frac{\omega}{k}$. Deploying our kinematic boundary conditions yields the general solution

$$\phi = \frac{a\omega}{k}\frac{\cosh k(z+H)}{\sinh kH}\sin(kx - \omega t) \tag{5.9}$$

which determines the velocity distribution throughout the fluid. We can also use it, along with 5.8 and 5.7, to find c in terms of λ and H:

$$c = \sqrt{\frac{g\lambda}{2\pi}\tanh\frac{2\pi H}{\lambda}} \tag{5.10}$$

This reveals a complicated relationship between the phase velocity of a water wave, the wavelength, and the depth of the medium. Some care is needed to see exactly what its significance is for our case of a rock dropped in an otherwise calm ocean.

We first consider the case where $H/\lambda \gg 1$, that is, the depth of the ocean is much greater than the wavelength. Consider, for example, the case where $H > 0.28\lambda$. Then $2\pi H/\lambda > 1.75$ and $\tanh(2\pi H/\lambda) > 0.94138$. More generally, as $x \to \infty$, $\tanh x \to 1$. As the square root of 0.94138 is 0.97, replacing $\tanh\frac{2\pi H}{\lambda}$ by 1 in equation 5.10 will produce a maximum error of 3% when $H > 0.28\lambda$. Using this calculation, Kundu and Cohen note that "waves are therefore classified as *deep-water waves* if the depth is more than 28% of the wavelength" (Kundu and Cohen 2008, p. 230). Deep-water waves include ocean waves generated by wind, where the typical wavelength is around 150 m, and the ocean depth ranges from around 100 m to around 4 km. They can thus always be treated by

$$c = \sqrt{\frac{g\lambda}{2\pi}} \tag{5.11}$$

Among other things, this allows us to understand the phenomenon of water wave dispersion noted earlier. See figure 5.5. Based on 5.11, we can see that the crests of waves with longer wavelengths will travel faster than the crests of waves with shorter wavelengths. If we treat the initial disturbance at t_1 as a more or less random superposition of deep-water waves, then waves of varying wavelengths will become dispersed or spatially separated. In addition, the leading waves will gradually decrease in amplitude because of the conservation of energy for the whole disturbance.[11] This result does not follow for 5.10 generally as can be seen by taking the other extreme case of shallow-water waves where $H < 0.07\lambda$ and

$$c = \sqrt{gH} \tag{5.12}$$

---

11. There are other features of water wave dispersion that I ignore here for simplicity of presentation.

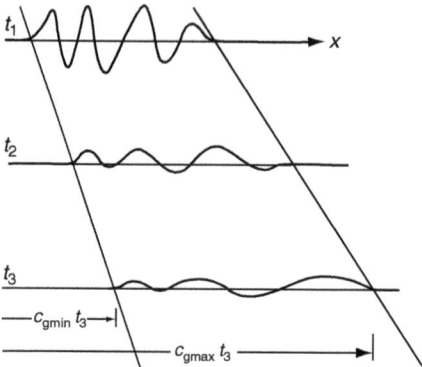

Figure 5.5: Wave Dispersion
Source: Kundu and Cohen (2008, p. 243)

This equation results from using the fact that $\tanh x \to x$ as $x \to 0$, so $\tanh \frac{2\pi H}{\lambda}$ is replaced by $\frac{2\pi H}{\lambda}$. Here the phase velocity is independent of the wavelength, and all wave crests travel at the same velocity. In such cases, an initial disturbance retains its irregular shape because the waves of differing wavelengths fail to disperse.[12]

Notice that 5.11 also results from 5.10 from thinking of $H \to \infty$. So we can treat the deep-water wave representation as corresponding to the idealization that the ocean is infinitely deep. This sort of assumption could also be used to justify part of our small-amplitude idealization that $a/H \ll 1$. This is in fact how Kundu and Cohen sometimes speak, as when they treat "Waves at a Density Interface between Infinitely Deep Fluids" (Kundu and Cohen 2008, p. 255). It is tempting to say that it is only at the limit where $H = \infty$ where $\tanh \frac{2\pi H}{\lambda} = 1$ and the equality $c = \sqrt{\frac{g\lambda}{2\pi}}$ holds. If we take this perspective, then it looks like the false assumption that the ocean is infinitely deep reveals something important about the behavior of real ocean waves. This can create various interpretive puzzles. One puzzle is that it looks like an agent that deploys the deep-water idealization has inconsistent beliefs. For in the idealization, she makes the assumption that $H = \infty$. But when pressed for her belief about the ocean depth, she will of course concede that the ocean has some finite depth. Indeed, this belief seems required for any subject who thinks coherently about the ocean. An ocean of infinite depth is about as physically impossible as it gets. So we appear to have a kind of deep or irreconcilable inconsistency at the heart of a successful scientific practice.[13]

I think the most fruitful approach to such cases is to question this starting point. We need an account of idealization which allows that a false assumption relating

---

12. A somewhat surprising fact is that actually the shallow-water idealization is used to understand the ocean in geophysical applications. See, for example, Kundu and Cohen (2008, ch. 14). This is because of the thermocline noted earlier in our brief discussion of ENSO. The bottom of the thermocline is often treated as the lower boundary of a layer of fluid.

13. For more on this approach see, for example, Maddy (1998, pp. 143–145): "we assume the ocean to be infinitely deep when we analyze the waves on its surface"; Azzouni (2004, p. 46), and Colyvan (2008, p. 117).

to the depth of the ocean is not part of the content of the representation itself, but is only the means to obtain the representation. The basic idea of this approach to idealization is that false assumptions take us from an interpreted part of a scientific representation to an idealized representation where this particular part is no longer interpreted. In particular, when we make the false claim that the ocean is infinitely deep, what we are really doing is shifting to a representation where H is decoupled from its former interpretation. This means that an agent can affirm the whole genuine content of the resulting representation, in this case the deep-water idealization. It also helps explain why this representation is only appropriate for certain purposes. Our idealization has literally nothing to say about what goes on in the bottom of the ocean, so if we are interested in these aspects of the situation, we need to shift to a different idealization.

This sort of proposal finds a natural place in the account of the content of a representation developed in chapter 2. When we deploy an idealization like $H = \infty$, we transform the representation so that it has merely a schematic content. The H no longer represents the depth of the ocean, and so our representation can be accurate no matter what the depth actually is. The genuine content is obtained by filling out the schematic content so that the appropriate scope of application is clarified. That is, which target system is the representation meant to accurately represent? This can be one of the trickiest aspects of scientific modeling. In this case, though, the informal argument that waves where $H > 0.28\lambda$ obey 5.11 gives us a good start on pinning down this genuine content. Using this argument to fix the scope of the representation produces a definite representation that can be experimentally tested with respect to the relevant kind of target system.

The characterization of deep-water waves is fairly ad hoc, and it would be better if there was some general theory that would allow us to clarify the scope of these highly idealized representations. This is where the perturbation theory mentioned by Lin and Segel becomes useful. Although this is a large area of ongoing research in applied mathematics, we can start to understand the importance of the theory by considering how it works in our water wave case. I discuss perturbation theory in two stages. In the first stage we treat the theory in terms of the scale of the features of the system. In the next stage, discussed in the next section, we see how this talk can be grounded more adequately in terms of series. For now, our goal is to use the parameters given in setting up the problem, such as the typical wave amplitude a and depth H, to form dimensionless variables. In our case, for example, x^* is a length variable, so a new dimensionless length variable x results from dividing all occurrences of x^* by a or by H. I adopt the convention for this discussion that the original variables are labeled with an asterisk while the dimensionless variables are without asterisks. If we carry this scaling through for each variable, we can rescale the equations by substituting the new variables in for the old variables while keeping track of the parameters. The goal is to do this in such a way that "each term is preceded by a combination of parameters that explicitly reveals the term's magnitude" (Segel 2007, p. 329) for the domain in question. If this procedure is successful, then I say that the set of scales is *adequate*. The existence of an adequate set of scales is not a purely mathematical claim, but is rather a claim about the features of a given physical system and their relative magnitudes. In those cases where we obtain such an adequate set of scales, the terms preceded by combinations of parameters that are orders of magnitude smaller than other terms can be safely ignored. This corresponds to the claim that the features tracked by these parts of the equation are irrelevant for our given representational purpose. From this perspective it should not be surprising that a set of scales is adequate only with respect to a given context. For in another context where we are trying to accurately capture other aspects of the system, a different array of features may be important and so a different set of scales is needed.[14]

Our two idealizations can be understood, then, as the application of two sets of scales that turn out to be adequate for certain water wave systems. In this case, the second set of scales results from adjusting the first set, so the two idealizations can be combined.[15] Consider, first, the small amplitude assumption and the elimination of problematic nonlinear terms like $u\frac{\partial u}{\partial x}$. On our scaling approach, what we are doing is not accurately described as assuming the contribution of this term is negligible or unlikely to produce a significant experimental error. Rather, we assume *relative to the linear terms* like $\frac{\partial u}{\partial t}$ the contribution of $u\frac{\partial u}{\partial x}$ is negligible. Consider a set of scales in terms of the amplitude a, period T, and wavelength λ (Segel 2007, p. 325):

$$t = \frac{t*}{T}, \quad x = \frac{x*}{\lambda}, \quad u = \frac{u*}{a/T} \tag{5.13}$$

Notice that there is a choice in scaling x. We employ the distance λ, but we could also have used any other distance given in the problem, for example, a. In making the choice we do, we claim that the relevant processes operate in the x spatial direction only on the order of λ. Similarly, we are claiming that the velocity in the x-direction of a fluid particle, u, is similar in magnitude to the ratio between the amplitude and the period of the waves in question. Accepting these scales, ordinary algebra entails that

$$t* = Tt, \quad x* = \lambda x, \quad u* = \frac{a}{T}u \tag{5.14}$$

Making the substitution into the term $u * \frac{\partial u*}{\partial x*}$ yields

$$\frac{a}{T}u\frac{\partial\left(\frac{a}{T}u\right)}{\partial(\lambda x)} = \frac{a^2}{T^2\lambda}u\frac{\partial u}{\partial x} \tag{5.15}$$

Similarly, making the substitution into $\frac{\partial u*}{\partial t*}$ gives

$$\frac{\partial\left(\frac{a}{T}u\right)}{\partial(Tt)} = \frac{a}{T^2}\frac{\partial u}{\partial t} \tag{5.16}$$

---

14. This kind of argument is needed to justify our link between dynamic similarity with respect to drag and the *Fr* and *Re* parameters from the last section. See Kundu and Cohen 2008, pp. 280–283.

15. This is not generally true, as we see later in section 5.6 in our discussion of boundary layer theory.

The ratio between the nonlinear term and the linear term is then given by

$$\frac{\frac{a^2}{T^2 \lambda}}{\frac{a}{T^2}} = \frac{a}{\lambda} \tag{5.17}$$

So, if this set of scales is adequate, then the magnitude of the nonlinear term is negligible when compared to the magnitude of the linear term on the assumption that $\frac{a}{\lambda}$ is much less than 1.

To complete the set of scales sufficient to produce the linear equations, we also need to specify what happens for the other variables z, w, p, and η. Our focus is on the status of the depth H. The only place where it comes in is with the boundary condition 5.5: "At $z = -H$, $\frac{\partial \phi}{\partial z} = 0$." As with the scale $x = \frac{x*}{\lambda}$, we adopt $z = \frac{z*}{\lambda}$. Repeating the sort of algebraic substitution given above, 5.5 becomes "At $z = -h$, $\frac{\partial \phi}{\partial z} = 0$." Here $h = \frac{H}{\lambda}$. To shift to the deep-water case, Segel considers what happens as $h \to \infty$ (Segel 2007, p. 335). This transforms the boundary condition into "As $z \to -\infty$, $\frac{\partial \phi}{\partial z} \to 0$". Notice that this claim involves the dimensionless variables. What we are saying is that terms preceded by h become orders of magnitude more important, or equivalently, that terms preceded by $1/h = \lambda/H$ become orders of magnitude less important. When h appears in the boundary conditions, this has the effect of moving the boundary as far from the system as possible. Thus for this set of scales to be adequate, boundary effects must be irrelevant to what we aim to represent.

To see the effects of this adjusted set of scales, consider what happens if we repeat the derivation of 5.9. We find that ϕ takes on the simplified form

$$\phi = \frac{a\omega}{k} \sin(kx - \omega t) \tag{5.18}$$

Using 5.18 instead of 5.9 when we deploy 5.8 and 5.7 gives us 5.11 immediately. So, if this adjusted set of scales is adequate, then the waves obey this simple relation between phase velocity and wave length.[16]

The upshot of this analysis of idealization is that we can replace the vague notion of an approximately true claim with the claim that for phenomenon p some set of scales s is adequate. In our case, the phenomenon that we are trying to capture is wave dispersion. So we can directly specify what we are trying to understand with our representation. Starting with equations that we take to represent the system fairly realistically, we might be frustrated in our attempts to capture p. This can prompt an investigation of one or more set of scales to see if transforming the equations using these sets reveals anything of relevance to p. In our case, the main feature of wave dispersion comes out directly on a particular choice of scales. Additional tests are needed to see if the representation does indeed accurately capture wave dispersion. On the assumption that it does, we have evidence that we have found an adequate set of scales.

---

16. The interpretive issues are discussed briefly at Segel (2007, pp. 340–341).

What we see, then, is a kind of trade-off between the completeness of a representation and its ability to accurately represent some phenomenon of interest. If we remain with the nonlinear Navier-Stokes equations or even with 5.10, then it is impossible for us to represent the observed pattern of dependence between wave length and phase velocity. This is not just a point about mathematical tractability, although our inability to solve a system of equations analytically or numerically is a symptom of the problem I allude to. The problem is basically that a complete representation of a complex system will include countless details, and these details typically obscure what we have selected as the important features of the system. By giving up completeness, and opting for a set of scales, we shift to a partial representation of the system that aims to capture features that are manifest in that scale. This partiality gains us a perspicuous and accurate representation as well as an understanding of features that would otherwise elude us.

5.5 PERTURBATION THEORY

The claim that a set of scales is adequate can be made more precise using concepts drawn from perturbation theory. This theory develops a network of interrelated techniques for solving mathematical problems that prove difficult to solve by more traditional means. As a textbook in this area summarizes things, "a perturbation procedure consists in constructing the solution for a problem involving a small parameter ϵ, either in the differential equation or the boundary conditions or both, when the solution for the limiting case $\epsilon = 0$ is known" (Kevorkian and Cole 1981, p. v). We can divide this procedure into three steps: (1) isolating the small parameter ϵ and recasting the problem in these terms, (2) considering a sequence called an asymptotic expansion of our unknown function that results from the first step, and (3) calculating the solution to the original problem using the sequence from the second step. In the simplest case, we are trying to find an $f(x)$ that satisfies some differential equations and boundary conditions. The approach to this case can be generalized to several functions with more than one independent variable. Given some ϵ that we take to be small for the domain in question, we first recast f as a function of both x and ϵ. The f we are after is then thought of as $f(x, \epsilon)$ where ϵ has some particular value, for example, close to 0.

The next step is to find what is called an asymptotic expansion of $f(x, \epsilon)$ to N terms. To clarify this, we need to introduce the notion of the order of a function. Consider two functions $\phi(x, \epsilon)$ and $\psi(x, \epsilon)$, where $\psi \neq 0$ for the domain in question. Suppose x is fixed and that as $\epsilon \to 0$, $(\phi/\psi) \to 0$. Then we say that $\phi = o(\psi)$ or "ϕ is of order ψ" or $\phi \ll \psi$. Using this terminology, we can posit a sequence of functions where as we go along in the sequence, the next function is of the order of the previous function. That is, we have $\phi_n(\epsilon)$ such that

$$\phi_{n+1}(\epsilon) = o(\phi_n(\epsilon)) \text{ as } \epsilon \to 0 \tag{5.19}$$

A standard example when $\epsilon < 1$ is ϵ^n: $1, \epsilon, \epsilon^2, \epsilon^3, \ldots$ (Kevorkian and Cole 1981, p. 3). Using these sequences, we then aim to isolate an asymptotic expansion of f. This sequence must satisfy

$$f(x, \epsilon) - \sum_{n=1}^{M} a_n(x)\phi_n(\epsilon) = o(\phi_M) \text{ as } \epsilon \to 0 \qquad (5.20)$$

for each $M = 1, \ldots, N$ (Kevorkian and Cole 1981, p. 3). The crucial feature of such an expansion is that the difference between our unknown f and the sum goes down quickly as we consider larger N.

To use the asymptotic expansion to solve the original problem, we first find f when $\epsilon = 0$. Then we use the features of the asymptotic expansion, the differential equations, and boundary conditions given in the original problem to fix the a_1 term. Based on this, we can calculate a first-order correction to f for $\epsilon = 0$. Additional corrections can be made by calculating a_2 or later terms. However, often no additional calculations are needed beyond the first and second terms. This is because we know that whatever error remains will be of order ϵ^3, and the assumption that ϵ is small tells us that this error will be very small.

In our wave dispersion case, we solved the problem after dropping the nonlinear term preceded by a/λ. Let $\epsilon = a/\lambda$. A perturbation theory approach allows us to consider an asymptotic expansion of our unknown function $u(x, t)$. Now, though, we consider it as $u(x, t, \epsilon)$. In the case where $\epsilon = 0$, we have the linear case that we can solve directly. Then this solution can be used as the input to the perturbation procedure just outlined. We can then calculate the first- or even second-order corrections. This will involve some adjustments to the original 5.11 obtained. However, we can be confident that if there is an asymptotic expansion of u that is valid for our entire domain, then these corrections will be tracked by the orders of magnitude of ϵ and ϵ^2. As we consider systems where ϵ gets smaller and smaller, we can see that these corrections will quickly pass below the level of experimental detection.

The point of this brief discussion of asymptotic expansions was to clarify what it means to say that a set of scales is adequate. The main advance that has been made is to go beyond the claim that a set of scales is adequate just in case it results in a set of equations where "each term is preceded by a combination of parameters that explicitly reveals the term's magnitude" (Segel 2007, p. 329) for the domain in question. The cash value of the notion of magnitude here is that carrying out the perturbation procedure in terms of an asymptotic expansion for our unknown functions will deliver an answer to any desired degree of accuracy.

5.6 MULTIPLE SCALES

A *regular* perturbation problem can be roughly characterized as a case where the solution of the original equations can be recovered by starting with the solution to the $\epsilon = 0$ case and adding corrections using an asymptotic expansion. The water wave dispersion case is of this type for reasons outlined in section 5.4. Unfortunately, the conditions of application for the regular perturbation theory are quite stringent. Problems arise when the $\epsilon = 0$ case is "different in qualitative character" (Lin and Segel 1988, p. 279) than the case when $\epsilon \neq 0$. An equation that can be used to illustrate this kind of qualitative shift is

$$\epsilon m^2 + 2m + 1 = 0 \qquad (5.21)$$

When $\epsilon = 0$, $2m + 1 = 0$, so $m = -\frac{1}{2}$. But if $\epsilon \neq 0$, then the equation has two roots and we cannot recover them by starting with the $\epsilon = 0$ case.[17] A symptom of this sort of failure of the regular perturbation approach is that the asymptotic expansion fails to remain valid for the entire domain under consideration. This can happen when the $\epsilon = 0$ case is not a limit of the small ϵ cases for some part of the domain. When this happens, we must shift to *singular* perturbation theory and try to develop an associated multiscale representation.[18] These techniques have a much wider domain of application, although a cost here is that the interpretive significance of a successful treatment is often much less clear. After reviewing what is perhaps the most famous success case, we turn to some more recent discussions of multiscale modeling. Throughout we will see that the failure of a single set of scales to be adequate corresponds to genuine features of the phenomenon that can be more accurately handled by a shift to two or more distinct sets of scales that are adequate in their respective domains. In the example discussed in this section, it turns out that spatial variation is central to the multiscale representation. Later, briefly in section 5.7, we will see another case where features operate on different temporal scales.

As my central example of singular perturbation and multiple scales, I discuss Prandtl's development of what is known as boundary layer theory. Its significance for our understanding of fluid mechanics is hard to exaggerate, and a leading textbook calls it "perhaps the greatest single discovery in the history of fluid mechanics" (Kundu and Cohen 2008, p. 851). Boundary layer theory refers not only to a part of fluid mechanics but also to a general technique of singular perturbation theory. To start to understand what Prandtl's theory amounts to in the fluid case, consider the flow of a fluid like air around an object. See figure 5.6. Here the vector U_∞ represents the left-to-right incoming velocity of the fluid. As usual, the general problem is to determine how the velocity and pressure values are arrayed around the object and the associated forces impressed on the object. Depending on the incoming velocity, the shape of the object, and the properties of the fluid, the object may experience both an upward force (lift) and a left-to-right force (drag). The lines labeled "$U(x)$" show the velocity in the x-direction as we approach the object for a particular value of y. Crucially, the velocity first increases and then dramatically decreases to 0.

We focus on steady flow, so none of the functions in question are time-dependent. We aim to discover for a given flow the velocity functions $u(x,y)$, $v(x,y)$, and the pressure function $p(x,y)$. u is the velocity component in the x-direction, and v is the velocity component in the y-direction. As shown in the figure, the x- and y-coordinates are not the regular Euclidean axes but are chosen so that the x-direction is always tangential to the surface of the body and the y-direction is always normal to the surface of the body.

---

17. Indeed, determining the roots directly, say, using the quadratic formula, yields values for m in terms of ϵ that are undefined when $\epsilon = 0$.

18. Singular perturbation theory deploys many different methods. In my discussion here I consider only the methods associated with multiple scales. For more discussion see Kevorkian and Cole (1981).

Scale Matters

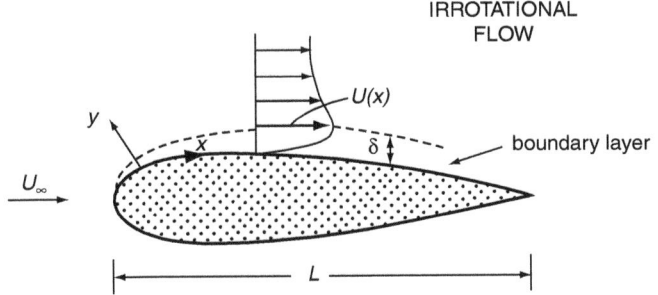

Figure 5.6: Boundary Layer Flow
Source: Kundu and Cohen (2008, p. 48)

Prior to Prandtl, there were two approaches to this sort of system. First, there were the Navier-Stokes equations. For our simplified situation the Navier-Stokes equations are

$$u\frac{\partial u}{\partial x} + v\frac{\partial u}{\partial y} = -\frac{1}{\rho}\frac{\partial p}{\partial x} + \nu\left(\frac{\partial^2 u}{\partial x^2} + \frac{\partial^2 u}{\partial y^2}\right) \qquad (5.22)$$

$$u\frac{\partial v}{\partial x} + v\frac{\partial v}{\partial y} = -\frac{1}{\rho}\frac{\partial p}{\partial y} + \nu\left(\frac{\partial^2 v}{\partial x^2} + \frac{\partial^2 v}{\partial y^2}\right) \qquad (5.23)$$

In addition, conservation laws tell us that

$$\frac{\partial u}{\partial x} + \frac{\partial v}{\partial y} = 0 \qquad (5.24)$$

For the case under consideration, additional boundary conditions are that

$$u(0, y) = U_\infty, \quad v(0, y) = 0 \qquad (5.25)$$

ρ here represents the density of the fluid. We treat ρ as a constant, which is only appropriate when the fluid is incompressible and homogenous, that is, the density is not changed by the pressure and the fluid's features do not vary. Neither assumption, of course, is correct for air, but for speeds well below the speed of sound these assumptions are widely employed.[19] ν in 5.22 and 5.23 represents the kinematic viscosity of the fluid. The viscosity of a fluid measures its tendency to absorb the momentum of adjacent fluid elements. A highly viscous fluid resists internal circulation, whereas a nonviscous fluid easily propagates internal disturbances. It may be represented by $\nu \equiv \frac{\mu}{\rho}$ or μ, known simply as viscosity. As with density, we treat viscosity as a constant. The units of ν and μ are cm^2/s and g/cm·s, respectively, and typical values are

---

[19]. The attentive reader will not be surprised to hear that the justification for this idealization also turns on scaling considerations. See Kundu and Cohen (2008, pp. 714–716).

water: $\nu = 10^{-2}, \mu = 10^{-2}$,
air: $\nu = 15 \cdot 10^{-2}, \mu = 2 \cdot 10^{-4}$ (Segel 2007, p. 87)

The appearance of the viscosity term makes the Navier-Stokes equations very difficult to solve, except for highly artificial contexts. The reason is that the viscosity terms are the only place where the second-order partial differential operators occur.

The second approach employed the Euler equations. The Euler equations result from the assumption that terms preceded by ν are small compared to the other terms. Neglecting these terms gives us

$$u\frac{\partial u}{\partial x} + v\frac{\partial u}{\partial y} = -\frac{1}{\rho}\frac{\partial p}{\partial x} \qquad (5.26)$$

$$u\frac{\partial v}{\partial x} + v\frac{\partial v}{\partial y} = -\frac{1}{\rho}\frac{\partial p}{\partial y} \qquad (5.27)$$

The conservation law and boundary conditions are unaffected. We can motivate this simplification by calculating a rough estimate of the relative size of the terms in 5.22. $u\frac{\partial u}{\partial x}$ is roughly of order U_∞^2/L, where L is the length of the body. By contrast, $\nu\frac{\partial^2 u}{\partial x^2}$ is roughly of order $\nu U_\infty/L^2$. The ratio between the two terms, then, is $(U_\infty^2/L)/(\nu U_\infty/L^2)$. If $U = 50$ km/h, $L = 10$ cm, and $\nu = 0.15$ cm²/s (the value for air), then the $u\frac{\partial u}{\partial x}$ term is about 100,000 times larger than the $\nu\frac{\partial^2 u}{\partial x^2}$ term (Segel 2007, p. 106). So the approximation procedure seems eminently reasonable.

This style of reasoning should be reminiscent of the justification of the small amplitude idealization given earlier in section 5.4. As with the wave dispersion case, we can associate the elimination of the viscosity terms with the claim that a certain set of scales is adequate for the domain in question. In this case the scales are

$$x = \frac{x^*}{L}, \quad y = \frac{y^*}{L}, \quad u = \frac{u^*}{U_\infty}, \quad v = \frac{v^*}{U_\infty}, \quad p = \frac{p^*}{\rho U_\infty^2} \qquad (5.28)$$

(Again, the original variables have asterisks and the dimensionless variables do not.) The scale for pressure may seem ad hoc, but it is chosen based on the units of pressure and so that the resulting $\frac{\partial p}{\partial x}$ term is of the same order of magnitude as the other terms. An important feature of this set of scales is that it treats both spatial variables and both velocity variables in the same way. As with 5.15–5.17 we can make the substitution to estimate the relative size of the terms of the equation. This produces:

$$\left(\frac{U_\infty^2}{L}\right)u\frac{\partial u}{\partial x} + \left(\frac{U_\infty^2}{L}\right)v\frac{\partial u}{\partial y} = -\left(\frac{U_\infty^2}{L}\right)\frac{\partial p}{\partial x} + \left(\frac{\nu U_\infty}{L^2}\right)\left(\frac{\partial^2 u}{\partial x^2} + \frac{\partial^2 u}{\partial y^2}\right) \qquad (5.29)$$

$$\left(\frac{U_\infty^2}{L}\right)u\frac{\partial v}{\partial x} + \left(\frac{U_\infty^2}{L}\right)v\frac{\partial v}{\partial y} = -\left(\frac{U_\infty^2}{L}\right)\frac{\partial p}{\partial y} + \left(\frac{\nu U_\infty}{L^2}\right)\left(\frac{\partial^2 v}{\partial x^2} + \frac{\partial^2 v}{\partial y^2}\right) \qquad (5.30)$$

which simplifies to

$$u\frac{\partial u}{\partial x} + v\frac{\partial u}{\partial y} = -\frac{\partial p}{\partial x} + \left(\frac{\nu}{U_\infty L}\right)\left(\frac{\partial^2 u}{\partial x^2} + \frac{\partial^2 u}{\partial y^2}\right) \quad (5.31)$$

$$u\frac{\partial v}{\partial x} + v\frac{\partial v}{\partial y} = -\frac{\partial p}{\partial y} + \left(\frac{\nu}{U_\infty L}\right)\left(\frac{\partial^2 v}{\partial x^2} + \frac{\partial^2 v}{\partial y^2}\right) \quad (5.32)$$

Here we see the presence of $\nu/U_\infty L$, which is a dimensionless quantity that is characteristic of a given flow. Its reciprocal is the Reynolds number noted earlier. For plausible values, we saw that this quantity is large, so its inverse is small. If we have chosen an adequate set of scales, then all the terms of the equation involving our functions are of the same order of magnitude. The $\nu/U_\infty L = 1/Re$ in front of some of these terms tells us that the product is of a different order of magnitude, and so they can be safely dropped.

In terms of asymptotic expansions, the claim that this procedure is adequate is equivalent to saying that our unknown functions have asymptotic expansions in the small parameter $\nu/U_\infty L = 1/Re$. The results of solving the Euler equations correspond to the starting point of the perturbation where this parameter is set to 0. But if we are right to assume that these expansions exist, we can be confident that additional corrections from higher-order terms will be relatively small in magnitude.

The crucial assumption here is that these are the right scales for the *whole domain* under consideration, including the region close to the object. The failure of this assumption is revealed by d'Alembert's paradox: for this sort of situation, the Euler equations predict that no drag will be felt by the suspended object.[20] This is easily refuted by experiment. Another way to see the limitations of the Euler equations is to note that they cannot be reconciled with a plausible and experimentally motivated boundary condition:

$$u(x, 0) = 0, \quad v(x, 0) = 0 \quad (5.33)$$

That is, at the surface of the object, the fluid should obey a "no-slip" condition. This requires that the velocity of the fluid elements decrease as we consider elements closer and closer to the object, in line with figure 5.6. By contrast, the Euler equations require that the velocity increase as we approach the object, as depicted in figure 5.7. Mathematically speaking, then, there is something wrong with our scaling procedure. It has transformed our presumably correct original equations into the demonstrably incorrect Euler equations.

The flaw in the procedure was the assumption that the scales would lead to terms of comparable magnitude in the entire domain. What we have just seen tells us that in some parts of the domain, presumably in close proximity to the object, the magnitudes of the terms dramatically change and our scaling becomes inadequate. Clearly, the terms originally paired with ν grow in relative significance. Must we then return to the intractable Navier-Stokes equations, or is there another alternative?

---

20. See Darrigol (2005) for some historical discussion of this problem.

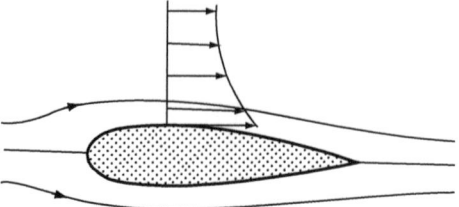

Figure 5.7: Ideal Fluid Flow

Source: Kundu and Cohen (2008, p. 166)

Prandtl's boundary layer theory opts to combine the Euler equations, restricted to the outer region where they work, with some new equations called the boundary layer equations, restricted to the thin region near the object called the boundary layer. These new equations are obtained by a different set of scales. These scales are picked using a new quantity δ, which we can think of as the width of the boundary layer. This reflects our recognition that when it comes to the y-direction, effects operate on a much smaller scale than L. We can estimate δ by first applying the following set of scales, and then seeing what δ must be for the resulting terms to balance:

$$x = \frac{x^*}{L}, \quad y = \frac{y^*}{\delta}, \quad u = \frac{u^*}{U_\infty}, \quad v = \frac{v^*}{U_\infty \delta / L}, \quad p = \frac{p^*}{\rho U_\infty^2} \tag{5.34}$$

This results in the equations:

$$\left(\frac{U_\infty^2}{L}\right) u \frac{\partial u}{\partial x} + \left(\frac{U_\infty^2}{L}\right) v \frac{\partial u}{\partial y} = -\left(\frac{U_\infty^2}{L}\right) \frac{\partial p}{\partial x} + \left(\frac{\nu U_\infty}{L^2}\right) \frac{\partial^2 u}{\partial x^2} + \left(\frac{\nu U_\infty}{\delta^2}\right) \frac{\partial^2 u}{\partial y^2} \tag{5.35}$$

$$\left(\frac{U_\infty^2 \delta}{L}\right) u \frac{\partial v}{\partial x} + \left(\frac{U_\infty^2 \delta}{L}\right) v \frac{\partial v}{\partial y} = -\left(\frac{U_\infty^2}{\delta}\right) \frac{\partial p}{\partial y} + \left(\frac{\nu U_\infty \delta}{L^3}\right) \frac{\partial^2 v}{\partial x^2} + \left(\frac{\nu U_\infty}{\delta L}\right) \frac{\partial^2 v}{\partial y^2} \tag{5.36}$$

Looking to 5.35, we want at least one of the ν terms to be of the same order of magnitude as the other terms and only $(\frac{\nu U_\infty}{\delta^2}) \frac{\partial^2 u}{\partial y^2}$ has our so far undetermined parameter δ. We can fix δ by requiring

$$\frac{U_\infty^2}{L} = \frac{U_\infty \nu}{\delta^2} \tag{5.37}$$

which simplifies to

$$\delta = \frac{L}{\sqrt{Re}} \tag{5.38}$$

Scale Matters

In turn, this allows us to rewrite (5.35) and (5.36) as

$$u\frac{\partial u}{\partial x} + v\frac{\partial u}{\partial y} = -\frac{\partial p}{\partial x} + \left(\frac{1}{Re}\right)\frac{\partial^2 u}{\partial x^2} + \frac{\partial^2 u}{\partial y^2} \qquad (5.39)$$

$$\left(\frac{1}{Re}\right)u\frac{\partial v}{\partial x} + \left(\frac{1}{Re}\right)v\frac{\partial v}{\partial y} = -\frac{\partial p}{\partial y} + \left(\frac{1}{Re^2}\right)\frac{\partial^2 v}{\partial x^2} + \left(\frac{1}{Re}\right)\frac{\partial^2 v}{\partial y^2} \qquad (5.40)$$

As Re is much larger than 1, $1/Re$ and $1/Re^2$ are much smaller than 1 and the corresponding terms can be dropped. This results in the boundary layer equations:

$$u\frac{\partial u}{\partial x} + v\frac{\partial u}{\partial y} = -\frac{\partial p}{\partial x} + \frac{\partial^2 u}{\partial y^2} \qquad (5.41)$$

$$\frac{\partial p}{\partial y} = 0 \qquad (5.42)$$

Again, from a series perspective, these equations represent the case when our small parameter is set to 0. Additional corrections could be obtained using the asymptotic expansions of the unknown functions, but for the domain in question we can be confident that they will not involve a significant correction.

Although 5.41 still remains second-order, we have removed one of the problematic terms. In conjunction with our earlier conditions and some additional boundary constraints, the problem becomes mathematically tractable. These new constraints arise from the general requirement that the flow within the boundary layer match the flow in the outer region along the upper edge of the boundary layer. By 5.42, we know that the pressure at the upper edge of the boundary layer remains constant in the y-direction and must agree with the pressure in the outer region. In addition to the no-slip condition, we impose the conditions

$$u(x, \infty) = U_\infty, \quad u(x_0, y) = u_o(x_0, y) \qquad (5.43)$$

where x_0 is some point upstream that can be used to determine the flow downstream in the boundary layer by matching it with the outer region u_o.

From a traditional mathematical perspective, there is a decided lack of rigor in these sorts of arguments. Discussions of scaling typically emphasize the empirical character of the choice of scales and, as we have seen, an apparently well-motivated choice of scales can prove inappropriate for certain purposes. In their discussion of scaling, for example, Lin and Segel offer six ways to arrive at the correct scales that include "utilize experimental or observational evidence," "make certain order of magnitude assumptions merely because the concomitant neglect of terms renders the problem tractable," "use a trial and error approach," and "employ the results of computer calculations" (Lin and Segel 1988, p. 222). Even after reviewing the manifestly successful example of boundary layer theory, Segel concedes that "formulation of the boundary layer problem is hardly a straightforward matter, and the critical reader may well feel uneasy at this point" (Segel 2007, p. 114). Although scaling is clearly a mathematical transformation of a given set of equations, its adequacy in a given case is not something accessible to the pure mathematician.

Another point of concern is the combination of two different scales and the resulting claim that they can be matched along an edge between the outer region and the boundary layer. Essentially what we are doing is considering two different asymptotic expansions for each of our unknown functions. We have tacitly appealed to both series expansions, which are adequate in their respective regions of the domain, and then matched them in a common, overlap region. This helps account for the peculiar procedure employed when we forced the two sets of equations to agree along the "edge" of the boundary. From our current perspective, what we are really doing is taking two series expansions and claiming that matching them yields an accurate solution of the original Navier-Stokes version of the problem.

The mathematical techniques used in this case provide considerable flexibility and can be deployed to manage a wide range of problems where regular perturbation theory fails. For example, when Fowler explains these singular perturbation methods in chapter 4 of his book *Mathematical Models in the Applied Sciences*, he notes that "They provide the analytic platform upon which much of the later discussion in this book is based" (Fowler 1997, p. 45). Lin and Segel motivate their discussion this way:

> It might seem that the practitioner of singular perturbation theory would find relatively few important problems to work on, but this is not at all the case. The abundance of such problems is at least partially explained by the fact that singular perturbation techniques seem to be appropriate for examining the behavior of solutions near their "worst" singularities. "Taking the bull by the horns" and examining functions near their worst singularities is the best way to obtain information about qualitative behavior, and it is the elucidation of this behavior that is very often the goal of a theoretician. In this lies the point of a remark once made about a distinguished applied mathematician, "All he can do is singular perturbation problems. But of course he can turn all problems into singular perturbation problems!" (Lin and Segel 1988, p. 277)

Here, then, is a powerful technique whose domain of application extends far beyond the fluid mechanics case I have focused on.

This example seems to have features that can be used to distinguish applied mathematics from pure mathematics more generally. In particular, the invocation of a thin boundary layer was motivated by the need to recover the details of a physical phenomenon, and not by any intrinsic mathematical interest. It is, of course, a mathematical discovery that the Navier-Stokes equations can be simplified in this way, but what made the innovation so important was its success in resolving various scientific problems having to do with fluid flow. Coinciding with these nonmathematical priorities, we find a failure of mathematical rigor. The disconnect between boundary layer theory as a practical tool for science and engineering and its status as a mathematical theory is well summarized by Nickel in his 1973 survey article "Prandtl's Boundary-Layer Theory from the Viewpoint of a Mathematician": "During the first 50 years of boundary-layer theory the fundamental mathematical questions could not be answered. It was not possible to establish a sound mathematical connection to the Navier-Stokes differential equations. There was no evidence of the existence, uniqueness, and well-posedness of a solution" (Nickel 1973, pp. 405–406).[21] Similarly, Segel quotes Carrier's 1972 evaluation of Prandtl's contributions:

> they have provided the basic foundations for the heuristically-reasoned, spectacularly successful treatment of many important problems in science and engineering. This success is probably most surprising to rigor-oriented mathematicians (or applied mathematicians) when they realize that there still exists no theorem which speaks to the validity or the accuracy of Prandtl's treatment of his boundary layer problem; but seventy years of observational experience leave little doubt of its validity and its value. (Segel 2007, pp. 77–78)

Though the situation had improved up to the 1970s and has no doubt improved since then, the failure of boundary layer theory as a rigorous mathematical theory is significant. It shows that the standards for admission as a tool of applied mathematics are different from what a pure mathematician would hope for. A sacrifice in intrinsic mathematical clarity is typically made if a consequent reward in physical understanding can be achieved.

A further feature of this application of singular perturbation theory is also worth noting. This is that the success raises interpretive questions that might be given an overly metaphysical answer. In this case, the edge between the boundary layer and the outer region is fundamental to the representation, but we do not take it to represent any genuine edge in the system itself. In fact, although we have pinned down the width of the boundary layer δ using considerations of scale, Kundu and Cohen note several inconsistent ways to estimate δ for different purposes (Kundu and Cohen 2008, pp. 346–348). This shows that a metaphysical interpretation of the features that arise in this sort of mathematical treatment is not warranted by the success of the mathematical representation. As we will see in the next section, the trade-off between accuracy and interpretive debate is a common result of singular perturbation techniques. In line with the more epistemic character of my approach, I argue that we have no general reason to conclude that the accuracy of these representations reveals new underlying metaphysical structure.

5.7 INTERPRETING MULTISCALE REPRESENTATIONS

In the last section we saw how a multiscale representation could be developed by dividing a system into two or more nonoverlapping domains. Each domain is then assigned its own set of scales, and the boundary conditions for each domain are adjusted so that they all agree on their respective boundaries. I turn now to another kind of multiscale representation. This occurs when the features or the processes are analyzed in terms of two or more scales that are appropriate for the same

---

21. This passage is noted by Vincenti and Bloor (2003).

domain. Somewhat surprisingly, two recent philosophy papers discuss examples of this kind and draw broader philosophical conclusions about their significance. After discussing each case, I conclude with my own take on the importance of multiscale representations.[22]

In "Reductive Levels and Multi-Scale Structure," McGivern focuses on the case of the damped harmonic oscillator (McGivern 2008, pp. 60–64). He chooses this example because we can represent such systems directly without recourse to a multi-scale analysis. This allows us to compare the benefits of the multiscale representation with what our most realistic representation tells us is really going on. From this starting point, we can then understand what is so useful about multiscale representations when more direct representations are not available. Recall that a linear harmonic oscillator is a system like a spring or pendulum that obeys

$$my'' + ky = 0 \qquad (5.44)$$

where m is the mass and k is a constant. Damping occurs when the linear motion is disrupted by a factor proportional to the velocity y'. This results in

$$my'' + cy' + ky = 0 \qquad (5.45)$$

A first attempt at scaling might assume that the damping effects were relatively small compared to the dominant processes. Adopting this perspective, we can rescale the equation using

$$t = \frac{t*}{\sqrt{\frac{m}{k}}} \qquad (5.46)$$

so that it becomes

$$y'' + \epsilon y' + y = 0 \qquad (5.47)$$

where

$$\epsilon = \frac{c}{(mk)^{1/2}} \qquad (5.48)$$

Assuming that $\epsilon \ll 1$, we can express our unknown y in terms of an asymptotic expansion involving ϵ. This yields

$$y(t) \sim \cos(t) + \frac{1}{2}\epsilon[-t\cos(t) + sin(t)] + \ldots \qquad (5.49)$$

where additional terms can be obtained by repeating our procedure for additional powers like ϵ^2 and ϵ^3.

---

22. See also Winsberg (2006, 2009).

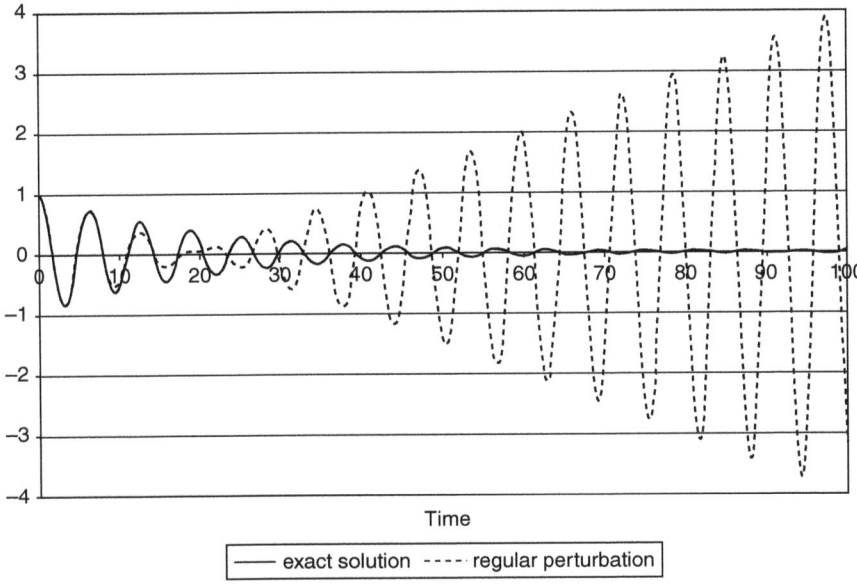

Figure 5.8: The Failure of a Single Scale

Source: McGivern (2008, p. 62)

As McGivern notes, this analysis in terms of a single time scale fails because as time grows the $-t\cos(t) + \sin(t)$ term grows in orders of magnitude no matter how small we assume ϵ to be. This leads to a dramatic failure in accuracy. See figure 5.8. The diagnosis of this failure is that we cannot assume there is a single process operating on a single time scale.[23] What we need to do is think of our y as a function of two time variables, $t_F = t$ and $t_S = \epsilon t$. When $\epsilon \ll 1$, $t_S \ll t_F$ so t_S corresponds to a "slow" time scale. Applying both scales produces

$$y(t) \sim e^{-t_s/2} \cos(t_F) + \ldots \quad (5.50)$$

This equation turns out to be incredibly accurate, thus vindicating our belief that two time scales are appropriate. See figure 5.9.

The success of this sort of multiscale approach can be extended into other cases where our original equations prove intractable. Even in these cases, experimentation can often reveal a match between the predictions generated by a multiscale representation and experimental data. Although the basic idea is similar to the boundary layer case, here the singular perturbation does not depend on dividing up the domain spatially. Instead, we have analyzed the phenomenon into two processes that operate on two different time scales.

Bishop's "Downward Causation in Fluid Convection" describes a different kind of case where a system requires an analysis on two scales (Bishop 2008). Consider

---

23. Notice how this is different from the boundary layer case where the assumption that there was no spatial variation in the processes proved false.

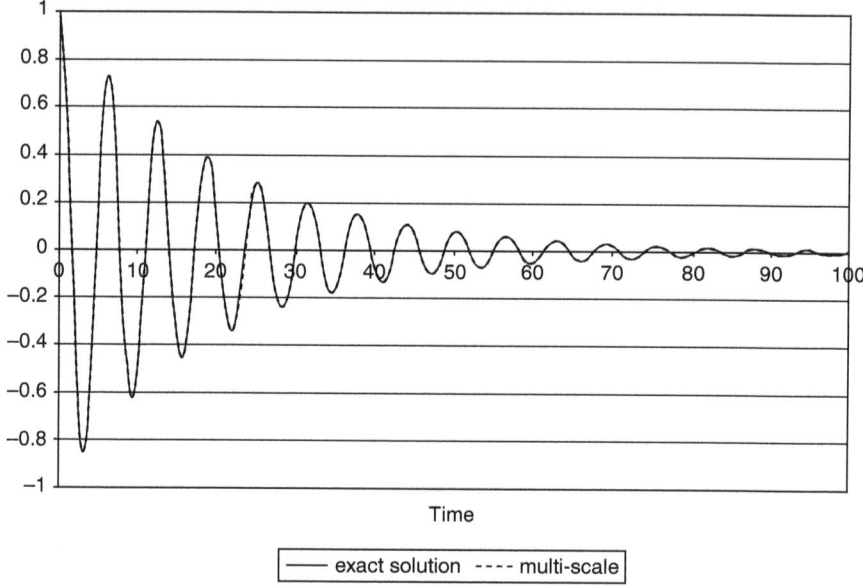

Figure 5.9: The Success of Two Scales

Source: McGivern (2008, p. 63)

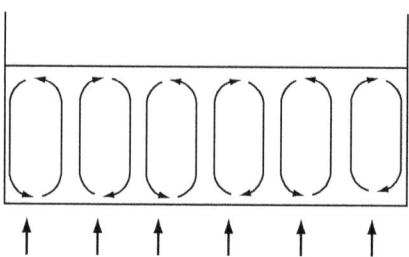

Figure 5.10: Bénard Cells

a fluid in a container where the bottom of the container is heated to a temperature above the temperature at the top of the container. The heating at the bottom decreases the density of the fluid at the bottom, and this creates an unstable system. Initially the difference in temperature can be handled through diffusion, where the lower elements heat up the higher elements and the heat eventually dissipates through the top of the fluid. However, when the difference in temperature ΔT passes a certain threshold ΔT_c, this dissipation is no longer fast enough. That is, the heating occurs in a shorter time scale than the diffusion process. A new, more stable configuration occurs when the fluid system develops interlocking convection cells known as Bénard cells.[24] See figure 5.10. These are large-scale features of the

---

24. These obtain in an interval $\Delta T_c < \Delta T < \Delta T_\alpha$. At some point beyond ΔT_α the flow becomes turbulent.

fluid system that influence how the small-scale fluid elements move. Bishop argues that a key to understanding such features is the relative size of the diffusive terms and the convective terms in the Navier-Stokes equations. Diffusive terms represent the processes that are responsible for the dissipation of heat noted above, and convective terms concern larger-scale fluid motions. Like our ϵ before, there is a parameter that can capture this difference: the Reynolds number, which has cropped up throughout our discussion of scaling and fluids several times. Here is how Bishop explains the significance of the Reynolds number for his case:

> The Reynolds number is a dimensionless parameter that roughly characterizes the effectiveness of convective versus diffusive processes. For small values, the slower time scale (shorter length scale) diffusive processes are dominant, while at high values the faster time scale (longer length scale) convective processes are dominant. The interplay between these two sets of time and length scales is important. If the values of the Reynolds number [are] too small, coherent long-lived structures typically do not form. On the other hand, if the value is too large, turbulence dominates the flow and coherent short-lived structures only form intermittently. It is in a range in between the extremes where coherent long-lived structures form. (Bishop 2008, p. 235)

Although the mathematical details are more complicated than in McGivern's case, the basic idea remains the same. A single-scale analysis would miss the role of these two types of processes and so make a false prediction. Because the interaction between these two types of processes is crucial to the formation and maintenance of Bénard cells, only a multiscale representation is appropriate.

Both McGivern and Bishop take their examples of multiscale analysis as challenges to certain popular metaphysical claims about how the physical world is structured. McGivern challenges micro-based physicalism, and Bishop presents his case as a viable instance of downward causation within the domain of the physical. Micro-based physicalism says that each property P of a large thing A is identical with a complex property of nonoverlapping small physical things $a_1, \ldots, a_n$ (McGivern 2008, p. 58). The complex property may make reference to the monadic physical properties of the a_i and their spatiotemporal or mereological relationships. McGivern complains that this position assumes a certain kind of spatial decomposition that is undermined by multiscale representations: "in the case of decomposition into fast and slow components, it is the same particles at the micro-level that realize both the fast and the slow components ... we appear to have a kind of structural property that doesn't fit with the standard micro-based account of reductive levels" (McGivern 2008, p. 65). Although he admits that there may be some way to rescue micro-based physicalism, McGivern notes that no advocate of this position has yet done the hard work on real scientific examples that would be necessary to effect this rescue.

A failure of micro-based physicalism entails the existence of a certain kind of emergent property, namely, those properties that are not micro-based. Such properties are not merely epiphenomenal because they are ascribed genuine causal and explanatory roles throughout the sciences. McGivern mentions, for example, "in oceanographic and atmospheric modeling, it is common to distinguish between the effects of fast-scale gravity waves and slow-scale Rossby waves ... and generalizations about such components can play distinct roles in explaining the behaviors of different atmospheres" (McGivern 2008, p. 70).[25] So unless the micro-based physicalist aims to reform scientific practice, she needs to reform or abandon her physicalism.

There are, of course, more generous conceptions of what counts as physical. One of the most well-worked-out approaches is Melnyk's realization physicalism (Melnyk 2003). Melnyk explicitly denies micro-based physicalism, but still insists that all instances of properties are physically realized. Roughly, this is the claim that any instance of a property P exists in virtue of a functional property instance, where the function is characterized in physical and broadly logical terms. It is hard to know if McGivern's properties pose a problem for realization physicalism because it is hard to know if their characterization in our multiscale representation fits with Melnyk's restrictions. Although these details are worth pursuing, it is probably Melnyk's intention to allow these techniques as they are central to physics, and it is physics, after all, that fixes what the physical amounts to.

Bishop, by contrast, claims that his convection case is an instance of downward causation within physics. If this is right, then we have emergent physical properties with novel causal powers that cannot be accounted for in more fundamental terms. That is, there would be causal powers that a property bestowed on its bearers that did not result from its more basic realization. In Bishop's case we have the Bénard cells, so the relevant property appears to be a feature of the spatially large and temporally extended region that is occupied by a chain of interlocking Bénard cells. At the same time, we have the spatially smaller fluid elements whose features are affected by the cells. Bishop suggests that this complicates any ordinary picture of realization:

> The properties individual fluid elements contribute to the cells are conditioned upon the properties instantiated by the cells. Furthermore the downward determination is instantaneous; that is, the properties of the fluid elements only contribute *when* they realize particular other properties such as those associated with the collective motion and large-scale structure of Bénard cells. (Bishop 2008, p. 243)[26]

All this occurs within the physical, so it cannot be meant as a counterexample to a realization physicalist who is only interested in physical–nonphysical interactions. Still, if there is this kind of emergence within physics, then it becomes reasonable to look for it in other cases, for example, in biology and psychology. Multiscale representation seems to be used in some areas of biology such as cell biology and ecology. So, if Bishop is on the right track, it appears that physicalism is in trouble.

---

25. In fact, Rossby waves are deployed in the delayed-action oscillator model of ENSO noted in section 5.2. Gravity waves include the deep-water waves discussed in sections 5.4–5.5.

26. Bishop qualifies the claim about instantaneous determination in his note 20. Still, the time gap here seems insufficient to interpret this case in terms of an ordinary feedback loop between scales.

Despite their differences, McGivern and Bishop both aim to draw metaphysical conclusions from their successful multiscale representations. I am wary of moving too quickly here for a simple reason. Most metaphysical discussion in contemporary philosophy is not informed by the complications of actual scientific practice. As a result, it is often easy to show that both sides of a metaphysical debate are undermined by examples from science. This seems to be the case with the ongoing debates about "levels" and their relationships. Both reductionists and their critics underestimate the difficulty in developing a workable representation of a complex system. So it is not too surprising that actual cases of successful science here fail to fit preconceived metaphysical positions. If this is right, though, we should use multiscale representations to help reform and sharpen the metaphysical positions, rather than using them to champion one or the other side.

Based perhaps on this disconnect between science and metaphysics, many philosophers of science express hostility to metaphysics and simply are not very interested in reforming metaphysics. The danger with this approach to multiscale representation is that it risks a slide into some kind of pragmatism or even instrumentalism about the features described by these representations. That is, rescaling with multiple scales is just a tool that allows us to handle otherwise intractable mathematical equations. We have learned how to work with this tool, but it need not reveal anything about the underlying features of the systems represented. Multiscale modeling, then, is just a piece of brute force mathematics whose larger significance can be ignored.

There is a third way, though, between a metaphysical interpretation and instrumentalism. This is to emphasize the epistemic benefits that multiscale representation affords the scientist. On this picture, a successful multiscale representation depends on genuine features of the system. We come to know about these features when a multiscale representation yields a successful experimental prediction. In this respect, it goes beyond the instrumentalist position. But it does not end up with a reductionist or emergentist metaphysical interpretation because the epistemic approach is consistent with a rejection of both reductionism and emergentism. To see why, consider Bishop's Bénard cells. When we represent a system in terms of both fluid elements and Bénard cells, we can accurately predict what will happen if, for example, we observe some dye traveling through the system over time. Talk of Bénard cells allows us to capture these features of the system and we would not be able to represent them directly in terms of the dynamics of the fluid elements alone. But this epistemic point about what we know based on our limited access to the details of the fluid dynamics need not entail any metaphysical conclusions about the existence of the Bénard cells in some more robust sense. Clearly, we can only understand the system by appeal to these larger-scale structures. But this may be a product of our ignorance and in the end may not correspond to what a fuller understanding of the system would reveal. So as with many metaphysical debates, we just do not know if Bénard cells have the features that Bishop ascribes to them or if some reduction along some lines is possible. Either metaphysical position seems premature, even if we can reform these views to accommodate this kind of case.[27]

---

27. For an attempt at reform see Wimsatt on "perspectives" (Wimsatt 2007, p. 228).

More generally, I suggest that we cannot move from the success of a multiscale representation to the conclusion that it reveals novel metaphysical features of the physical system. This metaphysical agnosticism is consistent with delineating a crucial epistemic role for the multiscale techniques in producing scientific knowledge.

5.8 SUMMARY

Our discussion over this long chapter has considered nine examples. In each case we can find an epistemic dimension to the mathematical contribution, although these contributions work differently as we change the examples. In section 5.2 we covered cases that involve scale separation either in the strict sense where all relevant features occurred at a single scale or in the extended sense where relevant features across scales were handled by other means. Section 5.3 discussed several senses in which systems can be similar and argued that the notion of scale can be fruitfully deployed to help us understand how we can successfully represent these families of systems. An important class of idealizations was our main concern in section 5.4 with the example of water wave dispersion. This set the stage for a survey of the central ideas of perturbation theory and how they link accurate representation, scaling and epistemic benefits in sections 5.5 through 5.7. A central theme here was the distinction between regular and singular perturbation techniques. Although regular cases are the most straightforward, they also require that special conditions obtain. This is that the $\epsilon = 0$ case can be the basis for a treatment of the $\epsilon \neq 0$ case. The regular cases are fairly rare. Instead, as we saw with boundary layer theory, unanticipated failures in the application of the regular approach can yield new insight into the processes operating in the physical system. Singular techniques are more flexible, but when they involve the invocation of multiple scales, there are associated interpretive risks. Although these risks were not that great for the boundary layer case, it turns out that at least for McGivern and Bishop there are supposed to be immediate metaphysical consequences from the successful use of such representations. I argued, though, that these metaphysical conclusions are often premature precisely because we can see that such representational techniques can operate successfully even when they do not track metaphysically fundamental features of systems. This does not stand in the way of my epistemic point, though. For the perturbation techniques have a wide domain of application and have allowed many accurate representations of complex systems that would otherwise be unavailable.

6

Constitutive Frameworks

6.1 A DIFFERENT KIND OF CONTRIBUTION

The last three chapters surveyed many different cases where mathematics seems to be making a contribution to the success of a scientific representation. Four kinds of cases have been described. First, we have concrete causal representations where the mathematics helps track causal features of a system. Second, there are the abstract acausal cases where mathematics contributes to the elimination of causal content. Third, we considered abstract varying representations where a family of different representations was linked by an overlap in mathematics. Finally, in the last chapter, we considered several different ways mathematics can specify the scale of a representation and thereby link representations across scales. Though very different, all of the examples so far have been similar in one important respect. They have all involved what I call derivative representations. By this I mean that these representations have been successful against the backdrop of other constitutive representations. The nature of the relationship between constitutive and derivative representations is difficult to pin down, and as we will see there is considerable disagreement on what it comes to or whether it even exists. To get the debate going, though, I will give the following preliminary characterization of what makes a representation derivative.

> A representation r_1 is *derivative* when its success depends on the success of another *constitutive* representation r_2.

This makes clear that our distinction is relative. A given representation r_2 may be constitutive with respect to another representation r_1, but then it may turn out to be derivative with respect to a third representation r_3. Beyond this, though, our proposal does little to clarify what sort of success is in question or why this sort of relationship might obtain between representations. It also leaves open the possibility that the success of a representation *qua* constitutive is different than the success of a representation *qua* derivative.

In line with the broader aims of this book, I restrict my focus to cases where mathematics makes a contribution to the success of the constitutive representation itself or to the relationship between a derivative and a constitutive representation. Eventually I argue that there is good reason to think that there are such cases, but to reach this conclusion some preliminary discussion is necessary. This is because the philosophical issues involved are fairly intricate. Among other problems, it does not appear that the best advocates for the derivative/constitutive distinction have given a satisfactory explanation of what the distinction comes to. I suspect that this weakness can be traced to a failure to focus on the status of the mathematics in these cases. Considering the mathematics and its function in the constitutive representations raises new problems for the whole approach, however. At the end of the chapter I argue that some form of a priori justification for pure mathematics is required if we are to defend an interesting derivative/constitutive proposal.

To clear the ground, I briefly review what I take to be the three most important attempts to defend a special role for constitutive representations. These are found in Carnap's "Empiricism, Semantics and Ontology," Kuhn's *Structure of Scientific Revolutions*, and Friedman's *Dynamics of Reason*. Carnap is best understood as emphasizing the normative role of his framework principles, although this is mixed with a more semantic conception of what the framework principles are doing. Kuhn, in turn, articulated a notion of scientific paradigm that incorporated both the normative and semantic functions of Carnap's framework principles, but also at times went further and included certain metaphysical components as well. More recently, Friedman has isolated what I take to be the defensible core of the constitutive approach: that these principles serve a crucial epistemic function in allowing the confirmation of derivative representations. Still, as with Carnap and Kuhn, the semantic picture that Friedman adopts obscures this epistemic point.

After reviewing these three proposals, I develop my own epistemic proposal and clarify its implications for mathematics and the status of the constitutive representations themselves. I do not claim any originality for this proposal, as it is identical in all important respects to Friedman's views, at least under one interpretation. Still, I hope my discussion can move the debates about Friedman's views forward by focusing more attention on the link between mathematics and constitutive representations. One shift from previous chapters is that the discussion here is more philosophical and involves fewer detailed examples. At the same time, the alleged cases of constitutive representations play a central part in our discussion. As we proceed, the differences between these examples and the earlier examples will show that we have a different kind of contribution from mathematics.

6.2 CARNAP'S LINGUISTIC FRAMEWORKS

Carnap's "Empiricism, Semantics and Ontology" (Carnap 1956) was written to clarify the role of abstract objects like propositions in his semantics, for example, in *Meaning and Necessity*. However, in the course of making these points, Carnap also offered some of his most direct statements about the status of his linguistic frameworks and the associated distinction between questions that are internal to a framework and those that are external. His basic idea is that the discussion of a new domain requires the introduction of new signs. For these signs to be meaningful, their use must be governed by rules, and so the introduction of these signs goes along with the specification of these rules. Finally, Carnap assumed that the rules in question would include at least some specification of how existence questions were to be resolved for the objects referred to by the new signs. This need not involve anything as rigid as verificationism, according to which the meaning of a sentence involving the new sign must be specified in terms of which observations would constitute conclusive verification of the truth of the sentence. Carnap allowed a much weaker link between new theoretical terms and observational terms. For example, a new theoretical term T_1 might have its meaning specified in terms of several "reduction sentences" of the form "$\forall x(O_i(x) \rightarrow T_1(x))$." Even weaker conditionals stopping short of a sufficient condition for something to be a T_1 seem acceptable if the language has a term for assigning varying degrees of confirmation.

Here is how Carnap explains his notion of linguistic framework:

> If someone wishes to speak in his language about a new kind of entities, he has to introduce a system of new ways of speaking, subject to new rules; we shall call this procedure the construction of a linguistic *framework* for the entities in question. And now we must distinguish two kinds of questions of existence: first, questions of the existence of certain entities of the new kind *within the framework*; we call them *internal questions*; and second, questions concerning the existence or reality *of the system as a whole*, called *external questions*. Internal questions and possible answers to them are formulated with the help of new forms of expressions. The answers may be found either by purely logical methods or by empirical methods, depending upon whether the framework is a logical or a factual one. (Carnap 1956, p. 206)

Carnap argues that mathematical entities always involve a logical framework. This means that all internal questions involving just mathematical and logical terms have a determinate answer given by the rules for the framework. An example of a factual framework would be a language for fluids of the sort we have developed in our earlier examples. Here certain internal questions could be resolved just using the rules of the framework, but other internal questions only have a determinate answer in conjunction with the acceptance of additional factual claims. For example, "inviscid fluids obey the Euler equations" might be given an answer by the rules of the framework, whereas "water is a viscous fluid" could require additional factual claims about water to link it to the definition of viscous fluid.

It is fair to summarize one aspect of Carnap's proposal as the claim that internal questions have answers that we have a reason to accept only because we have accepted a given linguistic framework. More generally, acceptance of a linguistic framework is a necessary condition on a given belief being rational. For certain beliefs and linguistic frameworks, the acceptance of a linguistic framework is also a sufficient condition for the belief to be rational. Science, like any genuine theoretical discipline, needs a linguistic framework as a basis for rational belief. Notice that Carnap has a very simple argument for this conclusion. It assumes that new signs must be given a meaning and that this meaning comes in the form of rules that will serve in part to resolve certain questions raised using the new signs. The attempt to ask a question using a sign independently of its linguistic framework is then futile because the sign will be without meaning, and so the question will be meaningless. This is how Carnap views traditional philosophical questions. These external questions do not take the rules of the linguistic framework for granted. But if these rules are in question, Carnap is unable to see what the content of an external question might be. Suppose, for example, that someone asked if fluids exist. If this question cannot be answered using the rules of the fluid framework, then it is intended as an external yet still theoretical question about the reality of fluids in some deeper sense. But Carnap responds that "An alleged statement of the reality of a system of entities is a pseudo-statement without cognitive content" (Carnap 1956, p. 214). If we take the rules for fluid talk for granted, then it is easy to establish the existence of fluids. But if we call these rules into question, then an investigation of the existence of fluids is no longer possible and the question is not something we could have a reason to answer either way.

Although Carnap rejects the possibility of external, theoretical questions, he allows that an external question can be given an answer if it is understood as a practical question of how best to achieve certain goals. If we are asked whether fluids exist, then this may be intended as a question about whether adopting the fluid framework would contribute to the goals we had set ourselves, for example, to design ships. There can be genuine disagreement here about whether this framework would be the best means to achieve this goal. But now, "the acceptance [of a framework] cannot be judged as being either true or false because it is not an assertion. It can only be judged as being more or less expedient, fruitful, conducive to the aim for which the language is intended. Judgments of this kind supply the motivation for the decision of accepting or rejecting the kind of entities" (Carnap 1956, p. 214). When theoretical reasons are no longer possible, then, Carnap allows that practical reasons may give a question some content.

We can fit this linguistic framework proposal into our talk of constitutive and derivative representations: a representation r_1 is derivative when its success depends on the success of another constitutive representation r_2. For Carnap, the rules of the framework exhaust the constitutive representations. These are the rules for the proper use of the signs that make up the language of the framework. For a given agent, the sentences of the language that are believed are then the derivative representations. The relationship between the constitutive and the derivative representations has two main facets, but the crucial one I emphasize is that adopting the rules of the framework allows an agent to have a reason to believe a sentence of the language. This relationship can also be described using the notion of presupposition. An agent's rational belief in a sentence of the language presupposes having adopted the rules of the framework. This is not the only kind of presupposition in question on Carnap's approach. There is also a tight semantic relationship between the constitutive and derivative representations. A sentence of the language only has a meaning in virtue of the rules of the framework. For Carnap, the semantic presupposition is the basis for the rational belief presupposition. But it should be clear that these are two different kinds of relationships, and other positions could drop one and try to retain the other.

It is trickier to see what Carnap thought the success of the constitutive and derivative representations came to, although it is obvious that he thought their success consisted in different things. He is reluctant to speak of the success of his derivative representations in terms of their correspondence with reality or any similarly metaphysical notion of truth. Instead, he seems to think the success of science amounts to the fact that we have beliefs that are rational, that is, either they derive their support from the rules of the framework or they are supported by the other beliefs that we have adopted on the basis of experience. On this reading, then, there is an immediate connection between the rational belief presupposition and the success of the derivative representations. The success of the constitutive representations is evaluated differently. Recall that Carnap insists that the reasons to adopt rules are different in kind than the reasons to believe a sentence of a language. For the constitutive representations, Carnap has instead his practical test: does adopting this framework contribute effectively to our goals? We must consider the practical implications of choosing the framework when we determine if the constitutive representations are successful.

As should be clear from the fluid framework example, mathematics plays a central role in Carnap's linguistic framework proposal. To begin with, mathematical vocabulary is introduced just like any other new set of terms with its own rules. As I noted, Carnap assumes that all wholly mathematical and logical sentences will be assigned a determinate truth-value by the rules of the linguistic framework. This provides the language with a stock of sentences that an agent has a conclusive reason to believe, even though Carnap grants that it may be very hard to find out which sentences have this feature. Additional terms can then be introduced into the language using new rules that make reference to mathematical and logical terms. This is of course how fluids are typically defined, as with the sentence "a fluid is a substance that deforms continuously under a shear stress." For Carnap, this sentence would be associated with a rule or series of rules involving the new term *fluid*. The derivation of the Navier-Stokes equations, for example, would involve the discovery of a set of sentences that are made true by the combined rules of the mathematics and the fluid framework. Additional experimental investigation would then be needed to determine which observed substances obey these equations or in which respects they fail to be fluids. For Carnap, then, mathematics plays a crucial role in making science as we know it possible. There is no deep account of why this or that successful framework uses mathematics, but just the observed historical fact that many successful frameworks are highly mathematical.

Though we have succeeded in clarifying Carnap's proposal and relating it to our derivative/constitutive distinction, the resulting picture of science and its success is highly problematic in a number of respects. The first issue I want to raise is Carnap's conception of meaning and its link to evidential support. We saw that Carnap's argument depended on two claims: (1) the meaning of a new term is given by rules for its proper use, and (2) the rules will relate, at least in part, to how sentences using that term can be supported. Neither claim seems correct. Against (1), it is now common to think that the meaning of some terms used in science is based on some causal or other kind of interaction between an agent and something in the world. This broadly externalist conception of meaning entails that the meaning of a new term is not exhausted by a set of rules that the speaker is aware of or even potentially aware of. This is the view I embraced in section 2.1 when I allowed that the content from a physical concept may go beyond what is reflected in the concept itself. To continue the fluid example, an original introduction of the term *fluid* could have occurred in terms of ordinary interactions with water, air, and so on. This can prompt a scientist to try to develop a rigorous theory of fluids, but the articulation of this theory does not confer a new meaning on the term. Rather, the term continues to refer to the substances that it originally referred to, and this allows us to evaluate new and old theoretical proposals in terms of the evidence available. By contrast, in Carnap's picture, different linguistic frameworks might use the same word *fluid*, but if their rules are different, the meaning of this term changes across frameworks. Even more problematically, a framework can only be criticized using the pragmatic question of how it has helped us achieve our goals. There is no framework-independent subject matter that we can hold fixed and use to evaluate scientific progress.

Semantic externalism has its limitations, especially when it comes to logical and mathematical terms whose reference is hard to imagine fixing through interactions with mathematical entities. Here it is considerably more plausible to consider the meaning of a new term as being fixed by the adoption of rules.[1] But, contrary to (2), these rules need not specify how claims can be supported. Instead, they might involve rules that reflect features of the objects in question. For example, in introducing the word *group* I may adopt certain rules for the use of this word. It is not an automatic consequence of adopting some rules, though, that my word winds up referring to anything. A plausible requirement is that these rules must correspond to the genuine features of the things I wind up referring to using the new word. Carnap, of course, does not recognize this restriction and would criticize my conception of the link between rules and reference as traditional metaphysics. My reply is that his picture makes meaning too easy to achieve and clashes with the way that mathematicians and scientists go about deciding when their words have successfully picked out something in the world. Even if the meaning of a new word in mathematics is given by rules, I can adopt these rules and still not be in a position to know if I am referring to anything. Similarly, in retrospect, we can look back at the work of a mathematician and understand what his or her claims mean. If the meaning is specified using rules, then we effectively adopt the rules for their terms. Still, we can come to see that their terms did not refer to anything, as with mathematical theories that we have come to see are inconsistent or defective in other respects.

What we need, then, is a different way to motivate the rational belief presupposition than Carnap's semantic presupposition argument. To accord with our working conception of scientific knowledge, this must go beyond a weak framework-relative notion of success for the derivative representations. When it comes to the constitutive representations, their status must receive more substance than Carnap's rules offered us.

6.3 KUHN'S PARADIGMS

Kuhn's picture of paradigms, normal science, and scientific revolutions has decisively influenced the philosophy of science and our conception of science more

---

1. I eventually reject this form of semantic internalism in chapter 13.

generally. It would be futile to attempt a summary of Kuhn's views here or the debates about their proper interpretation and correctness that followed the publication of *The Structure of Scientific Revolutions* in 1962. Instead, I present a selective reconstruction of those aspects of Kuhn's views that promise to shed light on our derivative/constitutive distinction as well as on the special place for mathematics in paradigms.

To start, we can introduce the two notions of paradigm that are at the core of Kuhn's picture of normal science. The first and "deeper" (Kuhn 1970, p. 175) notion concerns an exemplar of how a problem of a certain type should be handled. Exemplars are central to scientific training, as shown in their prominent role in textbooks, exams, and laboratory demonstrations. To become a scientist in a given scientific community, Kuhn argues, a person must master a stock of exemplars. These worked problems serve as the basis for the person's appreciation of the second and wider notion of paradigm as the entire "disciplinary matrix": "the entire constellation of beliefs, values, techniques, and so on shared by the members of a given community" (Kuhn 1970, p. 175).[2] Kuhn's view is that the mastery of a stock of exemplars allows the person to internalize the core beliefs, norms, and methods of the community. To appreciate the differences with Carnap, it is helpful to realize that this process is not mediated by rules of the sort Carnap places at the core of his linguistic frameworks (Kuhn 1970, p. 47). The learning does not proceed according to rules, and the mastery of the exemplars does not lead the person to act in line with some new set of rules. Instead, the person gains the capacity to approach new situations using the stock of exemplars provided in her training. This capacity involves new abilities to make comparisons and group cases, and these abilities are exercised in the course of normal science.

Given this conception of paradigms as the key qualification for entry to a given scientific community, it is not too surprising to find Kuhn insisting that normal science consists essentially of solving puzzles in line with the stock of exemplars that make up the core of the community's paradigm. This activity can involve a great deal of creativity, as it may require a novel or even controversial extension of an exemplar in a new direction to handle a new puzzle. For example, Kuhn speaks of the "articulation" of the Newtonian paradigm by Lagrange and others who completely transformed the mathematical framework of the paradigm and widely extended its scope. Still, Kuhn seems to think that normal science involves work on puzzles that the members of the community believe must have a solution in terms of the given paradigm (Kuhn 1970, p. 80). This gives the scientist the confidence to work on the problem using only the resources of the paradigm. Of course, some puzzles resist solution in these terms. These failures give rise to anomalies whose accumulation may prompt a sense of crisis in the scientific community. Only at this point are some motivated to develop a new paradigm via a new set of exemplars. When two paradigms confront each other, Kuhn argues that each will treat the evidence selectively so that its defenders will judge it to be superior. One way that this can happen follows directly from the central role of the different sets of exemplars. These worked

---

2. See also Kuhn (1979).

problems indicate not only how to solve a puzzle but also which puzzles are worth solving. If a new paradigm abandons some old exemplars and introduces new ones, then the associated standards of evaluation will be different.

Kuhn elected to describe this situation as a case of two "incommensurable" paradigms. The core sense of incommensurable is that there is no shared standard against which the achievements of the two paradigms can be judged, that is, "the incommensurability of standards" (Kuhn 1970, p. 149). But Kuhn also argued that the incommensurability of paradigms entails a difficulty in understanding that is analogous in many respects to translation: "Communication across the revolutionary divide is inevitably partial" (Kuhn 1970, p. 149). This seems to be because the meanings of words are determined by the stock of exemplars at the core of the paradigm. For example, Copernicans "were changing the meaning of 'planet' " (Kuhn 1970, p. 128) when they argued that the planets moved around the sun. Or in the case of the change from Newton to Einstein, "the physical referents of these Einsteinian concepts are by no means identical with those of the Newtonian concepts that bear the same name" as "Newtonian mass is conserved; Einsteinian is convertible with energy" (Kuhn 1970, p. 102). Based on the incommensurability of paradigms, there is no scientific basis for a member of one community to abandon his paradigm and adopt the new paradigm. As Kuhn put this point in one of his more convoluted slogans, "Though the historian can always find men—Priestly, for instance—who were unreasonable to resist for as long as they did, he will not find a point at which resistance becomes illogical or unscientific" (Kuhn 1970, p. 159). That is, as a scientist whose standards are fixed by his paradigm, Priestly is not unscientific for rejecting Lavoisier's chemical innovations. Unfortunately, Kuhn gives little substance to the sense in which Priestly was "unreasonable" for holding out for so long. He may mean that it was irrational for Priestly as a person to reject Lavoisier's proposals because this stance undermined Priestly's nonscientific instrumental goals, for example, making a living.

Kuhn's views on the role of mathematics in paradigms comes out mainly in his discussions of "symbolic generalizations" (Kuhn 1979, p. 298). These are implicity universal claims like $f = ma$ and $\nabla^2 \psi + 8\pi^2 m/h^2 (E - V)\psi = 0$. He claims that "no one will question that the members of a scientific community do routinely deploy expressions like these in their work, that they ordinarily do so without felt need for special justification" (Kuhn 1979, p. 298). These are the sorts of things scientists adapt from their stock of exemplars. Still, this simple agreement in using symbolic generalizations is quite weak and is analogous to agreement on "a pure mathematical system" (Kuhn 1979, p. 299). To move beyond this to something of scientific substance, the symbols must be associated with concrete problems and techniques for solving these problems. When this happens, $f = ma$ is transformed into $mg = md^2s/dt^2$ for the problem of free fall, and it is transformed in different ways to handle other stock problems like the simple pendulum or coupled harmonic oscillators (Kuhn 1979, p. 299). In this view, mathematics plays a crucial role in the presentation of many of the exemplars that are at the center of paradigms. A given mathematical statement and technique is presented as a model on which new attempts to solve new problems are to be judged.

With this brief summary of Kuhn's views, we can now ask what sort of derivative/constitutive relationship he envisions. My suggestion is that the constitutive representations are exhausted by the stock of exemplars on which Kuhn places so much emphasis.[3] The derivative representations are then all the other representations that a given scientific community comes to agree on. The connection between the constitutive and the derivative is that the adoption of a set of constitutive representations makes it possible for scientists to agree on the appropriate derivative representations. This community-wide agreement is central to Kuhn's whole account of normal science, and a breakdown in agreement shows that a revolutionary period has begun. He argues that mere exposure to some scientific subject matter is insufficient to generate agreement among scientists. This is because the contents of our experience are not univocally fixed by the stimuli that we receive. As Kuhn says several times, "the members of a group...learn to see the same things when confronted by the same stimuli" (Kuhn 1970, p. 193) by mastering the exemplars of the paradigm. So the evidence gathered by scientific experiments is already influenced by the stock of exemplars that are in play.

Kuhn clearly thinks that the sort of agreement on derivative representations that results from prior agreement on exemplars is a rational or scientific agreement. It is therefore inconsistent with an externally mandated agreement, for example, one enforced by social or political norms. The success of the constitutive representations in generating agreement should be tied to the ability of the exemplars to effectively solve puzzles and underwrite the prospects for the continued solution of puzzles that are within the purview of the community. Kuhn remains unclear on what the success of a set of derivative representations comes to, though. He definitely rejects the idea that the derivative representations of one paradigm can be said to be closer to the truth than another paradigm's derivative representations. Still, there is a sort of progress for science that he insists falls short of convergence on the truth: "the nature of such [scientific] communities provides a virtual guarantee that both the list of problems solved by science and the precision of individual problem-solutions will grow and grow" (Kuhn 1970, p. 170). This is consistent with a change in which problems are important and how these problems can be solved as old paradigms are replaced by new ones.

As with Carnap, the most problematic aspect of Kuhn's approach is his conception of meaning. Because there are two notions of paradigm at play, there are two associated claims that paradigms fix the meanings of scientific terms. To start, there is the view that the whole "disciplinary matrix" fixes the meaning of the terms. The result of this proposal is that almost any change in scientific context will affect the meanings of the terms employed by the scientist. This will make it difficult (if not impossible) to account for the widespread agreement in derivative representations throughout the period of normal science. It would also rule out any articulation of a paradigm of the sort Kuhn allows when he notes Lagrange's extension of the Newtonian paradigm. What Kuhn probably intends to defend, though, is the claim

---

3. Kuhn talks of a "constitutive principle" at Kuhn (1970), p. 133, but he does not say enough to make clear if he is using the word *constitutive* in our sense.

that the narrow set of exemplars at the core of the paradigm is responsible for the meaning of the terms. On this view, terms like *mass* or *force* are assigned meanings based on appearance in the set of worked problems that a student acquires in his scientific training. If this stock of exemplars is changed, then the meanings of these terms are suitably adjusted.

Although quite different in its details from Carnap's conception of rules and linguistic frameworks, Kuhn's theory of meaning seems equally flawed. There is no reason to think that the meaning of a scientific term is fixed in the simple way he suggests. More to the point, even if the meaning of a new scientific term is fixed at the time that a new set of exemplars is adopted by a scientific community, it does not follow that this meaning will change as these exemplars change. The analogous point has been made by semantic externalists who concede that the referent of a name may be fixed via description. Even if I introduce the name *Julius* to refer to the man who invented the zipper, I can still accept the claim that it is contingent that Julius invented the zipper. This is because the tools used to fix the meaning of a term need not be identified with the meaning of a term. Kuhn seems to think that a term like *mass* is introduced as the physical magnitude that would make the techniques displayed in the stock of exemplars viable solutions to the given problems. But even if we grant this, it does not follow that *mass* will change its meaning as the community shifts its stock of exemplars. An externalist conception of the meaning of scientific terms allows these terms to have a fixed meaning across scientific revolutions.

I reject, then, Kuhn's claim that the views of different paradigms are incommensurable in meaning. But there still remains his view that the paradigms have different standards of evaluation and this will frustrate any substantial evaluation of the two approaches. The argument for this sort of incommensurability need not turn on considerations of meaning, but only on the way the stock of exemplars makes agreement on the remaining derivative representations possible. Kuhn's idea is that the training of scientists into a given paradigm transforms their approach to a scientific domain by inculcating a set of standards of evaluation. Essentially, these standards are those implicit in the worked problems. They are extended and adapted to judge new solutions to outstanding problems. The problem with this proposal is that Kuhn never makes clear why a scientist must *adopt* a given set of standards to *understand* them. On his reconstruction, when a student is initially exposed to some new domain, she is unable to understand the proposed solutions to the problems at hand. When she works through her textbook and laboratory exercises, she comes to understand the worked problems at the core of the paradigm of the community. That is, she understands how to proceed if she is to continue to work in that paradigm. But Kuhn extends this fairly plausible point to include the claim that the student must also adopt the standards implicit in the exemplars. The adoption of these standards produces the incommensurability of standards across paradigms that are central to scientific revolutions. But why is it not possible for a student to understand the exemplars of a paradigm and yet reject the implicit standards? Many of us are familiar with exemplars of past scientific paradigms. This is clear for Newtonian mechanics, which is often introduced in basic physics courses with the proviso that it is not correct. Similarly, in the historical study of phlogiston or ether theories, historians and philosophers of science readily come to an understanding of what the exemplars of these paradigms amount to. But of course they refuse to endorse the standards of these communities. The reason for this is that they understand these standards and disagree with them.

There is a crucial gap, then, between the exemplars of a paradigm and the agreement on the derivative representations of that paradigm. The training of a scientist is supposed to acquaint them with the exemplars, and this is meant to explain the right kind of agreement on derivative representations that make up the rest of the paradigm. But I have argued that the exposure to the exemplars that is sufficient to get a student to understand them and how to "go on" in the scientific community need not result in an endorsement of those standards. Kuhn must explain where this additional endorsement comes from. His talk of "conversion" (Kuhn 1970, p. 204) does not inspire confidence that he has much to say. Furthermore, his proposal seems to flatly contradict the widespread ability to learn and yet not endorse past scientific paradigms.

6.4 FRIEDMAN ON THE RELATIVE A PRIORI

Finally, we turn to what I take to be the most promising attempt to clarify a defensible derivative/constitutive distinction. This is Friedman's notion of the relatively a priori parts of a scientific theory. The defensible core of this proposal is that some parts of a scientific theory must be endorsed for scientific testing to proceed. In this respect, Friedman can be seen to agree with Carnap and Kuhn. For all three there are constitutive representations that make rational agreement on the remaining derivative representations possible. This is not an accident, as Friedman explicitly aligns his project with aspects of Carnap's and Kuhn's views (Friedman 2001, p. xii). Unfortunately, as with Carnap and Kuhn, Friedman surrounds this commitment with flawed semantic views about the meanings of scientific terms. Once these semantic aspects of the relative a priori are set aside, we can turn to the epistemic angle concerned with rational agreement.

The conclusion that Friedman aims for is summarized by the claim that "the role of what I am calling constitutively a priori principles is to provide the necessary framework within which the testing of properly empirical laws is then possible" (Friedman 2001, p. 83). These principles are different from both the rules of Carnap's linguistic frameworks and the exemplars of Kuhn's paradigms. Some constitutive principles are purely mathematical claims, such as the theory of Euclidean geometry or Riemannian manifolds. Other constitutive principles are "principles of coordination," like Newton's three laws of motion. They are typically formulated in terms of the mathematics provided by the first kind of constitutive principle along with additional terms for basic physical magnitudes. Friedman insists that these principles must be in place for the framework to allow the testing of ordinary empirical laws like the law of universal gravitation. His argument for this turns on considerations of meaning: "Without a constitutive framework, the putatively empirical laws would have no empirical content after all, but would simply belong to the domain of pure mathematics" (Friedman 2001, p. 83). Or, more fully,

> To say that A is a constitutive condition of B rather means that A is a necessary condition, not simply of the truth of B, but of B's meaningfulness or possession of a truth value ... in our example from Newtonian physics, the law of universal gravitation essentially employs a concept—absolute acceleration—which has no empirical meaning or application (within the context of Newtonian physics) unless the laws of motion hold. Within the context of Newtonian physics, that is, the only way in which we know how to give empirical meaning and application to the law of universal gravitation is by presupposing that the laws of motion are true: if the latter principles are not true (in the sense that there exists a frame of reference in which they hold) then the question of the empirical truth (or falsity) of the law of universal gravitation cannot even arise. (Friedman 2001, p. 74)

What we have here is an ambitious semantic argument for an epistemic conclusion about the constitutive contribution from mathematics and coordinating principles. Although I think the conclusion is largely correct, I do not endorse this argument. This will leave the task of finding a more plausible argument for the same conclusion to the next section of this chapter.

To reconstruct Friedman's argument, let's fix the Newtonian framework using a traditional presentation of Newton's three laws of motion:

a. A body persists in its state of rest or of uniform motion unless acted on by an external force.
b. $F = ma$, that is, the net force on a body is equal to the mass times the acceleration.
c. To every action there is an equal and opposite reaction.

These laws employ mathematical concepts such as the straight line of Euclidean geometry in a and acceleration as the second derivative of position with respect to time in b. Additional nonmathematical concepts like force and mass appear in the laws. Friedman's argument, then, is that a–c must be presupposed for any empirical law of Newtonian mechanics to be true or false. In the absence of this presupposition, such a law would be merely "pure mathematics." The most famous empirical law of Newtonian mechanics is, of course, the law of universal gravitation. This says that for any two bodies with mass m_1 and m_2 separated by distance d, there is a force of attraction F_g along the line connecting the bodies given by

$$F_g = G\frac{m_1 m_2}{d^2} \qquad (6.1)$$

G is the gravitational constant that depends on the choice of units.

The crucial link between a–c and 6.1 is that 6.1 is meant to apply only to frames of reference that satisfy a–c. A significant part of the evidence in favor of 6.1 is the observed motion of the planets around the sun. But for these observations to be relevant to the correctness of 6.1, they must be presented in terms of positions and times that are marked out so that a–c are satisfied. For Newton this demand was cashed out with his assumption of absolute space and time. The accelerations

Constitutive Frameworks

invoked by b, then, were accelerations with respect to this absolute background. As Friedman explains, though, by the nineteenth century it was possible to clarify these presuppositions in terms of the existence of inertial frames of reference. This is a means of describing the relative motions of bodies using some fixed time scale. In Newtonian mechanics, more than one inertial frame exists. Still, if a law like 6.1 is true with respect to any single inertial frame, it will also hold in every other inertial frame. Simple transformations will allow the observations given with respect to one inertial frame to be presented with respect to another inertial frame. Essentially, taking a–c for granted and making certain plausible assumptions, Newton was able to argue that the frame of reference centered on the center of mass of the solar system is an inertial frame. This allowed him to marshal support in favor of 6.1 using the available observations of planetary positions over time.[4]

With this background we can consider four different presupposition relations of the sort suggested by Friedman as candidates for a viable constitutive/derivative relationship. The most ambitious position which Friedman seems to defend is that

i. a–c must be true for 6.1 to be either true or false.

A slightly less ambitious suggestion is that

ii. a–c must be believed to be true for 6.1 to be either true or false.

An even weaker position is that

iii. a–c must be believed to be true for a scientist to have a reason to believe that 6.1 is either true or false.

Finally, the weakest position consistent with what Friedman says is that

iv. a–c must be believed to be true for a scientist to have a reason to believe that 6.1 is true.

I refer to these last two proposals as epistemic interpretations of constitutive representations because they focus on what an agent has a reason to believe. By contrast, the first two ambitious positions concern truth and falsity. They are thus broadly semantic claims.

The most important consideration against Friedman's most ambitious semantic proposal i is that a–c seem to be false and 6.1 seems to be equally false as well. Newton was wrong to believe in absolute space and time. But this did not deprive 6.1 of a truth-value. It was false in Newton's time and is false today. The point is not affected by shifting to the notion of inertial frames. There are no frames of reference in which a–c hold. As Newtonian mechanics requires such frames,[5] it is clear that

---

4. For more discussion of inertial frames see DiSalle (2002).

5. Friedman (2001, p. 87, n. 21.).

Newtonian mechanics is false. Similarly, 6.1 requires the existence of inertial frames. Given that there are no inertial frames, 6.1 is false. It is no help to insist that 6.1 is some kind of tacit conditional of the form "In every inertial frame, 6.1." On this reading, 6.1 is trivially true, so it remains a counterexample to Friedman's semantic proposal i.

Presumably Friedman would respond to this objection by insisting that in such a situation "the putatively empirical laws would have no empirical content after all, but would simply belong to the domain of pure mathematics." That is, the 6.1 of Newtonian mechanics is neither true nor false of the physical world and so is merely a purely mathematical claim. To carry this defense through, though, Friedman must provide some reason to think that the meanings of the symbols making up 6.1 are somehow determined by which constitutive principles are true. One way to make this connection is by insisting that constitutive principles like a–c are responsible for the meaning of the nonmathematical terms like *force* and *mass* that appear in 6.1. Then the meanings of these terms will be fixed by the principles that happen to be correct. In the Newtonian case, though, it is hard to know what this clarification of the semantic proposal i would amount to. Based on our commonsense acceptance of our best contemporary science, there are no true constitutive principles that can be used to fix the meaning of the nonmathematical terms of 6.1. This suggests that 6.1 must be meaningless after all, and not merely lacking in a truth-value. My objection to this way of obtaining Friedman's conclusion is that it results in a totally unrealistic reconstruction of the history of science. Only an extreme positivist would insist that hundreds of years of scientific practice turned on debates about a meaningless claim. Friedman, like any other good historian of science, recognizes that Newton gave compelling arguments for his colleagues to accept 6.1. This requires that 6.1 be a meaningful claim.

Several remarks by Friedman suggest that he wants to use the beliefs of agents to fix the meanings of scientific terms. This is why he more or less endorses Kuhn's views on the incommensurability of meanings across frameworks: "there has indeed been a 'meaning change' in the transition from the old framework to the new: even if the same terms and principles reappear in the new framework, they do not have the same meaning they had in the old, for they may no longer function as constitutive" (Friedman 2001, p. 99, n. 37; see also p. 60). So, a–c are beliefs that an agent must have for 6.1 to have its Newtonian meaning. Then for agents who reject a–c and fail to replace them with any analogous principles, 6.1 turns out to be meaningless or else a purely mathematical claim. Finally, if an agent adopts an alternative set of constitutive principles, then 6.1 will be true or false but will have changed its meaning from what it had in the original Newtonian framework.

This is my most charitable interpretation of what Friedman's proposal actually is. It is ii from our earlier list. Given the link with Kuhn, though, it is not surprising that the same objections that we raised against him on meaning also apply to this version of Friedman's proposal. Simply put, there is no reason that I have to believe or endorse a–c to understand 6.1 in its original Newtonian sense. As we learn physics, we are constantly exposed to these sorts of cases where we are led to understand what the views of past scientists were while also being told to withhold assent or belief. It is precisely because I can understand a–c and 6.1 in their original sense that I can recognize that they are false.

As we have seen, though, there is a strand of Friedman's discussion that focuses on conditions that must be met for a potential law like 6.1 to have epistemic support. On our first epistemic proposal iii, a–c must be believed to be true for a scientist to have a reason to believe that 6.1 is either true or false. This is not a point about meaning and we can assume that all the claims in question have a fairly determinate meaning for the agents in question independently of their beliefs. Still, the epistemic proposal can be supported based on the point that empirical observations will only bear on the truth or falsity of 6.1 if those observations are presented with respect to an inertial frame. For an agent to believe that their observations bear on 6.1 one way or the other, she must believe in the existence of an inertial frame. But this belief requires a belief in a–c. So it follows that the empirical confirmation or refutation of empirical laws like 6.1 requires as a necessary condition that the agent believe a–c. For only with these prior beliefs can the agent believe that what she observes is with respect to an inertial frame, and this in turn is a necessary link between the data and the law.

In this view, there are constitutive representations that serve a necessary epistemic function. If agents believe these constitutive representations, they can take their observations to confirm or disconfirm their derivative representations. Otherwise, the observations will not bear the necessary connection to the derivative representations. So the success of the constitutive representations consists in their allowing the confirmation or disconfirmation of their associated derivative representations. One benefit of this approach to the derivative/constitutive distinction is that it is consistent with treating the success of derivative representations in ordinary scientific realist terms. That is, a derivative representation will be successful if it is true and well supported by the evidence. Based on our views of meaning, it is also possible to hold out the hope that we have arrived at true constitutive representations as well. Our confidence that this has occurred will tend to rise and fall as we come to confirm more or fewer derivative representations. But even though this is the ultimate hope for the constitutive representations as for the derivative representations, it remains that we judge the success of the constitutive representations in terms of their ability to allow the confirmation and disconfirmation of other representations.

At a few points in his discussion Friedman seems to reject this nonsemantic conception of the constitutive representations because it risks collapsing into confirmational holism of the sort defended by Quine. Friedman claims that before Einstein's proposed constitutive framework, his "new theory of gravitation was not even an empirical possibility" (Friedman 2001, p. 94). Continuing in a note, he cites this as "the crucial difference ... between standard cases of Duhemian empirical underdetermination and genuine scientific revolutions" because "what is here in question is the very notion of empirical justification or reason" (Friedman 2001, p. 94, n. 31).[6] On my purely epistemic proposal, the reason that an agent must believe a–c

---

6. See also the discussion against Quine at Friedman (2001, pp. 34–40.).

to confirm or disconfirm 6.1 is that it is only the conjunction of beliefs a–c with observations O that bear any evidential connection to 6.1. In the terminology of underdetermination, this is to insist that O by itself is insufficiently related to 6.1 to allow confirmation or disconfirmation. Our observations by themselves do not tell us that they are given with respect to an inertial frame. So we need to also have other beliefs that give us a reason to interpret O as given with respect to an inertial frame. Friedman seems to agree with these points as far as they go as he says that "neither space nor time nor motion has an unequivocal relation to our experience" (Friedman 2001, p. 76) for the Newtonian.

Confirmational holism is supposed to follow from this sort of underdetermination based on the additional claim that it is reasonable to adjust any of our beliefs in light of a conflict with observations. In our Newtonian case, Quine would argue that it is equally reasonable to adopt any of the following three responses. First, one might disconfirm 6.1. Second, one could reject the claim that O is given with respect to an inertial frame, and thereby insulate a–c and 6.1 from refutation. Finally, an agent could reject a–c and effectively conclude that there are no inertial frames of the sort required for 6.1. It is important to notice here that there is no need to resort to considerations of meaning to block the Quinean conclusion that each of these three responses is equally reasonable. It is sufficient to insist that prior to the collection of O the constitutive principles a–c had a high enough degree of confirmation that it would be unreasonable to reject them simply on the basis of a conflict between 6.1 and O. If we can establish this high degree of confirmation prior to testing for these constitutive representations, then we will have blocked Quinean holism. This would block the third response, but would still leave the first two options available depending of the specifics of the case.

Precisely here Friedman deploys his notion of the relativized a priori. He claims that the constitutive principles have a special a priori status. This may just mean that they help define the meanings of the basic scientific terms that appear in the derivative representations. If this is right, though, then there is no link between this semantic function and the traditional epistemic connotations of the a priori, that is, what can be justified independently of experience. Friedman appears to embrace a link between definitions and the relative a priori by endorsing Reichenbach's claim that relativized a priori principles are "constitutive of the concept of the object of [scientific] knowledge" (Friedman 2001, p. 30). The problem with this defense of the relativized a priori is that it fails to do the right kind of job against the Quinean position. As we have seen, the view that the constitutive principles fix the meaning of the words used to express the derivative representations just does not work. More to the point, even if these principles have this semantic function, this will not confer on them any special legitimacy or degree of confirmation. In the face of recalcitrant experience, the agent will have as much reason to reject the definitions they have adopted as to adjust their beliefs. So, even if a–c are definitions, the agent may take the third option in the face of a conflict between a–c, 6.1 and O and reject the definitions a–c. A statement being a definition does not automatically remove it from empirical scrutiny or insulate it from rejection for holistic reasons. To take a less controversial case, I may reject the definition of a mathematical concept if it can be shown to lead to contradictions, for example, the Russell class or the largest ordinal.

In the case of a–c, Friedman concedes that these definitions entail the existence of an inertial frame. Given a conflict with experience, Friedman has not explained why it is irrational to reject these definitions and so deny the existence of inertial frames. Until he rules out this possibility, he has not successfully countered his Quinean opponent.

6.5 THE NEED FOR CONSTITUTIVE REPRESENTATIONS

Our discussion of Friedman's proposal has isolated a potential epistemic contribution from constitutive representations, but we have not yet found a sufficient basis for their special role in our scientific knowledge. What is missing is what we need to block Quinean holism, that is, a special reason to believe a set of constitutive representations that will make it rational for an agent to hold such representations outside the normal domain of empirical testing. My suggestion for how to proceed out of this impasse is to resurrect the central idea of Carnap's linguistic frameworks. This is the claim that even though the absence of an agreed-on framework precludes theoretical justification for a scientific statement, it remains possible to assemble practical justifications for a certain course of action with reference to some agreed-on goals. But rather than focus on practical goals like designing ships or reliable bridges, I suggest that we consider the goal of finding out the truth concerning a given domain of investigation. This adjustment to Carnap's proposal, along with a rejection of his semantic account of constitutive representations, will lead to constitutive representations that are genuinely relatively a priori.

To see how this works, suppose that it really is the case that the only way for an agent to confirm or disconfirm an empirical law like 6.1 concerning the motions of the planets is for her to believe a–c. The agent is in a position to realize this if she knows the relevant mathematics, understands a–c and 6.1, and is able to make the relevant links in terms of the notion of an inertial frame. Based on this realization, the agent has two possible courses of action. She could elect to not believe a–c and so remain in the dark about whether 6.1 could be confirmed. Alternatively, she could decide to believe a–c and in the existence of inertial frames. This puts in place the conditions that are necessary for a genuine test of 6.1 to occur. For the test to be successful, though, a–c must be given a fairly high degree of initial confirmation. The justification for this assignment is practical. This is the only means available to the agent to find out the truth about the motions of the planets. If she forms her belief as a result of realizing this predicament, I would argue that her decision is fully rational and she should be able to convince other agents to follow her in her Newtonian investigations.

Given this starting point, there are then two possible outcomes. Either 6.1 and similar empirical laws come to be confirmed for those who have tentatively adopted a–c or else no empirical laws wind up being confirmed for the domain in question. If the former outcome occurs, then the practical decision to believe a–c is vindicated. Furthermore, the degree of confirmation of a–c can go up as it comes to look more and more like the scientists have hit on the truth about the domain in question, that is, there are inertial frames, one obtains around the center of mass of the solar system, and so on. By contrast, a failure to find any empirical laws that can be confirmed using a–c provides mounting evidence that the practical decision to believe a–c was incorrect. If this failure persists, it is reasonable for the agent to conclude that a–c are false and there are no inertial frames. This is, of course, my interpretation of our current epistemic situation. While it was practically rational for Newton to make the practical decision to believe a–c in hopes of finding out the truth, that hope has proven illusory. After Einstein's proposals, we can come to see that it would not be reasonable to return to a belief in a–c. That set of constitutive principles has been thoroughly explored by scientists since Newton's time and has come up short.

This last part of my proposal helps make clear how the constitutive representations can be both a priori and relative. They are a priori because the practical reasoning that gives rise to their adoption does not turn on any empirical evidence. Only the realization that some principles have to be believed for testing to proceed is responsible for this justification. At the same time, the history of science can provide empirical evidence that undermines this justification. In this respect I follow other recent writers on the a priori, such as Casullo, who allow that a priori justifications can be undercut by empirical evidence (Casullo 2003). The evidence that I invoke concerns the historical track record of investigation of a given set of constitutive principles. So given our stage in history, some sets of principles can no longer be reasonably provisionally adopted. But there will typically be several remaining candidates for any domain of investigation that the interested scientist can choose.

The discussion of this section forces us to retreat from the epistemic proposal iii that we have been working with so far to the weakest epistemic proposal iv noted before. So far I have defended the claim that constitutive representations are those that must be believed for the derivative representations to be confirmed or disconfirmed. The discussion of this section has shown a route to disconfirmation that does not proceed via the belief in the associated constitutive representations. This is fortunate as I have argued that 6.1 is false and also insisted that I can understand a–c without believing them. Our route to disconfirming 6.1 is historical. We look back at the history of science and recognize that the agents who adopted a–c were unable to confirm 6.1. So I can disconfirm 6.1 without believing a–c. The point remains, though, that agents would not be in a position to confirm 6.1 unless they believed a–c. The asymmetry noted here between confirmation and disconfirmation is important. Without it, it is not clear what sort of relatively a priori justification would be possible.

6.6 THE NEED FOR THE ABSOLUTE A PRIORI

To summarize the results of our discussion so far, there are constitutive representations that must be believed for other derivative representations to be confirmed. With this proposal in hand, we can try to determine if there is any special contribution that mathematics makes to these constitutive representations over and above the contributions surveyed in earlier chapters. In line with these earlier discussions, our aim is not to argue that mathematics is essential to these sorts of contributions. In our current constitutive case, it would be difficult to argue that the constitutive representations must involve mathematics. All that our discussion has assumed is a degree of remoteness from experience in our scientific hypotheses that exposes a gap between these hypotheses and the observations that we can collect. Still, it seems clear that mathematical claims are especially helpful in formulating constitutive representations. This is for at least two reasons. First, the generality of the mathematics lets the scientist consider a wide range of abstract structures that may be realized in the physical world. This is especially obvious for the different kinds of geometries that are deployed in different constitutive representations. Even the most restrictive Euclidean geometry allows a vast array of possible configurations of bodies in space and time. As Friedman emphasizes, this range of possibilities only increases as we shift to the non-Euclidean geometries countenanced by Einstein's general theory of relativity. Here, then, we have a flexible arena of possible physical states that the scientist can use to embed the physical systems that she encounters. A second reason that mathematics seems to be so useful in presenting constitutive representations, though, is the precision it affords. Recall that the function of these representations is to allow the confirmation of the derivative representations. A great aid in this process is a quantitatively formulated hypothesis that can be compared against the suitably presented observational data.

In the examples discussed in earlier chapters, it is often easy to identify the constitutive representations that have contributed to the success of the derivative representations which have been our primary focus. For example, in the many examples given involving fluids, we have taken for granted the correctness of the constitutive Newtonian representations. This is why, for example, we can take the failure to predict drag to undermine our confidence in the Euler equations when applied to airplane wings.[7] More generally, all our talk of confirmation has presupposed the correctness of Newtonian mechanics. If pressed to truly justify our conclusions about those earlier examples, the tacit role of these constitutive representations would have become manifest. Although we have not focused on many examples that turn on nonclassical physics or other areas of science, I think our discussion has made it plausible that constitutive representations of one kind or another would be found there. In this picture, then, the different sorts of epistemic contributions reviewed in chapters 3–5 for derivative representations require prior belief in some set of constitutive representations of the sort discussed here. The constitutive representations then involve a distinct kind of contribution of mathematics with its own broader significance.

The most pressing issue raised by the special role of mathematics in constitutive representations is the justification of the purely mathematical beliefs involved in these representations. The Quinean insists that mathematics is justified along with our scientific beliefs based on its role in our overall science. If we follow this line of thinking, then the mathematical claims are believed along with the remaining "coordinating" representations based on the practical consideration that only this family of constitutive representations would allow confirmation of the other derivative representations for the scientific domain at issue. Although this approach is hard to

7. See section 5.6.

completely close off, it seems to face the following problem. Recall the argument that practical reasoning could lead to a high degree of confirmation in the constitutive representations. If we scrutinize this reasoning, though, it seems to presuppose that the agent already believes in the relevant mathematics. How else could the scientist convince herself that the only way to confirm a given derivative representation was by believing the coordinating principles? If the agent does not believe in the truth of the relevant parts of mathematics, then it seems impossible for him or her to imagine the difficulties.

If these considerations are accepted, then in addition to the relatively a priori constitutive representations that have been our focus in this chapter, there needs to be some kind of absolutely a priori justification available for at least some part of the mathematics used in science. I am optimistic that this is a defensible conception of the epistemology of pure mathematics, but to fully defend it involves a series of delicate issues. Among these is an evaluation of so-called indispensability arguments, which argue that the indispensable role of mathematics in science by itself gives a scientist sufficient reason to believe in the existence of mathematical entities. If these arguments can be sustained, then there will not be any need to find a source of absolutely a priori justification for mathematics. An evaluation of these arguments must be postponed until chapters 9 and 10. As a preview, I argue that indispensability arguments are not successful. This will reinstate the demand for some a priori justification for mathematics. I start to meet this demand in chapter 14, where we will survey the conception of pure mathematics that results from our discussion of applications.

7

Failures

7.1 MATHEMATICS AND SCIENTIFIC FAILURE

So far our discussion has tried to survey some of the main ways in which mathematics can make a positive contribution to the success of science. A survey of the different ways in which mathematics contributes to the success of science risks being decidedly one-sided, however. If we just discuss the success cases, then we may overlook the negative aspects of what mathematics brings to science and so may misunderstand the character of its positive contributions. To appreciate the potential problems here, consider the success of those people who win millions of dollars in the lottery. A narrow study of these successful people might reveal some link between their character traits and their winning of the lottery, for example, a propensity to buy lottery tickets. But of course we recognize that the propensity to buy lottery tickets leads to failure much more often than it leads to success. A wider study of winners and losers is necessary to appreciate the link between success and character.

This sort of wider study is what I want to initiate in this chapter for the link between the use of mathematics and the prospects for scientific success. I draw on six sorts of scientific failures that correspond to the positive contributions discussed in earlier chapters. These are:

1. Acausal representations: An apparently reasonable way of storing stone columns proved unsuccessful.
2. Varying representations: The Black-Scholes model of options pricing contributed to the failure of the hedge fund Long-Term Capital Management in 1998.
3. Scaling (size): Bridge design failures.
4. Scaling (perturbation theory): The argument that flight is impossible.
5. Constitutive: Catastrophe theory as applied to the social sciences.
6. Confusion of contributions: Laplace's improved estimate of the speed of sound supported the incorrect view that heat was a fluid-like substance.

The sixth category here involves mistaking what is really an acausal contribution for a causal contribution. As we will see, in each case the mathematics involved contributed to these failures. Fortunately, the situation here is not as grim as with the lottery winners, where the very trait that was necessary to winning proves to be responsible for failure in the vast majority of cases. But something close seems to be the case. This is that the ways mathematics provides a positive contribution in some cases are the very same ways it makes a negative contribution in other cases. I conclude by discussing the implications of this result for the interpretation of scientific knowledge and suggest that a modest form of scientific realism remains viable. One lesson I draw from these failures is that the scope of our representations is just as important as the entities involved in the representations. This complicates traditional inference to the best explanation arguments for scientific realism. However, once we understand how mathematics can contribute to scientific failure, we can probe for these sorts of mistakes and thus allay any reasonable worries about scientific knowledge.

7.2 COMPLETENESS AND SEGMENTATION ILLUSIONS

I begin with a failure that is discussed by Galileo in "Dialogues Concerning Two New Sciences" and emphasized more recently by Henry Petroski in *Design Paradigms: Case Histories of Error and Judgment in Engineering* (Petroski 1994). The problem at hand was the best way to store a heavy stone column between the time it was manufactured and when it was to be transported to the site of a new building. As these columns tended to be quite long and relatively thin, there was a serious risk of damage during the storage period. The most common storage method deployed two supports to prop the column above the ground (see figure 7.1). Thought of mathematically, this makes perfect sense. We can represent the column as a straight line and realize that the weight of the entire column will be more or less evenly distributed based on the distance between the two supports and how far these supports are from the ends of the column. The problem with this arrangement is that the stored column would often crack in the middle, presumably indicating the region that was placed under the greatest strain by this method of storage. See figure 7.2. One way to respond to this failure would be to try to estimate what strain a given column could sustain and how this strain could be minimized by the placement of the two supports. Though I do not want to engage with the details of how these calculations would go, suffice it to say that they take place within a thoroughly *static* setting. That is, we take it for granted that the column can be placed on the supports without incident. The question then becomes how the weight will be distributed within the column as it rests on the supports and whether it will surpass a given threshold. We do not care when or how the column will crack, but only whether it will crack. This permits a representation that lacks much causal content to be acceptable for our purposes.[1]

It should be clear that we have an instance of what I call an abstract acausal contribution. This is because our representation of the column-support system has

---

1. A similar case noted earlier is Batterman's example of the buckling strut in section 3.5.

Figure 7.1: Column Storage

Figure 7.2: Two-Support System

erased any information about how the column's shape evolves over time or indeed any other causal information tied to the constituents of the system. All we have left is a means to calculate the weight distribution across the column. Again, even when our representation predicts that the column will crack, we do not have any means of representing where it will crack or how a small initial crack will expand through the column. This sort of abstraction is, of course, incredibly useful. It allows us to focus our meager understanding of the system on the point of interest and, when suitably supplemented by experimental information, it would allow an effective understanding of where to place supports and when these supports would be successful.

Here, then, is a limited scientific success tied to an abstract acausal representation of a physical system. This success suggested a certain kind of extrapolation, though. As Galileo relates,

> A very large column of marble was laid down, and its two ends were rested on sections of a beam. After some time had elapsed, it occurred to a mechanic that in order to insure against its breaking of its own weight in the middle, it would be wise to place a third similar support there as well. This suggestion seemed opportune to most people. (Galileo 1974, p. 14)[2]

In retrospect, we can see this extrapolation as entirely natural given the static representation used to evaluate the two-support system and the sorts of problems it had. What is more reasonable than to add more supports to the system, especially in the middle region where our cracks tended to appear? See figure 7.3. Clearly, the sorts of calculations that we used to evaluate the two-support systems can be easily extended to help us distribute the three supports, and we will predict that our new means of column storage is more widely applicable than the old approach.

Unfortunately, this innovation in column storage proved to be a failure. In fact, the three-support method was less successful than the traditional two-support

---

2. Given at Petroski (1994, p. 49).

Figure 7.3: Three-Support System

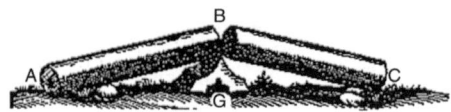

Figure 7.4: The Failure of the Three-Support System

approach! All too often, the columns stored using three supports cracked as shown in figure 7.4. It may be initially hard to see why this is. When I read Petroski's account, it seemed that the problem must be that the supports were not all built to be the same height. Perhaps the failure of the three-support system was tied to the faulty execution of an otherwise appropriate storage strategy. The source of the problem is more fundamental, though. What we have ignored in our representation is the fact that the supports may sink into the ground over time. Let us assume that the three supports sink at independent rates. Then two-thirds of the time one of the end supports will sink more quickly than the middle support. When this happens a greater strain will be placed on the point of the column above the middle support than if we had used a two-support approach. As Galileo notes, "abandoned by the support at the other end, its excessive weight made it do what it would not have done had it been supported only on the two original [beams], for if one of them had settled, the column would simply have gone along with it" (Galileo 1974, p. 15).[3] Even if the two supports of the two-support system sank at different rates, this would not produce such a dramatic change as would tend to occur in the three-support system.

Galileo uses this scientific failure to show the need for his new science. The specific point of this case is that an understanding of the basis of the strength of materials is necessary to avoid these sorts of failures. Galileo explains the cracking of the column with reference to his account of when a beam fixed to a wall will break under its own weight. Assuming that a column of length l will break under its own weight when fixed to a wall, a column of length $2l$ will break when supported only in the middle. This allows us to see that a two-support system will support a column of length less than $2l$.[4] Petroski, in his discussion, uses Galileo's example to focus on those engineering failures that can result from changing something that undermines an otherwise successful design: "it is the responsibility of the redesigner

---

3. Given at Petroski (1994, p. 51).

4. Galileo (1974, pp. 130–131). Heyman (1998, pp. 9–10).

to consider how the alteration interacts with the rest of the existing system and how the alteration satisfies the original objectives of the design problem" (Petroski 1994, p. 54). What I would like to suggest, though, is that the successful application of an abstract acausal representation often makes it extremely difficult to appreciate the domain in which the representation will remain accurate. The reason for this is quite simple. When we remove causal information or represent a system from an acausal perspective, we may drop details that are essential to the accuracy of the abstract representation. The partial character of the abstract representation cuts both ways, then, when it comes to accuracy, confirmation, and practical success. By representing a column as a static straight line supported by fixed supports, we make our representation manageable and can make and confirm countless successful predictions. But these successes depend on something that we have factored out, namely, the rate at which the supports sink into the ground. In the two-support system this rate of sinking is, in fact, irrelevant, but it becomes highly relevant when we add additional supports. The abstract nature of our representation hides this dependence, though, as nothing in our calculations for the three-support system suggests that there will be any problems.

The mathematical character of the representation, then, can be squarely blamed for this failure. We can distinguish two closely related ways such acausal representations may lead us to think we have established accuracy in a wider domain than is actually the case. First, there is what I call the illusion of *completeness*. If we have a successful mathematical representation that allows the calculation of some physical magnitude, then it is very tempting to conclude that our representation includes everything that is relevant to that magnitude. That is, the mathematical character, say, in the form of an equation with certain inputs and parameters, takes on an air of finality and comprehensiveness. But when this representation is acausal, it is also highly partial because we have abstracted away from the causal processes operating in the system. This creates a tension and often leads a scientist to overestimate the scope of their representation. The second, though related, source of these failures is what I call the *segmentation* illusion. This is the mistake, noted by Petroski, of thinking that a change in one aspect of a system will not result in a change in another aspect of the system. Again, when our mathematical representation is acausal, we may have failed to include the causal connections between the two aspects. Perhaps a change in one aspect within certain bounds fails to affect another aspect, and so our abstract representation is successful within these bounds. But this limitation may not be explicit in our representation, and so we may misjudge the interconnected character of the system in question.

Although this is just one case, I think it reveals something quite general about the limitations of acausal representations, especially those that are highly mathematical. The most important lesson of the column case is that even when a particular causal process is irrelevant for the accurate representation of a target system of kind A_1, that process may become highly relevant for the accurate representation of the closely related target system of kind A_2. In our case, nothing about the stone columns has changed and the means of support remains identical except for the addition a new beam for support.

7.3 THE PARAMETER ILLUSION

The heat equation is one of the most important partial differential equations for the history of science.[5] To recall its main target system, consider a metal bar whose temperature at point x at time t is given by $u(x, t)$. Assuming that an initial temperature distribution and appropriate boundary conditions are given, we can find the resulting changes in temperature over time in the bar by solving:

$$u_t = \alpha^2 u_{xx} \qquad (7.1)$$

Here α is a constant determined by the features of the metal bar. This equation relates changes in temperature at a point over time to the distribution of temperature in the x-direction. Intuitively, a concentration of higher temperature in one region will spread out or diffuse throughout the bar over time.

The heat equation and its higher-dimensional extensions have proven extremely successful for the study of temperature in certain domains as well as for other physical phenomena that share the core diffusive features of heat (Narasimhan 1999). But these successes can be tied to an important kind of failure. An extension of the mathematics used here to financial economics is a useful example of a failure tied to the abstract varying contribution from mathematics. The case from financial economics that I have in mind is the Black-Scholes model for the pricing of options:

$$V_t + \frac{1}{2}\sigma^2 S^2 V_{SS} + rSV_S - rV = 0 \qquad (7.2)$$

A (call) option gives the owner the right to buy some underlying asset like a stock at a fixed price K at some time T. Clearly some of the factors relevant to the fair price of the option now are the difference between the current price of the stock S and K as well as the length of time between now and time T when the option can be exercised. Suppose, for instance, that a stock is trading at $100 and the option gives its owner the right to buy the stock at $90. If the option can be exercised at that moment, the option is worth $10. But if it is six months or a year until the option can be exercised, what is a fair price to pay for the $90 option? It seems like a completely intractable problem that could depend on any number of factors including features specific to that asset as well as an investor's tolerance for risk. The genius of the Black-Scholes approach is to show how certain idealizing assumptions allow the option to be priced at V given only the current stock price S, a measure of the volatility of the stock price σ, the prevailing interest rate r, and the length of time between now and time T when the option can be exercised. The only unknown parameter here is σ, the volatility of the stock price, but even this can be estimated by looking at the past behavior of the stock or similar stocks. With the value V computed using this equation, a trader can execute what appears to be a completely risk-free hedge. This involves either buying the option and selling the stock or selling the option and buying the stock. This position is apparently risk-free because the direction of the stock price is not

---

5. We have already encountered it twice. See sections 2.3 and 5.3.

part of the model, so the trader need not take a stand on whether the stock price will go up or down.

Equation 7.2 is solved using the known terminal value of $V(S, T)$, that is, the curve of fair prices for the option at the time T when the option can be exercised given the stock price S. The mathematical relationship between 7.1 and 7.2 becomes clear once we realize that there is a transformation of any well-posed terminal value problem for 7.2 into a well-posed initial value problem for 7.1.[6] Analytic techniques exist for solving 7.1 in many cases, and even when these prove intractable numerical techniques can be used to find the function with little computational effort. The Black-Scholes model and the heat equation, then, are mathematically interchangeable for all practical purposes. But the interpretation of the two equations is completely different, as are the assumptions necessary to derive them. This is then a particularly striking example of two representations that are linked in a family of abstract varying representations by mathematics.

The nature of the hedging strategy implied by 7.2 and its risks comes out clearly in the derivation of 7.2.[7] I summarize the basic ideas here, and a more detailed derivation can be found in Appendix B. The basic assumption underlying the derivation of 7.2 is that markets are efficient so that "successive price changes may be considered as uncorrelated random variables" (Almgren 2002, p. 1). The time interval between now and the time T when the option can be exercised is first divided into N-many time steps. We can then deploy a *lognormal* model of the change in price δS_j at time-step j:

$$\delta S_j = a\delta t + \sigma S \xi_j \qquad (7.3)$$

The ξ_j are random variables whose mean is 0 and whose variance is 1 (Almgren 2002, p. 5). Our model reflects the assumption that the percentage size of the random changes in S remains the same as S fluctuates over time (Almgren 2002, p. 8). The parameter a indicates the overall "drift" in the price of the stock, but it drops out in the course of the derivation.

We calculate V by first describing a *replicating portfolio* made up of some combination of the underlying stock and cash that "will yield exactly the same eventual payoff as the option in all possible future scenarios" (Almgren 2002, p. 4). The value of this portfolio is given by

$$\Pi(S, t) = D(S, t)S + C(S, t) \qquad (7.4)$$

where D is the number of shares of stock held and C is the amount of the portfolio held in cash. Over time Π is adjusted by either using cash to buy more stock or else selling stock to add to the assets held in cash. Using a preliminary estimate of the value of V as a function of S and t, it is then possible to represent the *change* in the

---

6. Wilmott (2007, p. 160). In this chapter Wilmott also lists other ways of solving the Black-Scholes equation. See also Almgren (2002, pp. 7–8).

7. I follow Almgren (2002) closely.

value of a *difference portfolio* $\delta(V - \Pi)$ which buys the option and offers the replicating portfolio for sale. The expression for $\delta(V - \Pi)$ contains two problematic terms. One reflects the change in the value of S over our time step. But it can be set to 0 if we assume

$$D = V_S \quad (7.5)$$

That is, at each time step the number of stock held in the difference portfolio is adjusted. The remaining problematic term is due to δS^2, that is, the change in the square of the stock price. This term can be eliminated over the whole time period Δt using the lognormal model. The basic idea is that as we increase the number of time steps in our time interval to infinity, the limit of the sum of the squares of the random fluctuations ξ_j will be 1 precisely because the ξ_j have mean 0 and unit variance. This gives us an expression for $\Delta(V - \Pi)$ that involves only V and the interest rate r. The last major step is to deploy the efficient market hypothesis again and assume that $\Delta(V - \Pi)$ is identical to the result of investing $V - \Pi$ in a risk-free bank account with interest rate r. That is,

$$\Delta(V - \Pi) = r(V - \Pi)\Delta t \quad (7.6)$$

Simplifying and rearranging terms leads to 7.2.

Now the hedging strategy suggested by 7.6 can be made clear. If 7.2 fails to match the trading price of the option, then either $\Delta(V - \Pi) > r(V - \Pi)\Delta t$ or $\Delta(V - \Pi) < r(V - \Pi)\Delta t$. In the former case, we should borrow the money at interest rate r necessary to purchase the difference portfolio. In the latter case we should sell the difference portfolio and lend out the profits from this sale at interest rate r. Over time, the hedge should be adjusted continuously as the trading prices of the option and stock change. But at each time step our dynamic hedging strategy has eliminated the risk associated with the fluctuations in the stock price. In the limit of continuous trading with infinitely many time steps and no transaction costs, we are guaranteed a risk-free profit (Almgren 2002, p. 7). This, in essence, is what hedge funds do, although their pricing equations involve more sophisticated representations of the volatility of the assets they invest in than what we have considered.

I have spent the time to explain the Black-Scholes model to illustrate its role in the failure of the hedge-fund Long-Term Capital Management (LTCM, see figure 7.5) in 1998. Though it would be a stretch to blame this failure completely on the Black-Scholes model, I want to argue that the mathematical character of this representation, combined with its early successful application, were crucial factors in the overconfidence of the traders working for the hedge fund. A further amusing feature of this case is that Scholes and Merton, two of the three original economists who developed the model, not only worked for LTCM but also won the Nobel Prize for Economics for their model shortly before the fund's quick collapse. As Lowenstein explains in *When Genius Failed: The Rise and Fall of Long-Term Capital Management* (Lowenstein 2000), in the four years leading up to its failure, the fund reported "returns of more than 40 percent a year, with no losing stretches,

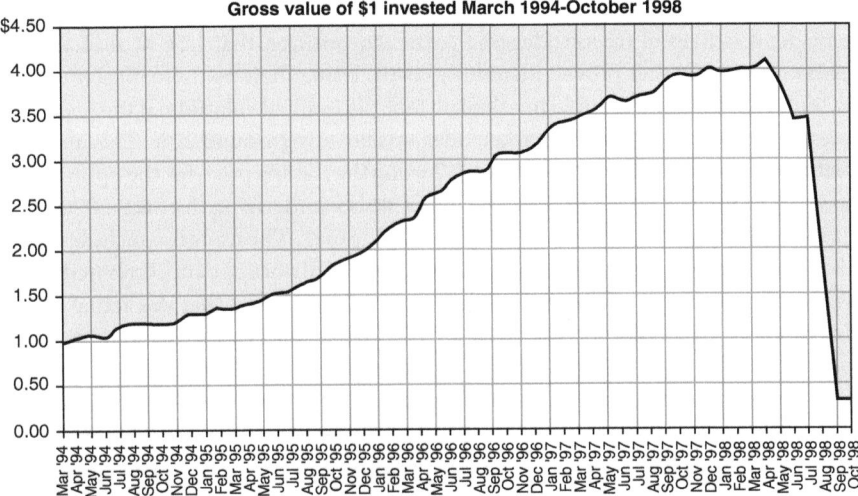

Figure 7.5: LTCM Return on Investment

no volatility, seemingly no risk at all" (Lowenstein 2000, p. xix).[8] The fund began the year with investments of over $4 billion (Lowenstein 2000, p. xx). At the time of its collapse the fund had mostly borrowed assets of $100 billion, but these had been leveraged into an exposure of more than $1 trillion (Lowenstein 2000, p. xix). However, unanticipated instability in the markets set off largely by international financial turmoil quickly wiped out these assets, and the associated lack of liquidity in the markets prevented the fund from recovering cash from its investments. In a pattern that is now all too familiar, the Federal Reserve was forced to step in and organize a bailout involving a consortium of 14 major banks.[9]

In the wake of this collapse, another Nobel Prize–winning economist, Merton H. Miller, asked

> whether the LTCM disaster was merely a unique and isolated event, a bad drawing from nature's urn; or whether such disasters are the inevitable consequence of the Black-Scholes formula itself and the illusion it may give that all market participants can hedge away all their risk at the same time. (in Lowenstein 2000, p. 143)[10]

Lowenstein seems to endorse the link between the formula and the illusion mentioned by Miller, going so far as to speak of the "Mertonian view of markets as efficient machines that spit out new prices with all the random logic of heat molecules

---

8. Lowenstein draws on Lewis (1999b), Kolman (1999).

9. For a quick summary, see Wilmott (2007, pp. 537–540).

10. Lowenstein cites an address called "A Tribute to Myron Scholes" from January 1999, but it appears that this paper has not been published.

dispersing through a cloud" (Lowenstein 2000, p. 123). To see how this link could be made, it is helpful to consider one particular position that LTCM took at the beginning of 1998 that proved incredibly costly. Using their mathematical models, the traders at LTCM were able to calculate that the implied volatility of the five-year options on the S&P 500 stock market index was hovering around 20%. This implied volatility can be easily found by solving 7.2 using the trading price for the options for the index to fill in V, the current value of the stocks underlying the S&P 500 index to supply S, and then calculating the implied value of σ. The traders concluded that this meant the options were overpriced because the historical data combined with their statistical model of the stock index, that is, 7.3, showed that the actual value of σ was only 15% (Lowenstein 2000, p. 124). Implementing the hedging strategy noted, they then sold the options that they took to be priced too high and bought the underlying stocks. Their position did not involve any risk that the stocks would go down because if that happened their options investments would provide a corresponding profit. That is, they had hedged perfectly against the risk associated with the fluctuations in the stock prices. However, what the traders were assuming is that the implied volatility of the stock options would drop from 20% to the historical level of 15%. When, instead, the implied volatility continued to rise, they expanded their position because now the hedging strategy looked to be even more of a sure thing. By the end of August 1998, the implied volatility had risen to 30% (Lowenstein 2000, p. 158). In September 1998 the implied volatility continued to go up, peaking at 38% (Lowenstein 2000, p. 191). This position alone lost the fund $1.3 billion (Lowenstein 2000, p. 234).[11] Eventually this position, combined with several others of the same sort, worked to undermine the confidence that LTCM's banks had in the continued viability of the fund. They were no longer able to finance their positions and adjust their various portfolios in line with their models, but any liquidation of their assets would have incurred huge losses and wiped out the equity of the fund.

Now, what exactly was the mistake that LTCM's traders made, and how can it be tied to the Black-Scholes model? It is tempting to blame the idealizations that went into the derivation of 7.2. For example, it seems obvious that markets are not efficient, traders do not operate in continuous time, and transaction costs are not 0. So, perhaps it was foolish for anybody to believe that 7.2 would ever be applicable to an actual market where its idealizing assumptions failed to apply. My suggestion, though, is that this criticism misdiagnoses the failure of 7.2 because it has an overly restrictive conception of when we can reasonably apply an idealized representation. It is too much to ask that we apply such representations only to systems where the idealizations are "approximately true" or where we can prove that only "negligible errors" are introduced. To see why, consider the applications of the heat equation 7.1. The idealizations concerning continuity appear to be just as extreme as the view that markets are efficient. For example, in a standard derivation of 7.1 we treat the iron bar as if it were continuously divided and as if temperatures are well defined down to the smallest distance and time scales. But as I have said, scientists have

---

11. The long-term error here becomes clearer when we note that the implied volatility was still 30% in September 1999 (Lowenstein 2000, p. 229).

applied this representation to understand temperature dynamics and have extended it to other kinds of "diffusive" phenomena. This suggests that scientific practice does not require that we check the validity of an idealization using any notion of approximate truth. In chapter 2 and again in chapter 5, I argued that a more reasonable test is to see whether the idealizations preserve the correctness of the representation for those aspects of the phenomenon we are focused on. For the heat equation, our focus is clearly on the larger scale temperature changes, and experimental testing confirms the accuracy of the representation within appropriate bounds.

What happens when we extend this conception of an adequate idealization to 7.2? One might make the new complaint that the traders at LTCM were foolish to apply their representations to actual markets because it is impossible to test 7.2 along the lines I have just outlined for 7.1. I disagree. In fact, the representation was deployed extensively since its publication in the 1970s. The main traders at LTCM had extensive experience with its use prior to setting up their hedge fund. The substantial returns they earned before setting up the fund convinced them and their investors that the idealizations central to 7.2 did not undermine the accuracy of their representation on the key question of the pricing of options. Furthermore, the fund traders had four years of successful trading for LTCM itself. The very pattern of this success, as given in figure 7.5, shows that 7.2 had passed very stringent empirical tests. For example, in that four year period, they never had two consecutive months of losses. In these sorts of circumstances, it seems eminently reasonable to continue to apply 7.2.

Still, the common situation where we have a highly successful though highly idealized representation is a dangerous one. I want to argue now that the mathematical character of the representation was partly responsible for the failure of LTCM for two reasons. First, the scope with which the representation could be reliably applied was hidden by the mathematics. Crucially, after long restricting their focus to bonds and options for bonds, LTCM moved into stocks and options for stocks. This overextension was a critical factor in the disastrous equity position. Second, the traders made incorrect assumptions about the ways in which their positions would interact with each other and with other market players. In particular, their assessment of the risk of losses of their combined positions was hopelessly distorted. On both counts, the very success of the highly mathematical representation generated the overconfidence in 7.2 that proved its undoing.

Both reasons can be traced to the abstract varying character of the representation. This not only allowed the link between 7.1 and 7.2, but also allowed the representation to be variously interpreted in terms of stocks, bonds, or other commodities. I want to group these reasons under the heading of *parameter illusions* because the presence (or absence) of a certain parameter in the representation as it is reinterpreted can be tied to the failures. The scope of reliable application will be overestimated as the flexibility of the mathematics increases and our understanding of why a core application has been successful in the past decreases. Perhaps there are relevant differences between bonds and stocks that are responsible for the success in applying 7.2 to bonds. Among other factors that could be important here are the predictable rate of return that a bond will pay in interest compared to stocks. That is, there may be absent parameters that reflect the differences between bonds and stocks. A related point is the viability of the associated representation of price fluctuations captured by our lognormal model 7.3. The main assumption encoded in 7.3 is that the percentage fluctuations reflected by σ are constant over time. This is what allows us to use past market data to fix σ and also what underwrites the calculation of implied volatility using current prices. It may be the case that this model is applicable to bonds and not to stocks. But as the role of 7.3 drops out in our derivation of 7.2, this possibility seems to have been largely discounted by the traders. A related worry is that our tests of 7.2 are restricted to a period when prices of a particular commodity may be more stable than over the long term. Because σ is a constant and this representation proves successful, it seems like we have evidence that this parameter is a constant. Again, the abstract representation of the underlying markets, then, may blind us to possible sources of future failure.

A deeper problem, though, is suggested by the way 7.2 represents risk itself. As we have seen, the derivation involves constructing a "risk-free" difference portfolio. This is what allows us to estimate V by considering the return on the money invested to buy the difference portfolio in a truly risk-free bank account. The problem is that the only risk we have taken care of is due to the price fluctuations of our underlying asset like the stock or bond. Our representation does not include any parameters tracking the risk that 7.3 is the wrong model or that σ has been poorly estimated using past market data. Put differently, we have explicitly factored in a premium for risk associated with price changes, but we have ignored any premium for risk associated with shifts in the level of volatility itself. This introduces a bias into the way 7.2 prices options. In a period where market participants worry about an increase in volatility itself, say, due to an international crisis or political instability, positions like the equity position will seem attractive because the price for stock options will be too high according to 7.2. The risks associated with these sorts of positions are systematically ignored because they find no place in the mathematical representation.

We see, then, some of the effects of a parameter illusion, which is more subtle than the illusion of completeness encountered in the column case discussed by Galileo. Risk is explicitly considered in the derivation, but only a certain kind of risk. Other sorts of risks are not represented, so the traders seemed to have thought they were absent. This comes out in the way that LTCM estimated the chances of a profit or loss in their entire fund over time. They seem to have treated each of their positions as independent, so that a loss in one position would not be correlated with losses in other positions. As a result, they estimated that a "loss of 50 percent of their portfolio" would occur roughly every 10^9 to 10^{30} days (Kolman 1999). This basic error can be traced, in part, to the way 7.2 works. It treats each position independently of any other positions and does not attempt to represent the interactions between prices for different commodities. However, it does not take a Ph.D. in economics to worry that a shift in volatility of one commodity could affect the volatility in others. In particular, as we have seen more recently, turmoil in one market can generate panic and turmoil in another.

There is a further sort of failure that LTCM's traders encountered based on the actions of other market participants. Once losses in a given position that LTCM had taken became known, this gave other participants an incentive to trade against the position that LTCM had taken, as well as any other position that LTCM had that was known. This is because LTCM clearly needed to generate cash by closing out some of its open positions. In the short term, this sort of counteractivity induced further losses and put even more pressure on LTCM to liquidate some assets, even at a substantial loss. At the extreme, this pressure gave investment banks like Goldman Sachs a reason to enter the market in a way that decreased LTCM's equity. This then made it easier to intervene and "rescue" the fund by acquiring its assets at a discount. Considering these sorts of problems, we see that the past success of 7.2 depended on some very specific circumstances. The merely academic point that markets are not exactly efficient does not do justice to the problems that LTCM ran into. Indeed, the actions of other market participants is a fully predictable way losses could accumulate and sink the fund. Of course, there is no aspect of 7.2 that corresponds to the actions of other market participants, and it seems that LTCM's estimates of potential loss failed to reflect it.

My suggestion, then, is that the successful application of highly idealized mathematical representations like 7.2 carries substantial risks of failure. These risks are not explicit in the representation itself, and the parameter illusion may suggest that these risks are absent. In the wake of LTCM's collapse, Merton suggested that "The solution...is not to go back to the old, simple methods. That never works. You can't go back. The world has changed. And the solution is greater complexity" (in Lewis 1999b). Although this is no doubt one sort of response, I suggest that more complicated mathematical representations may not fix the problems encountered by LTCM's traders. In fact, by being more complicated and purporting to incorporate additional sources of risk, the parameter illusion may be compounded and lead traders to be even more overconfident of the cogency of their trading strategies. The near-collapse of the credit markets in the fall of 2008 suggests that the models developed between 1998 and 2008 have not overcome these basic limitations.

7.4 ILLUSIONS OF SCALE

In chapter 5 we considered several different contributions related to scaling. A crucial distinction made there was between systems that were dynamically similar and systems that were merely geometrically similar. While pairs of systems that are dynamically similar behave in the same way, there is no guarantee that merely geometrical similar systems will behave in analogous fashion. This is an especially pernicious problem in structural engineering, where "scale effects" are often to blame for the collapse of structures like bridges or buildings. The problem, essentially, is that calculations carried out and implemented for small structures lead to a functioning structure that is stable, but when the same calculations are applied to larger structures, some kind of structural failure results. This is not because the original representations are wholly acausal, as with the static representation of a column. The representations consider, for example, the load that a given support will be forced to bear. The structure is then built to handle several times that load to guarantee a high margin of safety. Nevertheless, structural failures from scale effects can still arise when new aspects of the structure come to dominate as the structure is made larger and larger. The most famous example of this is the Tacoma Narrows Bridge, which collapsed in 1940 (Petroski 1994, ch. 9).

The Tacoma Narrows Bridge was a suspension bridge, like the Golden Gate Bridge, which supports its load primarily using cables above the bridge span. These cables need to be strong enough to withstand the changes in load as cars pass across the bridge. A focus on the materials of these cables and the associated struts making up the framework of the bridge led engineers to conclude that a sufficiently stiff structure could support the fluctuations in the load. But as Petroski notes, "in concentrating on the stiffness under the traffic load, designers were overlooking the role of stiffness in suppressing oscillations of the bridges in the wind" (Petroski 1994, p. 158). As the length of suspension bridges increases, these oscillations become increasingly important although this fact was not revealed by the mathematical representations of the bridges. The effects of wind were ignored completely, and this proved an effective simplification for shorter bridges. As the spans lengthened, though, the torsional oscillations induced by the wind proved to be sufficient to destroy the bridge. The dramatic oscillations of the bridge were captured on film and are continually used to remind engineers of the unanticipated sources of failure.[12]

For our purposes, the Tacoma Narrows Bridge failure is an instance of flawed mathematically supported scaling. The solution to the failure involved reinforcing the structure of the bridge to prevent the torsional oscillations induced by the wind. There was of course no reason the engineers designing the bridge could not have considered this possibility earlier and so avoided the failure. But I think the segmentation illusion noted in the column storage case played a role in directing the engineers' attention away from this possibility. The central problem was making sure that a suspension bridge that was this long was stable enough to support the cars that would pass over it. Once the representation of the cables and girders predicted that the threshold had been met for this purpose, no further investigations seemed necessary because worries about torsional oscillations were absent. These calculations, combined with the past successes of suspension bridges, gave the engineers the illusion that they had accounted for all reasonable sources of failure.

A second sort of contribution related to scaling was the resolution of d'Alembert's paradox using Prandtl's boundary layer theory.[13] Recall that the Euler equations for fluid flow result from setting viscosity to 0. This seemed to be a reasonable idealization based on the small viscosity observed in the air. As a result, it was possible to show that such a fluid would not induce any drag on an object as it moved through the fluid. This conclusion was so flatly contradicted by ordinary experience that it seemed a kind of paradox. Nevertheless the success of the Euler equations in representing fluid flow in other cases seemed to many to reinforce their validity. D'Alembert's paradox was conveniently forgotten, and the Euler equations led "many nineteenth century experts to glumly predict the impossibility of heavier than air flight" (Wilson 2006, p. 186). As Darrigol relates in his *Worlds of Flow: A History of Hydrodynamics from the Bernoullis to Prandtl* (Darrigol 2005), much of

---

12. See http://www.youtube.com/watch?v=3mclp9QmCGs.

13. See section 5.6.

the history of our understanding of fluids can be told in terms of two traditions. One tradition deployed highly mathematical representations in the study of fluids. A second, more practical approach engaged in engineering and design tasks proceeded more by ad hoc empirical adjustments. By the end of the nineteenth century there were several theoretical attempts to account for the role of viscosity in fluid flow, but none of them led to any real possibility of lift being generated by an object moving through a fluid like air. The crucial insight that viscosity induced a certain kind of circulation around an object like a wing eluded the best theoreticians. As Darrigol puts it, experts like "Rayleigh, Lamb and Kelvin knew too much fluid mechanics to imagine that circulation around wings was the main cause of lift" (Darrigol 2005, p. 305). The misplaced confidence in these mathematical representations led Kelvin, for example, to write "I have not the slightest molecule of faith in aerial navigation other than ballooning or of expectation of good results from any trials we hear of" (Darrigol 2005, p. 302).

Although we will not delve into the mathematical representations that came before Prandtl's boundary layer approach, it is still possible to see this sort of failure as the converse of the Black-Scholes failure. Again we have an abstract varying family of representations, but this time the confidence in the extrapolation of the representations across cases led some scientists to be sure that a certain task was impossible. The pernicious effects of these sorts of arguments should not be underestimated. A scientific argument that something is impossible does much to discourage the investigation of the phenomenon in question and can greatly set back the pace of scientific and technological progress.

The sort of scaling that produced these representations was quite reasonable given the state of fluid mechanics at the time. The successes of these techniques involved the boosts in confirmation that I have repeatedly tied to the positive contributions of mathematics. With the argument that flight is impossible, though, we see how this positive contribution can become a negative contribution. Because the scientists did not understand why the Euler equations or their innovations were so successful, they also misunderstood the scope of application of these representations. This is a problem that can be tied to the mathematics involved. With retrospect, using Prandtl's boundary layer theory and the associated mathematics of singular perturbation theory, we can of course see that this failure was avoidable. But as with the Tacoma Narrows Bridge collapse, this does not refute the point that the use of mathematics in this way brings costs as well as benefits.

7.5 ILLUSIONS OF TRACTION

The last kind of positive contribution that we have seen is in developing constitutive representations. On the view of these representations that I defended in chapter 6, belief in a constitutive representation is what allows an agent to gather evidence that will provide positive support to a family of derivative representations. As a result, normal science can proceed along Kuhnian lines once a sufficient number of agents adopt a given set of constitutive representations. The arguments in favor of believing in these constituent representations are largely practical. These beliefs are the only means to achieve a shared end, so it is rational to consider this or that belief, at least until such time as their practical value has been shown to be minimal. From this perspective it might seem easy to find failed constitutive representations. We can simply look at the history of science and find constitutive representations that have been discarded. This seems unfair, though, because these constitutive representations are not really failures if probing their associated derivative representations convinced scientists to move on to other constitutive representations. What is needed, then, for a genuine failure are cases where the constitutive representations at least initially seemed to many to be promising avenues for confirming derivative representations, but where this promise proved illusory. That is, it seemed like believing these representations was a good means to the end of confirming some derivative representations, but this was a mistake. Again, our interest is in finding cases where this illusion was tied to the mathematical character of the representations.

I develop a case that seems to be like this: catastrophe theory as applied to the social sciences. Catastrophe theory was proposed in the 1970s by Thom with great fanfare and subsequently developed by Zeeman and others as a novel "mathematical method for dealing with discontinuous and divergent phenomena" (Zeeman 1976, p. 65). It has the generality and apparent precision of other constitutive representations, but was quickly criticized based on its lack of success in its core applications in the social sciences. We consider one such failure and see how the mathematical character of the representations was responsible for the illusion of success. Although many of the results associated with catastrophe theory have been incorporated into contemporary science, the program has failed to offer any new constitutive framework: "catastrophe theory was propelled on a wave of hype and enthusiasm during the mid-1970s only to die out in bitter controversies by the end of the decade. Caught in fierce debates, the movement nearly vanished from the scene of science" (Aubin 2004, p. 96).[14]

At the heart of the theory is a mathematical theorem that is often referred to simply as Thom's theorem.[15] It involves a classification of surfaces of a certain sort. For our purposes, we restrict most of the discussion to surfaces given by functions $f(\alpha, \beta, x)$. We treat α and β as the control variables and x as the dependent variable. One sort of *equilibrium equation* for f is obtained by setting $\frac{\partial}{\partial x} f(\alpha, \beta, x) = 0$. The set of triples (α, β, x) that satisfy this equation is called the *critical set* of f. We can also isolate some points in the critical set of f as *singular* if they also satisfy $\frac{\partial^2}{\partial x^2} f(\alpha, \beta, x) = 0$. Finally, we can consider the pairs of values of the control variables (α, β) for which there is at least one x such that (α, β, x) is singular. The set of these points is the *catastrophe set* or *bifurcation set* of f.

The significance of these definitions comes out clearly in a few examples. Consider first

$$f(\alpha, x) = \frac{x^4}{4} - \frac{x^2}{2} - \alpha x \qquad (7.7)$$

---

14. I draw on Zeeman (1976), Sussman and Zahler (1978), Truesdell (1984, §12) and Aubin (2004).

15. I follow the very accessible exposition of Sussman and Zahler (1978, §5).

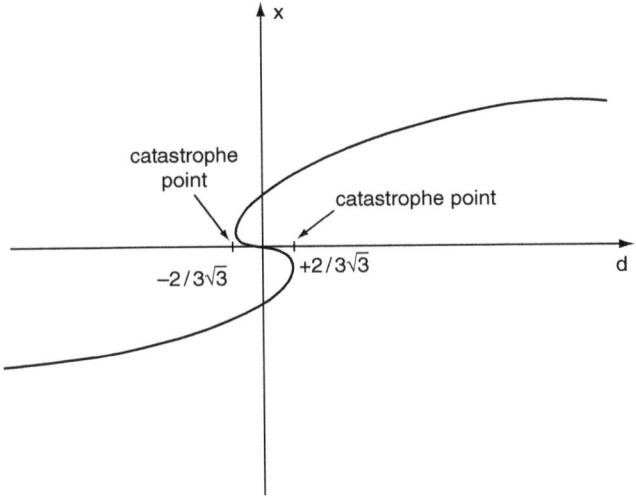

Figure 7.6: A Fold Catastrophe
Source: Sussman and Zahler (1978, p. 146)

The equilibrium equation for this function, whose solutions are f's critical set, is

$$\frac{\partial}{\partial x} f(\alpha, x) = \frac{\partial}{\partial x}\left[\frac{x^4}{4} - \frac{x^2}{2} - \alpha x\right] = 0 \qquad (7.8)$$

$$x^3 - x - \alpha = 0 \qquad (7.9)$$

The catastrophe set of f is obtained by solving

$$\frac{\partial^2}{\partial x^2} f(\alpha, x) = \frac{\partial}{\partial x}\left[x^3 - x - \alpha\right] = 0 \qquad (7.10)$$

$$3x^2 - 1 = 0 \qquad (7.11)$$

This equation is satisfied just in case $x = \pm\frac{1}{\sqrt{3}}$. Using the equilibrium equation, we can then determine that these values for x obtain only when $\alpha = \pm\frac{2}{3\sqrt{3}}$. These two values for α are the members of the catastrophe set. They mark the points at which the critical set changes in character. As a graph of the equilibrium equation shows, when α is either less than $-\frac{2}{3\sqrt{3}}$ or greater than $\frac{2}{3\sqrt{3}}$, then there is only one value for x (figure 7.6). However, in the interval $-\frac{2}{3\sqrt{3}} < \alpha < \frac{2}{3\sqrt{3}}$, x has three admissible values. At the points $\alpha = \pm\frac{2}{3\sqrt{3}}$ there is a discontinuous change in the structure of the critical set. This is an instance of what is called the *fold catastrophe*.

The catastrophe set can have a more interesting structure when our function f involves two control variables, as with

$$f(\alpha, \beta, x) = \frac{x^4}{4} + \frac{3\beta x^2}{2} - \alpha x \qquad (7.12)$$

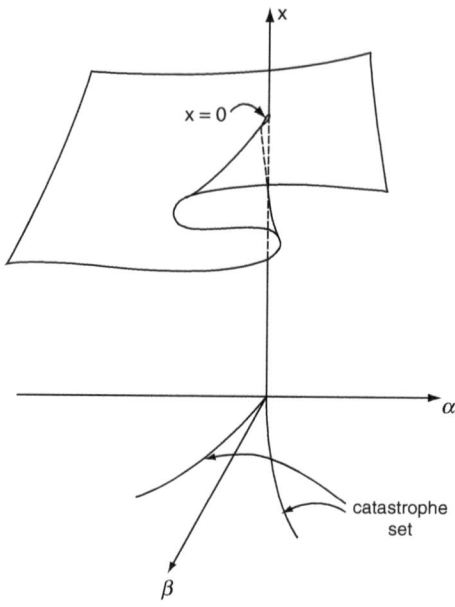

Figure 7.7: A Cusp Catastrophe

Source: Sussman and Zahler (1978, p. 149)

Proceeding as before, we have the equilibrium equation

$$\frac{\partial}{\partial x}f(\alpha, \beta, x) = \frac{\partial}{\partial x}\left[\frac{x^4}{4} + \frac{3\beta x^2}{2} - \alpha x\right] = 0 \quad (7.13)$$

$$x^3 - 3\beta x - \alpha = 0 \quad (7.14)$$

The catastrophe set is found by considering

$$\frac{\partial^2}{\partial x^2}f(\alpha, \beta, x) = \frac{\partial}{\partial x}\left[x^3 - 3\beta x - \alpha\right] = 0 \quad (7.15)$$

$$3x^2 - 3\beta = 0 \quad (7.16)$$

For this equation to be satisfied, $x = \pm\sqrt{\beta}$. Plugging this value for x into the equilibrium equation yields that $\alpha^2 = 4\beta^3$. That is, whenever the two control variables α and β are such that $\alpha^2 = 4\beta^3$, we will have a discontinuous change in the structure of the critical set. This can be visualized by plotting the surface of the critical set using the axes α, β, and x (figure 7.7). The catastrophe set then appears as a projection of this surface onto the α-β plane. As can be seen from the figure, the curve of the catastrophe set corresponds to the edges of the folds in the surface of the critical set. This sort of case is called the *cusp catastrophe*.

Thom's theorem concerns the surfaces that result from starting with an infinitely differentiable function f with one to four parameters $\alpha_1, \ldots, \alpha_4$ and with either one or two dependent variables x_1, x_2. The critical set of this function

Failures

is then almost always made up of ordinary points or points that exhibit one of seven elementary catastrophes like the cusp catastrophe.[16] In the case where f is a function of two control variables and one dependent variable, there are only two kinds of catastrophes that can arise, namely, the fold and cusp catastrophes that we have seen. Crucially, though, the theorem does not say for an arbitrary surface how many different catastrophes there will be, for example, several cusps bunched together.

A physical interpretation of the cusp catastrophe follows from thinking of x as standing for some quantity of interest for a system that is constrained to seek a local minimum on the surface given by the critical set. For certain evolutions of the α-β, the system will exhibit a discontinuous jump as it moves from the higher surface of the fold at the edge of the cusp. The conditions under which this jump occurs can be seen as a result of the structure of the catastrophe set. The program of applying catastrophe theory proceeds in the opposite direction. Given some jump or discontinuous phenomenon that we have observed in nature, we attempt to embed that phenomenon in an appropriate surface. If we can find cases where the quantity of interest can be represented as our x and it is feasible to isolate two control parameters for our α and β, then Thom's theorem seems to find direct application. We can be assured, it seems, that any surface responsible for the discontinuous phenomenon must be built out of folds and cusps. Indeed, there are qualitative features of the phenomenon that permit us to narrow our focus to surfaces with a cusp. In the cusp case (and not in the fold case), we have the potential for a kind of divergence that we may observe and wish to account for. This happens when systems that begin arbitrarily close together on the α-β-x surface show different behavior as the system seeks its local minimum. As in figure 7.8, we can account for this using the cusp catastrophe if we imagine that the systems 1 and 2 start in the region of the surface "beyond" the cusp and diverge when one system proceeds to the higher surface and the other proceeds to the lower surface.

This is essentially the method employed by Zeeman when he describes a "war model" that can be used to model the decision of one country to attack another based on the factors of cost and threat.[17] A similar model is used to account for dog attacks based on the control variables of rage and fear. Zeeman even held out the hope that this approach could also find application in the management of prison riots. In each case, the challenge is to come up with hypotheses that allow the phenomenon to be embedded in a surface with a cusp. The success of this endeavor seems to be guaranteed by Thom's theorem. For if almost all surfaces fit the conditions of Thom's theorem, it seems reasonable to assume that the surface corresponding to our phenomenon will have catastrophes corresponding only to Thom's elementary catastrophes. It then seems to be a fairly straightforward matter

---

16. For five control variables there are 11 elementary catastrophes and "For six or more parameters there exist infinitely many nonequivalent singularities" (Sussman and Zahler 1978, p. 152). The mathematical meaning of "almost always" is discussed at Sussman and Zahler (1978, p. 185).

17. Zeeman (1976, p. 76). A more complicated surface with four control variables is also considered at p. 80.

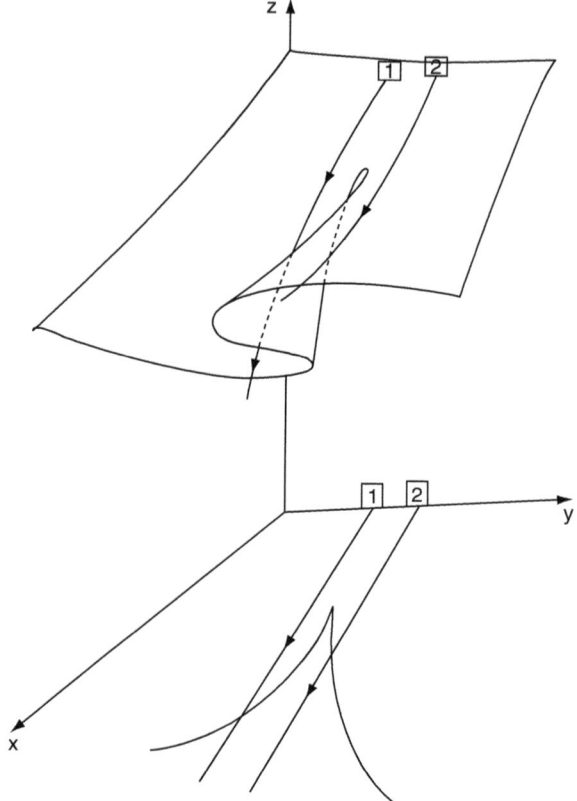

Figure 7.8: Divergent Trajectories
Source: Sussman and Zahler (1978, p. 125)

to isolate the relevant qualitative features of the jumps in the dependent variables and extract the features of the surface that are driving the system.

In their extended discussion, Sussmann and Zahler raise many objections, but what I want to focus on is the potentially constitutive function of catastrophe theory and its failure in this respect. The problem is that it is possible to embed any phenomenon in some surface if no restrictions are placed on the features of the surface. The proponents of catastrophe theory used this flexibility to apply Thom's theorem. But in practice they also assumed that the surfaces in question had several kinds of special features. The need for this becomes clear when we recall that Thom's theorem applies only for functions with up to four control variables and two dependent variables. It is not obvious that the phenomena in question can be accurately represented with these impoverished resources. However, even granting that the conditions of application of Thom's theorem are met, there is a further kind of simplicity that must be assumed to apply catastrophe theory in the way its proponents did. This tactic is reflected in Zeeman's claim that "whenever a continuously changing force has an abruptly changing effect, the process must be described by a catastrophe" (Zeeman 1976, p. 80). That is, a single catastrophe was used to account for the features of the phenomenon. In virtue of this stipulated sort of simplicity, the results of applying catastrophe theory seemed more substantial than they actually were. To see the significance of this move, notice that if our infinitely differentiable function $f(\alpha, \beta, x)$ is randomly chosen, then we have good reason to apply Thom's theorem and conclude that the points in the critical set will either be ordinary or else correspond to folds or cusps. But this gives us no reason to think that the surface in question will have exactly one cusp. If we assume this, then we will get interesting predictions out of our model, but these predictions are not the result of the mathematics being used.[18] Misleading remarks like "Through catastrophe theory we can deduce the shape of the entire surface from the fact that the behavior is bimodal for some control points" (Zeeman 1976, p. 67) only further muddy the waters.

This brief discussion of catastrophe theory has allowed us to see why it could not function as a viable constitutive representation. Simply put, it only allowed for a substantial confirmation of its associated derivative representations if it included some additional principles of simplicity that would constrain the choice of control variables and the topology of the resulting surfaces. But these further principles are not naturally related to the core mathematics of Thom's classification results. The appearance of a viable constitutive framework can be tied, then, to the flexibility of the mathematics and the lack of recognition of the tacit simplifying assumptions needed to get the approach off the ground. Again and again it seemed like phenomena from very different areas of science could be fruitfully modeled using the framework of catastrophe theory. But the illusion of *traction* here was based on misunderstanding the points at which this framework hooked on to the systems in question. The contrast with Newton's three laws is quite dramatic. Newton's laws, when supplemented with one or more force laws, quickly resulted in very specific constraints on physical systems, but the mathematical framework of catastrophe theory never placed enough limits on what could happen for a meaningful test to occur. Despite the persistence of Thom, Zeeman, and a few others, most scientists quickly realized that it was not a viable means to confirm any derivative representations, and so it dropped from the scene. It remains a useful case to study, though, for it is far from clear that scientists are immune to the illusion of traction that gripped the advocates of catastrophe theory.

7.6 CAUSAL ILLUSIONS

Now I turn to a final source of scientific failure that can be tied to mathematical contributions to science. This source is not related to any single category of contribution, as with our previous examples. Instead, the source involves confusing one category with another. In principle, this could involve any of the categories I have discussed, but by far the most dangerous kind of mistake is treating what is really an abstract acausal contribution as a concrete causal contribution. I call such cases

---

18. Sussman and Zahler (1978, §12).

causal illusions. When they occur, a scientist believes that her mathematics reflects genuine constituents and causal relations which prove illusory. A well-known example of this is the interpretation of the mathematics used to represent heat that led to the identification of heat with a fluid-like substance known as caloric.

In the late eighteenth and early nineteenth centuries, scientists had developed a successful representation of temperature that involved viewing temperature as a manifestation of an underlying physical quantity, which was labeled "heat." Various theories of heat were developed, but the most successful was the identification of heat with a fluid-like substance that was dubbed caloric. Different accounts of caloric were proposed. I discuss the one developed by Laplace. Crucial assumptions of the theory are that caloric is a conserved substance and that it exists in several different states. For example, there is free caloric that is responsible for the temperature of the air. But there is also a freely moving "intermolecular caloric" (Chang 2004, p. 71) within bodies whose density is identified with the temperature of the body. Finally, Laplace posited a "latent caloric" (Chang 2004, p. 71), which remains bound within the molecules of a body but could be released by chemical reactions or other disturbances. Equipped with this conception of heat, Laplace was able to use tools developed in the study of fluids like water to clarify changes in temperature. The most celebrated achievement of this approach was Laplace's correction of Newton's calculation of the speed of sound c. The correction marked a shift from $c^2 = \frac{p}{\rho}$ to

$$c^2 = \frac{\gamma p}{\rho} \qquad (7.17)$$

Laplace's basic insight is that the compression of air that occurs during the propagation of a sound wave involves a change in temperature that is significant enough to affect the speed of the wave. According to Laplace this change in temperature is due to the conversion of latent heat in the air into heat that raises the temperature of the air. This accounts for the additional factor γ. For an outline of the mathematics of this derivation, see Appendix C. For our purposes, the crucial feature of Laplace's argument is that it took for granted the earlier derivation of the speed of a wave-like disturbance. In one dimension this wave equation for disturbance $u(x,t)$ is

$$\frac{\partial^2 u}{\partial t^2} = c^2 \frac{\partial^2 u}{\partial x^2} \qquad (7.18)$$

The speed of the wave is given by c. The derivation of this equation involves many simplifications, but it had proven accurate for several wave systems, such as water waves.

A common view of the relationship between the belief in caloric and Laplace's successes is Chang's claim that "the Laplacian derivations did rely on particular ontological assumptions about the nature of caloric" (Chang 2003, p. 904). There are dicey historical issues here connected with the history of Laplace's discovery and the role of theoretical assumptions in motivating them. Setting these issues aside for the moment, let us try to reconstruct the actual contribution from the mathematics. Based on our current understanding of the link between heat and the kinetic energy of the molecules, it is clear that we have an abstract acausal contribution. The correction factor γ reflects large-scale features of the air that bottom out in the molecular interactions of the genuine constituents of the air. Laplace was able to get an improved result because his derivation aggregated these causal effects in a way that preserved the accuracy of the representation for the relevant features of interest while ignoring the causal details. So, the mathematics reflects the large-scale changes in air over time, but it does not do this by accurately capturing the causal connections between the genuine constituents of the air. The situation here is more intricate than what we had with the static representation of the column, but the basic story is the same. The representation involves conditions that fail to reflect genuine causal information concerning what is going on in the air.

This point about the actual contribution of the mathematics to the success of the heat equation is consistent with other sorts of merely apparent contributions. In particular, we can see how in the absence of a kinetic theory of heat, it might appear that the contribution the mathematics was making here was what I have been calling a concrete causal contribution. To see why, recall the analogous treatment of water waves. The mathematics treats the water as a continuous substance, but this idealization could be reconciled with a causal interpretation of the wave equation and its solution in terms of small particles interacting according to unknown laws. So, too, it might reasonably appear that the representation of sound waves was grounded in a causal basis of the particles of air and caloric whose causal interactions are unknown. Of course, we explain the success here quite differently by emphasizing the acausal contribution from the mathematics. My point is that precisely because mathematics can contribute in both causal and acausal ways, the successful use of mathematics in science may lead to a misunderstanding of why this or that contribution works. In this case, the successful use of mathematics prompted a causal interpretation that is incorrect. In retrospect, this causal interpretation can be seen to be unwarranted when compared to a more modest acausal interpretation of how the mathematics relates to the features of the physical system.

7.7 FINDING THE SCOPE OF A REPRESENTATION

We have seen six examples of scientific failures that can be blamed, at least in part, on the mathematics. The way mathematics contributes positively in some cases is the same way it contributes negatively in others. On this approach the contributions of mathematics to science take on a somewhat tragic air. Just as the strengths of a tragic hero prove to be his undoing, so, too, the flexibility of the ways mathematics can contribute to science accounts for both sorts of contributions.

Let us conclude our discussion by turning to the implications of our conclusions for the proper interpretation of our scientific knowledge. The example of caloric has been much discussed in the context of the pessimistic meta-induction against scientific realism. If Chang is right and "the Laplacian derivations did rely on particular ontological assumptions about the nature of caloric" (Chang 2003, p. 904), then the empirical success of Laplace's proposal supports the existence of caloric based on standard inference to the best explanation arguments. Given that caloric does not exist, the scientific realist seems forced to provide features of contemporary science that distinguish it from past science. It seems that we must then retreat to some kind of constructive empiricism that restricts our scientific knowledge to what is observable.

Our discussion has raised a much more disturbing possibility, though. This is that the use of mathematics has the potential to give rise to systematic failures via the various illusions that we have seen. These include the illusion of completeness and segmentation in the case of acausal representations and the parameter illusion with our varying representations. Other sorts of illusions arise in connection with scale illusions, the problem of traction for constitutive representations, and the causal illusion in the case where we mistakenly think an acausal representation is actually causal. If these illusions are pervasive, then not only scientific realism but even constructive empiricism is in trouble. This is because the problems we encountered were not restricted to the unobservable. The cracking of columns or the bankruptcy of hedge funds are all too observable phenomena.

My suggestion for how to proceed is to rethink the basic realist strategy of arguing that the best explanation for the success of science is that its theories are true or approximately true. We see the problems with this strategy even with Psillos's "divide and conquer" response to the pessimistic meta-induction. His proposal is to restrict inference to the best explanation to what could be called minimal explanations. These explanations appeal only to what is essential to getting the explanation to work. As Psillos puts it, "Theoretical constituents which make essential contributions to successes are those that have an indispensable role in their generation" (Psillos 1999, p. 110). This allows us to criticize a scientist like Laplace for appealing to an explanation that invokes caloric. As "Laplace's account does not explicitly rest on any particular representation of heat, although he happened to be an advocate of the caloric theory" (Psillos 1999, p. 120), the minimal explanation would appeal to some features of the situation, but not the existence of caloric. In this case, weaker assumptions about the links, for example, between heat and the compression of the gas, are sufficient to get the derivations to work, so the stronger assumptions found in the caloric theory are dispensable and not supported by the experimental success.

Chang responds by insisting that "the only plausible explanation ... was to understand it as the disengagement of caloric from ordinary matter, caused by mechanical compression" (Chang 2003, p. 904). This is because the kinetic theory had not been sufficiently developed to account for this kind of heating. Presumably Psillos thinks that it was open to Laplace to characterize the contribution of the mathematics in acausal terms. That is, the success of the derivation showed that there was some underlying features of the air responsible for this heating, but that it need not be anything like caloric. So there is an explanation here of the success of the mathematics, but it is not necessarily a causal explanation, as Chang seems to assume. Interestingly enough, Chang is well aware of this approach to the mathematical contributions of successful scientific representations. In *Inventing Temperature* he relates how "post-Laplacian French science" exhibited "a loss of nerve in theorizing about unobservable entities" (Chang 2004, p. 96). A central example of this is Fourier who "remained noncommittal about the ultimate metaphysical nature of heat" (Chang 2004, p. 96). This did not stop Fourier from improving our understanding of temperature dynamics. In fact, a lack of concern for the causal grounding of his derivations seems to be a crucial factor in emboldening Fourier to his groundbreaking mathematical innovations. These include the derivation of the heat equation 7.1 discussed earlier. Fourier's more free-wheeling approach to the link between mathematics and its causal interpretation was initially viewed with some suspicion by Laplace and the members of his school. But the accumulation of experimental success where the more rigid Laplacian school failed was adequate to convince many scientists that Fourier's attitude was superior.[19] To get a sense of the importance of Fourier's mathematical innovations, we can note Wilson's point that "applied mathematicians often declare that, of all the discoveries in mathematics, the ones most prized are due to Fourier" (Wilson 2006, p. 247).[20]

What we see, then, is that scientific realists should grant the possibility of different sorts of contributions from mathematics in the service of combating the pessimistic meta-induction. Part of the history here is that scientists came to appreciate these different contributions only after an overly causal or mechanical approach led them astray. But if the scientific realist says only what Psillos has said, there is a danger that some kind of positivism or instrumentalism will wind up looking very attractive. This is because we never seem forced to conclude that the mathematics is tracking causes and may always retreat to an acausal interpretation of the mathematics. No particular causal interpretation ever seems "essential" or "indispensable." We can rest more weight on the purely mathematical character of our representations and eschew any further interpretive risk. This is, after all, one way to make sense of the rise of positivism in the nineteenth century.[21]

This point serves to expose the weakness of an inference to the best explanation argument for scientific realism or even for constructive empiricism. What these arguments miss is the need to focus more narrowly on the appropriate scope of applicability of a given mathematical representation and what this tells us about the accuracy of the representation. Many of the failures that we have seen are consistent with limited success, but the failures arose from not realizing how this success occurred or where it would end. This is most clear with the completeness and segmentation illusions for the column storage problem and the parameter illusion for the Black-Scholes model. Notice furthermore how the interpretive question here did not turn on the existence of any entities, but only on the extent to which a given aspect of the representation would remain accurate if we adjusted the target system. The same worries arise for the scale illusions. As we build larger bridges, the scientific realist must allay the worries that her representations miss out on features that become important at that different scale. Finally, in addition to the caloric case, the traction illusion of catastrophe theory shows the possibility for a radical sort of failure where an entire research program is misconceived.

We see, then, that the scientific realist should focus less on the existence of entities that may be involved in the success of our best representations and more on the

---

19. Fox (1974), noted in Chang (2004).

20. Wilson quotes Lanczos from Prestini (2004).

21. See Fox (1974, p. 129) for the Fourier-Comte link.

scope of applicability that this or that representation has. If this scope is sufficiently wide and varied, then its success may be best explained by the existence of a certain kind of entity that operates across the target systems in question. But prior to this there is the important task of clarifying where the representation can be reasonably applied and where there is some reason to think its success will end. By reorienting the scientific realist toward this sort of task, the argument for realism becomes much more of a case-by-case affair where it may turn out that some of our best representations are successful in such a narrow domain that we must concede in the end that we lack any reason to believe in the entities posited. Or, as with catastrophe theory, it may turn out that our apparent confirmation of a family of derivative representations was simply a mistake. Still, armed with the list of potential illusions that we have developed in this chapter, and certainly others that can arise independently of the contributions of the mathematics, the scientific realist is sure to find some cases of genuine success. It is true that this local sort of strategy will not rule out the possibility that some sort of error has crept into our reasoning or that some success is purely fortuitous. But the realist was never engaged with the philosophical skeptic. We can concede the abstract possibility of complete error, while arguing that it remains reasonable to think that none of the errors we are likely to make has been committed in some particular case. At the moment, this sort of optimism is simply a program for how some form of scientific realism can be reconciled with the mathematical character of our best scientific representations. A fuller discussion of this proposal must be postponed until chapter 13 and the discussion of Wilson's more sustained challenge to my approach to representation.

PART II

Other Contributions

8

Discovery

8.1 SEMANTIC AND METAPHYSICAL PROBLEMS

When it appeared, Mark Steiner's *The Applicability of Mathematics as a Philosophical Problem* (Steiner 1998) was the only sustained attempt to articulate a problem of philosophical interest related to applications besides the indispensability argument for platonism. The prevailing attitude toward such a problem in the philosophy of mathematics is reflected in Steiner's remark that his book "is my final attempt to persuade my advisor and mentor, Professor Paul Benacerraf, that there really *is* a philosophical problem about the applicability of mathematics in natural science" (Steiner 1998, p. vii). Later Steiner notes that the influential anthology edited by Benacerraf and Putnam (1983) lacks any discussion of the issue (Steiner 1998, p. 14, n. 8). Steiner's response to this situation is to distinguish several different problems that arise in connection with the applicability of mathematics. He goes on to argue that a problem associated with the role of mathematics in scientific discovery is especially important and that the most promising solution to this problem undermines various kinds of naturalism. In this respect, he dwells on the significance of applications for our knowledge outside of mathematics: "Following the great physicists ... I would like to explore its implications for our view of the universe and the place in it of the human mind" (Steiner 1998, p. 2). The way Steiner develops his problems and the solutions he proposed have been a great aid in helping me formulate the central challenge of this book, namely, to account for the contributions that mathematics makes to the success of science. But for reasons that I hope to make clear in this chapter, Steiner's problems and solutions do not provide much help in resolving our main question. Much of what he says is consistent with my epistemic proposal for how mathematics makes its contributions, so it is important to review his discussion and see where our two positions have fruitful points of contact.

The first two chapters of Steiner's book present four different problems related to four different aspects of the applicability of mathematics. These are (1) the need for a uniform semantic interpretation of mathematical language when it appears in scientific reasoning, (2) the clarification of the metaphysical relationship between mathematical and physical objects that allows applications, (3) the explanation of how a particular mathematical concept can be used to describe a feature of a target system, and (4) the explanation of how mathematical concepts in general can be used to describe features of the physical world.

The semantic problem arises even in simple cases of counting. Steiner presents the following argument:

1. $7 + 5 = 12$.
2. There are seven apples on the table.
3. There are five pears on the table.
4. No apple is a pear.
5. Apples and pears are the only fruits on the table.
   Hence,
6. There are exactly twelve fruits on the table. (Steiner 1998, p. 16)

The problem is then how to relate the substantival occurrence of the numeral "7" to the adjectival occurrence of the word "seven." The former word seems to be a name that picks out an object, whereas the latter word seems to function semantically as a term for a property like "red" or "heavy." As Steiner summarizes things, "the problem is to find a constant interpretation for all contexts—mixed and pure—in which numerical vocabulary appears" (Steiner 1998, p. 16). In the next three pages, he argues that "Frege addressed this semantical problem, and solved it" (Steiner 1998, p. 17). The solution consists in the claim that all occurrences of number-words in scientific reasoning can be handled using the substantivalist reading. For example, premise 2 in the argument can be read as the sentence "The number of apples on the table is seven," and this has the form "The number of Fs is m," where m is a name for a number. Steiner then concludes his discussion of the semantic problem by noting that the right arguments can be made valid by adding premises such as "For all Fs and Gs, if the number of Fs is m and the number of Gs is n and no F is a G, then the number of F or Gs is $m + n$." For Steiner these general truths are not necessarily wholly logical, but our uniform semantic reading of the premises of an argument allows the argument to become valid by adding premises whose justification comes from our logical and mathematical knowledge.

There are at least two puzzling features of Steiner's brief discussion of the semantic problem and its Fregean solution. The first problem concerns the viability of the Fregean solution. As Steiner clearly recognizes, the scope of the problem is much wider than the simple counting case he discusses. So, even if Frege's strategy works in this sort of case, it is not immediately clear how to extend it to the sorts of examples reviewed earlier in this book. A related point concerns the assumptions necessary to vindicate the Fregean solution. To start, it seems to presuppose a platonist interpretation of mathematics. Even assuming this, though, the Fregean solution must also make strong claims about the semantics of ordinary natural languages like English or the semi-technical languages used in the sciences. The solution is only acceptable, I argue, if the correct logical form of all these sentences really does bottom out in substantival occurrences of all mathematical terms. But neither Frege nor Steiner provides evidence for this sort of conclusion. For example, Frege merely says that adjectival uses of number-terms "can always be got round" (Frege 1980, §57). This is a merely technical point about how a language could be devised and fails to meet up with our starting point of our languages as they are now used.

A second puzzling aspect of Steiner's discussion is his note that grants the option of solving "the 'semantic' problem of the applicability of mathematics with a theory according to which all numerals are really predicates" (Steiner 1998, p. 17, n. 21). Here he mentions only Hellman's book *Mathematics without Numbers* (Hellman 1989). As we explore in more detail in chapter 9, Hellman develops a modal-structuralist interpretation of mathematics according to which all mathematical claims can be taken to be about merely possible structures. Hellman, then, does not intend to commit himself to the actual existence of any abstract objects, although the price for this sort of rejection of platonism is an undefined modal operator. Steiner takes this to be an adjectival strategy because the purely numerical claims like "$7 + 5 = 12$" come out as general claims about the features of any number structure: "It is necessary that in any number structure S, adding$_S$ the 7_S to the 5_S results in the 12_S." This claim is not about any particular domain of objects, but is relativized to any possible number structure.[1] Simple applications can then be seen as instantiations to an actual domain. Even in the counting case, some care is needed to carry out this strategy because the fruit on the table will not instantiate the entire natural number structure, but only an initial segment. Nevertheless, Steiner clearly imagines a non-Fregean approach to the semantic problem that achieves a univocal reading of mathematical terms by using a modal-structural interpretation of the mathematical language.

As with the Fregean approach that Steiner defends, it is not clear if a modal-structuralist is entitled to their modal assumptions or the assumption that natural language sentences have logical forms that match their interpretation. Hellman, for one, is not committed to this strong assumption as he is not trying to solve the semantic problem as Steiner presents it. But given Steiner's apparent admission that this sort of adjectival strategy could succeed, the burden is then on Steiner to explain how his substantival solution is superior. If one says that they have a solution to a problem, this usually means that they think their solution is the best one, and not just that they have a strategy for proceeding that might ultimately be successful. Unfortunately Steiner has not given us any reason for preferring his Fregean strategy over an adjectival strategy. Until he does this, I insist that the semantic problem remains open.

As a last point about Steiner's semantic problem, it is worth briefly relating it back to our discussion of the content of a scientific representation and its role in scientific reasoning in chapter 2. My discussion there presupposed that agents were able to refer to mathematical objects as well as the relevant physical properties of the target system. These assumptions were then used to clarify the content of a scientific representation. But unlike Steiner I did not relate the content of a representation to any sentence or collection of sentences that the agent might utter in the course of scientific reasoning. This does not mean that I think these representations are nonlinguistic or somehow accessible to agents independently of their understanding of some natural language. Still, I have deliberately bypassed Steiner's semantic problem by refusing to take a stand on how natural language sentences should be interpreted or how their contents relate to the contents of our scientific representations. Typically many sentences will be involved in the specification of a

---

1. Hellman adds, for any mathematical domain, the additional assumption that a structure for that domain is possible.

given scientific representation, and I am not optimistic that there is a simple story about how this works. At the same time, I have assumed that once the contents of these representations are given, scientific reasoning can proceed more or less as we reason with sentences. That is, an acceptable scientific inference will preserve truth or accuracy. Using this starting point, then, we can account for the role of scientific representations in prediction and use successful predictions as a basis for an increase in confirmation.

Steiner's metaphysical problem turns on an alleged metaphysical gap between mathematics and its applications. Many think there is "a gap between mathematics and the world, a gap that threatens to make mathematics irrelevant" (Steiner 1998, p. 19). This gap is especially clear for platonism, and Steiner notes Dummett's complaint that platonism "leaves it unintelligible how the denizens of this atemporal supra-sensible realm could have any connection with or bearing upon conditions in the temporal, sensible realm that we inhabit" (Dummett 1993, p. 430, at Steiner 1998, p. 20). Hartry Field took this problem to undermine our epistemic access to mathematical objects; he also claimed that it called into question the scientific acceptability of theories that deployed mathematics. He argued that reference to mathematical objects brought in matters extrinsic to the physical system, but that scientists prefer intrinsic explanations in terms of the objects making up the system itself (Field 1980, p. 43). Although Steiner does not make this point, it should be clear that the metaphysical problem arises for almost any interpretation of mathematics. For example, Lewis's megethology interprets mathematics in terms of the part/whole relationships between concrete objects (Lewis 1993). Even this interpretation faces a gap between the subject matter of mathematics and its applications. The theory of hyperbolic differential equations is about some dispersed collection of mereological wholes, and this intricately structured entity seems just as extrinsic to fluid systems as any platonic entity.

Steiner again appeals to Frege to solve his metaphysical problem. Noting Frege's slogan from *Foundations* that "The laws of number... are, however, applicable to judgments holding good of things in the external world; they are laws of the laws of nature" (Frege 1980, §87), Steiner claims "mathematical entities relate, not directly to the physical world, but to concepts; and (some) concepts, obviously, apply to physical objects" (Steiner 1998, p. 22). As with the semantic case, Steiner describes only how this will work for the natural numbers and then generalizes this strategy to the rest of mathematics. Frege's strategy for the natural numbers involved a relationship between first-order concepts that apply to objects. Let two concepts F and G be *equinumerous*, abbreviated as $F \approx G$, just in case the objects in their extensions can be put in one-one correspondence. Then Frege adopted a definition of numbers that entailed, in the context of second-order logic, the principle:

$$\forall F \exists!x \forall G(G\eta x \leftrightarrow F \approx G) \tag{8.1}$$

The intended interpretation of η is the relationship between a concept G and object x that obtains just in case the number of G's is x.[2] This principle entails another

---

2. $\exists!x$ means that there exists exactly one x.

principle known as Hume's principle: the number of Fs is identical to the number of Gs just in case $F \approx G$. Steiner relates how Boolos showed that this theory is consistent (relative to Peano arithmetic) and strong enough to recover ordinary arithmetic.[3] So even though the logical system Frege deployed in *Basic Laws* was inconsistent, we can still adopt the key insight behind his approach to the natural numbers. The basic idea of this approach to number theory is that the numbers can be defined in terms of their relationship to the concepts that are relevant to the application of the numbers in counting. This then closes the gap between the numbers and their canonical application and accounts for the relevance of numbers to their application.

Steiner then extends this strategy to the rest of mathematics and its applications:

> Nor is this insight limited to arithmetic, since mathematicians have modeled all classical mathematics in set theory, ZFC. To "apply" set theory to physics, one need only add special functions from physical to mathematical objects (such as the real numbers). Functions themselves can be sets (ordered pairs, in fact). As a result, modern—Fregean—logic shows that the only relation between a physical and mathematical object we need to recognize is that of set membership. And I take it that this relation poses no problem—over and beyond any problems connected with the actual existence of sets themselves. (Steiner 1998, p. 23)

We start with a set theory that includes physical objects as individuals. Then we have impure sets at various levels of the cumulative hierarchy postulated by the axioms of ZFC. In addition we also have all mathematical objects appearing as sets. Relationships between sets of physical objects and the sets that are the mathematical objects will be new sets, for example, sets of ordered pairs where the first positions are occupied by the physical objects and the second positions are occupied by the mathematical entities.

There is an important change, though, when we move from the natural number case to the rest of mathematics. The Fregean proposal for the natural numbers did not involve just any relation between the numbers and what is counted, but instead turned on the very relation that is relevant to counting, namely, one-one correspondence. The set-theoretic strategy, by contrast, merely establishes some elaborate set-theoretic relation between the real numbers and their application in science, for example, via the theory of hyperbolic differential equations. If this set-theoretic solution is deemed adequate, then it shows how low the bar is set to solve the metaphysical problem. Steiner raises the issue by alluding to a gap between mathematics and the physical world. A bridge across this gap need only show that mathematics is related in some way to the physical world. But as Steiner's own set-theoretic response indicates, there is no requirement that the bridge do anything else. It need not illuminate what, as we have continually put it, mathematics contributes to the success of science. To know that every physical object is a member of a variety of

---

3. See Boolos (1998). The idea goes back to Parsons (1983) and Wright (1980).

sets, including ordered pairs that are members of sets identified with functions of a certain sort, is totally trivial once we have adopted ZFC with these objects as individuals. So, this solution to Steiner's metaphysical problem does not address Field's worry about extrinsic explanation. My impression is that Steiner thinks Field has mixed up the metaphysical problem with a distinct descriptive problem, which Steiner also discusses. We turn to that problem shortly.

Given that the metaphysical problem admits of such an easy solution, it is not surprising that there are other viable approaches to bridging the metaphysical gap. As with the semantic problem, these other solutions at least put the burden on Steiner to explain why his Frege-inspired solution is the best or most promising. Anyone who believes in mathematical objects will find many metaphysical relationships between these objects and the physical world. Presumably, though, in their account of how applications work, only some of these relations will be salient. Even if we believe in set theory and Steiner's set-theoretic relations, it may be that these sorts of relations are not relevant to applications. As an example of an alternative approach to this problem, I note only Wilholt's *Zahl und Wirklichkeit [Number and Reality]* (Wilholt 2004). Wilholt raises a worry similar to Steiner's problem about the relevance of mathematical entities to their domain of application. He solves this problem by proposing that some mathematical entities are really properties or relations, and these properties or relations are possessed by physical objects. For example, whole numbers are properties of aggregates of causal processes, whereas positive real numbers are relations involving ratios of extensive magnitudes.[4] So here we have an alternative to Steiner's solution and it is not immediately clear which solution is superior.

Some may object that Wilholt's proposal turns on dubious claims about what these mathematical entities are. But Steiner also faces this objection because there are strong arguments against identifying mathematical entities like the real numbers with sets. Here I have in mind Benacerraf's arguments in "What Numbers Could Not Be." If we take these arguments seriously, then we arrive at a structuralist interpretation of mathematical entities so that "in giving the properties (that is, necessary and sufficient) of numbers you merely characterize an abstract structure—and the distinction lies in the fact that the 'elements' of the structure have no properties other than those relating them to other 'elements' of the same structure" (Benacerraf 1983, p. 291). The structuralists must appeal to a different sort of relation between their mathematical structures and target systems. They are, in fact, the sorts of structural relations I introduced back in chapter 2 when I proposed an account of what the content of a mathematical scientific representation is. If we refuse to identify these structural relations with sets, then we have yet a third proposal for overcoming Steiner's metaphysical problem. Unsurprisingly, I take this third proposal to be the correct one as it invokes relations that are relevant to applications as they are actually made. A full vindication of this structuralist approach to applications must await further discussion in chapter 14. There I engage with the arguments by neo-Fregeans in support of the Fregean strategy that Steiner endorses.

---

4. These views are discussed in more detail in Pincock (2005).

Discovery

For now I add only a brief discussion of Dummett's worries about Steiner's Fregean strategy. Steiner quotes Dummett's worries about platonism but then goes on to complain that Dummett must be assuming that "mathematical objects do not participate in causal (or spatio-temporal) relations" (Steiner 1998, p. 21). However, this is not really Dummett's worry about the platonist proposal. As his discussion in "What Is Mathematics About?" makes clear, Dummett objects to the impredicative character of the associated conception of the natural numbers or real numbers. The conception is impredicative because the definition of the mathematical domain presupposes that the extent of the domain has already been clearly circumscribed. This is what an advocate of a nonclassical constructive mathematics calls into question (Dummett 1993, p. 439). Dummett concludes his article by advocating a constructive recasting of our applications of mathematics. Logicists like Frege

> were defeated by the problem of mathematical objects because they had incompatible aims: to represent mathematics as a genuine science, that is, as a body of *truths*, and not a mere auxiliary of other sciences; to keep it uncontaminated from empirical notions; and to justify classical mathematics in its entirety, and, in particular, the untrammelled use of classical logic in mathematical proofs ... I have urged the abandonment of the third [aim]. (Dummett 1993, p. 445)

It is not clear that this is a worry that Steiner needs to take seriously, but his discussion of the Fregean solution to the metaphysical problem does nothing to address it. Dummett is after an account of the semantics of our scientific language that will address both Steiner's semantic and metaphysical problems, and the Fregean approach is undermined based on a competing conception of what semantics can assume.

8.2 A DESCRIPTIVE PROBLEM

I turn now to Steiner's descriptive problems. Although they are different than the metaphysical problem, the picture of how applications work that Steiner deployed in solving the metaphysical problem helps shape how Steiner will introduce his descriptive problems. Notice that concepts that have physical objects in their extensions appear as sets in the impure set theory Steiner works with. Then these physical concepts are paired up with mathematical concepts via his functions. This is why it makes sense for Steiner to ask "Can we say—in *non*mathematical terms—what the *world* must be like in order that valid arithmetic deductions should be effective in predicting observations?" (Steiner 1998, p. 24). To explain the requirement that the account be nonmathematical, Steiner adds a note which says that the motivation is not the same as Field's concerns, but only "in explaining, in nominalistic language, the conditions under which a mathematical concept will be applicable in description" (Steiner 1998, p. 24, n. 2). This question can be asked for each mathematical concept, and if we restrict our focus to individual concepts or small families, then we have what I call a *particular* descriptive problem. But we might also ask how mathematical concepts quite generally serve in descriptions. Then the question "concerns the applicability of mathematics as such, not of this or that concept" (Steiner 1998, p. 45), and we have what I call a *general* descriptive problem.

Steiner argues that we can solve the particular descriptive problem for some mathematical concepts but not for others. He draws on measurement theory to conclude that "it is not hard to set down conditions, in nonmathematical language, for a magnitude to have an additive structure" (Steiner 1998, p. 29). This allows us to account for applications of arithmetic in predicting which weights will balance a scale as these weights, both realized and unrealized, will have the additive structure that ensures they can be "parametrized" by the natural numbers. Similarly, "the two-dimensional geometrical structure of the plane" has the multiplicative structure of the natural numbers, and this accounts for our application of multiplication to tiling a floor with square tiles (Steiner 1998, p. 29).

A further success case is supposed to be the mathematical concepts of an equation being linear or nonlinear. Steiner first argues that linear equations correspond to the operation of a "Principle of Superposition ... joint causes operate each as though the others were not present" (Steiner 1998, p. 30). But he also allows that there can be other cases where linear equations are used where superposition of causes fails because "the nonlinear may often be approximated by the linear" (Steiner 1998, p. 30). This approximation is "valid if the curve is smooth, or at least has smooth pieces" and this smoothness is "certainly a physical property" (Steiner 1998, p. 31). He then goes on to discuss the use of fractals to describe phenomena that are not smooth. He concludes his discussion by saying:

> Linearity is applicable to the extent, and only to the extent, that the Principle of Superposition holds, and to the extent that nature operates in a smooth, or at least piecewise smooth, manner. Whatever we are to say about this question, we can at least conclude this: there is no mystery concerning the applicability of linearity; the mathematical property of linearity can be reduced to physical properties which nature may either exhibit or not exhibit. (Steiner 1998, p. 32)

Though other cases are discussed, they all involve the search for this sort of parallel between a mathematical concept and a physical analogue that can be specified in nonmathematical terms. Unsurprisingly, in certain cases Steiner is unable to find such a parallel physical concept. For example, analytic functions are applied "to fluid dynamics, to relativistic field theory and to thermodynamics" (Steiner 1998, p. 37). They are functions of complex variables of the sort treated in the complex potential example in section 4.2. As we saw there, they can be defined as the functions that obey the Cauchy-Riemann equations. But Steiner notes that they also have other useful mathematical properties, such as being infinitely differentiable and being representable by an infinite power series (Steiner 1998, p. 36). His fluid dynamics case is the same as our complex potential case. After considering two other uses of analytic functions, he concludes that "we have no one property that corresponds to analyticity in all applications" (Steiner 1998, p. 38). So, to a certain degree, a mystery remains for the descriptive applicability of this mathematical concept.

Now we can ask whether it is reasonable to think that the mathematical concepts that figure centrally in our scientific representations have these parallel physical concepts and whether a failure to find such a parallel concept suggests a certain mystery in the success of that representation. I think the answer to both of these questions is clearly "no." Given the material developed in part I of this book, we can see that Steiner is making the unwarranted assumption that the contribution mathematics makes to a successful representation is always descriptive. Furthermore, he is assuming that when a given mathematical concept functions descriptively in two representations, this concept must be describing the same physical property in both cases. All the mathematics extrinsic to a given representation is obviously functioning nondescriptively. This is because the mathematics does not figure directly in the specification of the content of the representation, but only in linking the representation to other representations as in a derivation or solution of a given mathematical problem. These mathematical links are clearly applications of mathematics, but they are not reasonably forced into Steiner's descriptive category.

It may be that Steiner is implicitly restricting his focus to what we have called the intrinsic mathematics of the representation, that is, the mathematics that figures directly in fixing the accuracy conditions that the representation imposes on a given target system. Even here there are problems with tackling the issue as Steiner does. As we have seen, a given piece of mathematics can be variously interpreted or else not interpreted at all in physical terms, and this is at the heart of many positive contributions mathematics makes to the success of science. Let us see how this plays out in the case of the use of complex potentials in fluid dynamics and electrostatics that we discussed in section 4.2. As Steiner recognizes, this involves two mathematical concepts, namely, linear equations and analytic functions. Our presentation of this case causes problems for Steiner's interpretation of the descriptive use of linear equations. It is true that there is a relevant mathematical property of linear equations. This is that two solutions can be combined to yield a third solution. We saw that this mathematical feature of the equations allowed the use of the inverse method, where simple solutions are combined together to yield more complicated solutions. This sort of superposition is not easily reconciled with Steiner's superposition of causes. The simple solutions do not correspond to causal processes that are operating in the system described by the more complicated solution. For example, we combined the uniform flow with a source flow to find a third solution. But this third solution involves the flow around a symmetric object. There is no source here, and it is hard to make sense of the idea that the uniform flow is operating here as a cause. What we see, then, is that we can exploit the linearity of the equations in a way that is independent of any superposition of causes. At the same time, there is not much of a mystery about what is going on here. We are accurately representing this or that fluid flow, but that representation does not exploit the linear character of the equation in the way Steiner suggests.

Steiner also emphasizes the "smoothness" that allows us to treat nonlinear equations using linear equations. It is not clear what this comes to or how it is to be characterized in nonmathematical terms. Recall our argument that the potential function was well defined and that it satisfied the Laplace equation involved the claim that the fluid is irrotational. This requires that its viscosity is 0. Now in fact, any fluid like air or water that we represent in this way has some nonzero viscosity, and if we bring in these terms then the potential is not well defined and we are forced back on the nonlinear Navier-Stokes equations. This was discussed in some detail in section 5.6. So in those cases where an irrotational fluid representation is accurate in respects relevant enough to capture the phenomenon of interest, it is fair to say that we have a case where "the nonlinear may ...be approximated by the linear." But I must reject Steiner's view that this bottoms out in some physical feature of "smoothness." From a mathematical perspective, we can see the irrotational fluid equations as the result of letting the viscosity go to 0 in the nonlinear Navier-Stokes equations. This mathematical transformation does not correspond to any physical feature of the fluid flow, though. Steiner may be thinking of the relatively small contribution that the nonlinear terms make in the cases where the irrotational representation is viable. This is not a feature of the fluid system, though, as if we consider the flow close to an object, then we see that a boundary layer develops. Here the viscosity terms are extremely important. In short, smoothness is not a necessary condition on the application of linear equations.

Related problems arise in connection with Steiner's discussion of analytic functions. As Steiner emphasizes, analytic functions have a number of important mathematical properties. In our complex potential case, we directly exploited the fact that such functions satisfy the Cauchy-Riemann equations. There is no reason this feature has to relate to the other mathematical features of these equations. For example, our particular application did not rely on the expansion of the functions into infinite power series. Other applications of analytic functions may depend on this feature. This would be crucial to a case where the power series was used to find a function that approximately describes the system. Some of the perturbation theory cases noted in chapter 5 could be cast in terms of analytic functions, although we did not do this in our presentation. What we see, then, is that we should not expect a single physical property to be tracked by the use of this mathematical concept. This feature of analytic functions is not unusual. Typically a fruitful mathematical concept will involve a number of interesting mathematical properties, and it is sufficient for an application to exploit just one in an accurate scientific representation. The way Steiner frames his descriptive question, then, makes it harder to answer than it should be and suggests a mystery where none exists.

This is not to say that we always understand how a mathematical concept functions in an accurate scientific representation. Often the interpretive status of a given component of the mathematics is unclear. The whole situation receives a natural diagnosis if we return again to the account of content outlined in chapter 2. There I assumed that we have referential access to mathematical objects and physical properties. A content is assigned to a representation when we indicate which features of the mathematical objects stand for which physical properties and what sort of structural relationship must obtain. This assignment may involve just the simplest sorts of mathematical and physical properties. For example, we may say that our mathematical function u stands for the velocity of the fluid in the x-direction. However, these simple referential assignments can wind up imposing very elaborate conditions on the target system in virtue of the additional mathematics in play and the requirement that a given structural relationship obtain. In our complex potential case, we saw how the intricate mathematical structure associated with the analytic functions and the complex plane became relevant to the fluid flow in virtue of the requirement that the flow be irrotational. This is an unexpected and highly fruitful association. To understand it we need to review the mathematics of the case and see how the content arises from the initial referential assignments and structural associations. With this background in place, we can see clearly how this apparatus can contribute to an accurate description of such phenomena, but there is no reasonable demand that all mathematical concepts in play be "reduced," as Steiner puts it, to physical analogues.

In the absence of a reductive association between a given mathematical concept and single physical property, it is true that we can encounter some interpretive uncertainty with our best mathematical scientific representations. This uncertainty is ignored by those who are overly confident of the scope of their representations, and this can lead to the sorts of failures discussed in chapter 7. For example, Laplace did not understand why he was able to improve on Newton's estimate of the speed of sound and wrongly took his success to support the existence of caloric. These risks of failure go hand in hand with the positive contributions mathematics makes to science. But this situation is not mysterious. It is an instance of the more general fact that we do not know everything and that complete certainty is achieved only by refusing to believe anything.

8.3 DESCRIPTION AND DISCOVERY

In addition to the particular descriptive problem that asks how a given mathematical concept describes a target system, Steiner also considers the general descriptive problem of how mathematical concepts quite generally succeed in describing features of the physical world. This problem is closest to the one I have pursued in this book. But as with the particular case, Steiner makes the unwarranted assumption that the presence of mathematical concepts must be tied to their descriptive role in picking out physical concepts. This overlooks the wide range of potential contributions that mathematics has for this or that scientific representation. In his discussion of the general case, though, Steiner also makes the unfortunate decision to follow Wigner's aesthetic characterization of mathematics. This further distorts Steiner's pursuit of the general descriptive problem and paves the way for his interest in the distinct discovery problem and its philosophical significance.

Steiner introduces the general descriptive problem by saying that it is

> the most profound. It concerns the applicability of mathematics as such, not of this or that concept. It is therefore an *epistemological* question, about the relation between Mind and Cosmos. It is the question raised by Eugene Wigner about the "unreasonable effectiveness of mathematics in the natural sciences." (Steiner 1998, p. 45)

The Wigner paper in question here is Wigner (1960). In that paper, Wigner distinguishes the role of mathematics in reasoning from the use of mathematics to formulate successful scientific theories: "the laws of nature must already be formulated in the language of mathematics to be an object for the use of applied mathematics" (Wigner 1960, p. 6). This procedure is surprisingly successful for Wigner because the resulting laws are incredibly accurate and the development of mathematics is largely independent of the demands of science. As he describes it, "Most advanced mathematical concepts ... were so devised that they are apt subjects on which the mathematician can demonstrate his ingenuity and sense of formal beauty" (Wigner 1960, p. 3). When these abstract mathematical concepts are used in the formulation of a scientific law, then, there is the hope that there is some kind of match between the mathematician's aesthetic sense and the workings of the physical world. One example where this hope was vindicated is in the discovery of what Wigner calls "elementary quantum mechanics" (Wigner 1960, p. 9). Some of the laws of this theory were formulated after some physicists "proposed to replace by matrices the position and momentum variables of the equations of classical mechanics" (Wigner 1960, p. 9). This innovation proved very successful, even for physical applications beyond those that inspired the original mathematical reformulation. Wigner mentions "the calculation of the lowest energy level of helium ... [which] agree with the experimental data within the accuracy of the observations, which is one part in ten millions" and concludes that "surely in this case we 'got something out' of the equations that we did not put in" (Wigner 1960, p. 9).

Interpreters have wondered over two aspects of Wigner's discussion. To start, it is not clear what he means by "unreasonably effective." There seems to be some request for an explanation. The second interpretive question concerns what the failure of this explanation is supposed to imply. Wigner presents our lack of explanation as a profound mystery we cannot solve: "The miracle of the appropriateness of mathematics for the formulation of the laws of physics is a wonderful gift which we neither understand nor deserve" (Wigner 1960, p. 14, at Steiner 1998, p. 13).

It is possible to take Wigner's explanatory question to just be after an explanation of why certain physical claims are true. It is surprising that these claims are true partly because they involve highly abstract mathematical concepts. If a scientist in the nineteenth century was considering the future development of physics, he probably would not have anticipated that quantum mechanics would have arisen as it did. Still, one can respond to this version of Wigner's worries by noting that we cannot explain everything. Some physical claims can be explained, but we need to use other physical claims to do this. There is no mystery in this, and it is hard to see what special mystery there is relating to the mathematical character of truths that we have no explanation of. A related interpretation takes Wigner to be asking Steiner's particular descriptive question. However we have also seen that this sort of reductive demand for an account of particular descriptive success is at times unwarranted.

Steiner takes Wigner to be asking the general descriptive question and considers two possible arguments that Wigner might be after for his conclusion that mathematics is unreasonably effective in science. The first argument merely notes that many concepts are unreasonably effective, and that these concepts are also all mathematical. But Steiner objects that this does not support the conclusion that "mathematical concepts are unreasonably effective" as the premises do not answer the question "is their effectiveness related to their being mathematical?" (Steiner 1998, pp. 46–47).

The second argument that Steiner considers is more substantial as it takes a stand on a distinctive feature of mathematical concepts:

1. Mathematical concepts arise from the aesthetic impulse in humans.
2. It is unreasonable to expect that what arises from the aesthetic impulse in humans should be significantly effective in physics.
3. Nevertheless, a *significant* number of these concepts are *significantly* effective in physics.
4. Hence, mathematical concepts are unreasonably effective in physics.
(Steiner 1998, p. 46)

Steiner focuses on the third premise and objects that it is unclear when a significant number of mathematical concepts have been considered or what a significant level of effectiveness would amount to.

In both attempts to make sense of Wigner's discussion Steiner assumes he is raising a descriptive worry. But Steiner goes on to formulate a problem in terms of a strategy for discovering new scientific claims. It is possible that Wigner was also raising this discovery problem, although only Steiner carefully lays out any argument for a philosophically interesting conclusion. Crucial to Steiner's argument is premise 1 from Wigner's argument, which states that mathematical concepts are essentially aesthetic. Because this is a claim I reject, it is worth examining it carefully here. Then, in the next section, we will see the rest of Steiner's argument, his evidence in support of the remaining premises, and his conclusions.

Wigner's argument for 1 is to simply state that premise: "Most advanced mathematical concepts ... were so devised that they are apt subjects on which the mathematician can demonstrate his ingenuity and sense of formal beauty" (Wigner 1960, p. 3). Steiner goes further and presents passages from von Neumann and Hardy that link the mathematical and the aesthetic. He also cites a survey that "asked readers to rank mathematical theorems in order of their beauty" (Steiner 1998, p. 65). This is all taken to support a premise like 1, which Steiner also gives as the claim that "the aesthetic factor in mathematics is constitutive" (Steiner 1998, p. 65). This appears to mean that a necessary condition on being a properly mathematical concept is that it satisfies our sense of what is beautiful. As Steiner describes it, even though mathematics began with concepts that were directly inspired by physical concepts, modern mathematics now develops according to internal norms and priorities. Among these norms are aesthetic criteria: "Concepts are selected as mathematical because they foster beautiful theorems and beautiful theories" (Steiner 1998, p. 64). This allows us to make sense of why a theory of chess is not a mathematical theory. Though a theory of chess would share some features with genuinely mathematical theories, such as being highly abstract and largely structural, it remains that the theory of chess fails the aesthetic test set by mathematicians: "the distinction between mathematics and chess is a predilection of mathematicians, rather than a logical distinction" (Steiner 1998, p. 63).

Unless we are willing to ascribe massive error to mathematicians, it is quite reasonable to claim that mathematicians study mathematical concepts. This does not suggest that what makes a concept mathematical is that mathematicians study it or that we should cash out mathematical concepts in terms of the features of the mathematicians who study them. Analogously, we admit that philosophers study philosophical concepts, but at the same time deny that what makes a concept philosophical is that philosophers study it. This point is not sufficient to block Steiner's premise because it leaves us with two additional tasks. First, we need to understand why mathematicians might make a link between aesthetics and mathematics, say, in the context of a question about chess. Second, we should offer a characterization of mathematical concepts that does not appeal to aesthetic factors as constitutive. On the first task, my suggestion is that mathematicians are apt to confuse the question "what makes this concept mathematical or nonmathematical?" with "what makes this concept a good or bad mathematical concept?" Aesthetic criteria are of course central to answering the second question, but we should not let this fact influence our answer to the first. A mathematician discussing complex analysis and the Riemann hypothesis once insisted to his class that compared to the Riemann hypothesis, the conjecture known as Fermat's last theorem was pointless and not worth pursuing. By this he did not mean to relegate number theory and Fermat's last theorem outside of mathematics, but was instead expressing his conviction that complex analysis is a better or more important area of mathematics than number theory. These judgments about the relative importance of this or that area of mathematics are central to mathematical practice as they provide the mathematician with the motive to pursue his or her specialized area of research. On this interpretation of the relationship between aesthetics and mathematics, Hardy's claim that "beauty is the first test [of mathematical ideas]: there is no permanent place in the world for ugly mathematics" (at Steiner 1998, p. 65) is saying that only good mathematics should be pursued by mathematicians. Similarly, if a mathematician dismisses chess as "not mathematics" because it is "not beautiful," then what she is really saying is that it is not good or important from a mathematical perspective.

It is hard to go beyond this negative point and provide competing criteria for what mathematics is or when mathematicians are justified in their judgments about what is good or important mathematics. On the platonist approach to mathematics that I have been assuming throughout our discussion, pure mathematics is properly defined as the study of a certain domain of abstract objects and their properties. But mathematics as a discipline should also include what is loosely called "applied mathematics." Here the relationships between these abstract objects and certain physical systems are also studied in a thorough if not always fully rigorous fashion. This sort of characterization is not that helpful, though, as it fails to engage with what mathematicians have studied over the history of mathematics. I would argue that this history of the development of mathematics gives us the best handle on what mathematics is and when it is important according to internal mathematical standards. Unfortunately, there has not been much sustained focus on these questions by philosophers of mathematics. An important exception is Kitcher (Kitcher 1984) and many of the essays in Mancosu (2008c).[5] From a nonhistorical perspective, Maddy has carefully reviewed the norms that set theorists follow in their evaluation of new set-theoretic axioms (Maddy 1998). These studies would be the starting point of my own characterization of mathematics and good mathematics. What they emphasize is the way new mathematics arises out of old mathematics based on what are arguably objective features of the mathematical entities themselves. The theory of complex numbers and the analytic functions that appear there are not arbitrarily selected based on the preferences of this or that mathematician. Instead, there is a clear sense in which they represent the "natural" or "correct" extension to pursue or the "proper setting" in which a mathematical question should be answered. This is again just an indication of how I believe this question would be resolved, but I think it is sufficient to show a viable alternative to Steiner's proposal. We return to these issues again in chapter 13 when we consider their significance for Mark Wilson's account of conceptual evaluation.

8.4 DEFENDING NATURALISM

Steiner devotes the four remaining chapters of his book to a philosophical problem related to the role of mathematics in the discovery of new scientific representations. He emphasizes the difficulty that scientists encountered toward the end of the nineteenth century once it became clear that we could not arrive at the correct physics simply by extrapolating features of what we observe down to the very small or up to the very large. Peirce is credited with an awareness of this issue. As a result, Steiner argues, scientists adopted the following strategy for developing new representations. They started with a scientific representation that was already a going concern for scientists and then altered it via a mathematical analogy. Steiner calls such analogies "Pythagorean" when at the time t when it is employed it is "a mathematical analogy between physical laws (or other descriptions) not paraphrasable at t into nonmathematical language" (Steiner 1998, p. 54). A mathematical analogy is "formalist" when it is "one based on the syntax or even orthography of the *language* or *notation* of physical theories, rather than what (if anything) it expresses" (Steiner 1998, p. 54). However, in light of his aesthetic conception of mathematics, Steiner argues that the strategy of discovery via mathematical analogies involves a kind of anthropocentrism. This is because aesthetic judgments are not objective, but only "species-specific" (Steiner 1998, p. 66): "To say that the mathematical sense reduces to the aesthetic is to deprive the aesthetic of the only argument for its objectivity—namely, that the aesthetic sense is based on the objectivity of mathematical form, as the Pythagoreans in fact argued" (Steiner 1998, p. 66). This leads to an argument in favor of anthropocentrism:

I. Both the Pythagorean and formalist systems are anthropocentric; nevertheless,

---

5. See also Aspray and Kitcher (1988) and Mancosu (1997).

II. Both Pythagorean and formalist analogies played a crucial role in the fundamental physical discoveries in this century. (Steiner 1998, p. 60)

The success of these strategies supports anthropocentrism and so has negative implications for the rejection of anthropocentrism or what Steiner calls "naturalism." That is, "The weak conclusion is that scientists have recently abandoned naturalist thinking in their desperate attempt to discover what looked like the undiscoverable," whereas "the apparent success of Pythagorean and formalist methods is sufficiently impressive to create a significant challenge to naturalism itself" (Steiner 1998, p. 75). Theism is one version of anthropocentrism, but other nontheist options are available, including a Pythagorean metaphysics that "identifies the Universe or the things in it with mathematical objects or structures" (Steiner 1998, p. 5). Starting with the role of mathematics in scientific discovery, then, Steiner extracts a significant conclusion about the relationship between humans and the natural world.

Premise I of Steiner's argument is really the conclusion of another argument that we can reconstruct using the claims that a necessary condition on being a mathematical concept is that it conform to the aesthetic judgments of mathematicians and that the aesthetic judgments of mathematicians do not reflect objective features of the world. In the last section I argued against the first claim and suggested that all Steiner could establish was the weaker point that

> A necessary condition on being a *good* mathematical concept is that it conform to the aesthetic judgments of mathematicians.

It is arguable, though, that this is still sufficient for Steiner's anthropocentric argument to work, as it turns out that the Pythagorean and formalist discovery strategies seem to have restricted themselves to the good mathematical concepts. As I emphasized, mathematicians choose which areas to work on based on their judgments about what is good or important. So, these areas tend to be articulated to the point where they can be usefully deployed in mathematical analogies between old and new representations. If we take this route on Steiner's behalf, then the second claim in the argument for his premise I becomes

> The aesthetic judgments of mathematicians about which mathematical concepts are good do not reflect objective features of the world.

Unfortunately Steiner does not adequately support this claim.[6] His whole argument for it boils down to the point that the only route to grounding the objectivity of aesthetic judgments is by reducing these judgments to mathematical judgments. But there is no reason to accept this point. Any number of other accounts of our aesthetic judgments allow them to track objective features of the world. For example, we may be realists about aesthetic properties and argue that our minds have insight into these features that we express in aesthetic judgments. Even an argument based on natural selection seems imaginable according to which our tendency to

---

6. See Bangu (2006) for additional critical discussion.

make aesthetic judgments is an adaptation precisely because these judgments track objective features of our environment. Other routes are no doubt available to ground the objectivity of aesthetic judgments. So at the least, Steiner must engage more with these alternatives if he is to support I.

To counter Steiner here, in fact, we need not develop a theory of the objectivity of our aesthetic judgments in general, but only the restricted class of judgments by mathematicians about mathematics. Here, I think, there is a promising route to grounding the objectivity of such judgments based on the way we have linked these aesthetic judgments to judgments about mathematical importance and goodness. Suppose, first, that there is an objective fact of the matter about which developments in mathematics are important and good. Then it would be reasonable to think of all the people out there, the mathematicians are best equipped to grasp these objective matters in their judgments. For the mathematician is very familiar with many areas of mathematics, and her training has shown her how to evaluate a given development in mathematics. This preparation gives her the expertise to determine mathematical importance and value. On this approach we can claim that the aesthetic judgments of mathematicians about which mathematical concepts are good *do* reflect objective features of the world, namely, which concepts figure in important and valuable mathematical developments. Of course, the burden of this approach is to establish the objectivity of these features of mathematics, and I have not done this. The issue would turn on the careful consideration of the historical development of mathematics alluded to in the last section. My point here has been simply that we have alternative approaches to these issues that conflict with Steiner's premise I and he has done little to address them.

My concerns about premise I do not undercut the many examples Steiner develops in support of his premise II. His discussion of Pythagorean cases distinguishes six different kinds of Pythagorean strategies and includes, by my count, 15 different examples. One strategy assumes that all solutions to a given equation are similar in virtue of the mathematical link established by the equation (Steiner 1998, p. 77), whereas another strategy involves formulating a new equation based on the mathematical form of an old one (Steiner 1998, p. 94). Steiner's test for when an instance of these strategies is Pythagorean involves the reductive conception of description that was implemented when he raised his descriptive problems. For example, a decision to formulate a new equation so that it is linear could be defended based on some prior awareness that the causes in play obeyed a principle of superposition. But even where the mathematical analogy involves a concept whose descriptive success has been understood in some cases, Steiner still demands that the scientist have some reasonable belief in the physical correlate of this mathematical analogy. Otherwise, the instance is deemed Pythagorean.

In our discussion of the descriptive problem, I argued that this reductive test is unreasonable and that it ignores other kinds of well-understood contributions from mathematics to the success of a scientific representation. I believe that a review of Steiner's examples would show that some of them are classified as Pythagorean based on this overly demanding reductive test. For example, Steiner discusses how Maxwell assumed that there was a displacement current based on his attempt to develop a unified mathematical representation of electromagnetic phenomena

(Steiner 1998, p. 77). This displacement current was added to the equation that contained an expression for the "real" current, which had been experimentally tested. But Steiner claims that only a mathematical analogy grounded Maxwell's assumption that a displacement current was possible even when this real current was 0. Strangely, though, Steiner also notes one motivation that Maxwell had beyond the mathematics: "By tinkering with Ampère's law, adding to it the 'displacement current,' Maxwell got the laws to be consistent with, indeed to imply, charge conservation" (Steiner 1998, p. 77). It seems to me that the demand for charge conservation was eminently reasonable for someone in Maxwell's epistemic position. This assumption is also somewhat mathematical, and Maxwell may not have been in a position to see what his displacement current would come to. Still, in the context of trying to discover a new, more unified representation, Maxwell had more to go on than the mathematics.[7]

Another example that has been given a different interpretation by another commentator is the discovery of isospin. As Steiner presents it,

> Heisenberg reasoned that the nucleus of the atom is invariant under $SU(2)$ transformations, those which describe the spin properties of the electron; and that there had to be, therefore, a new conserved quantity, *mathematically* analogous to spin. This quantity is today called isospin, and its discovery launched nuclear physics. Both the neutron and proton, states of the same particle, are called "nucleons." (Steiner 1998, pp. 86–87)

The basic idea here is that we track the spin of an electron using a vector. Spatial rotations affect this vector through the application of a member of the group $SU(2)$. Heisenberg extended this apparatus to the protons and neutrons that make up the nucleus of the atom and posited an analogous property of isospin, which could be tracked by a similar vector using the same group. Without citing Steiner, French (2000) gives an extended discussion of Heisenberg's reasoning. For French, these sorts of applications of group theory have a "dual role" noted by Wigner himself: "the establishment of laws—that is, fundamental symmetry principles—which the laws of nature have to obey; and the development of 'approximate' applications which allowed physicists to obtain results that were difficult or impossible to obtain by other means" (French 2000, p. 107). In the isospin case, French emphasizes the crucial role of experimental results in licensing the idealizations necessary to get this particular application of group theory to work. These included the observed fact that protons and neutrons had nearly identical weights. As a result, for

> the construction of isospin ... the effectiveness of mathematics surely does not seem quite so unreasonable, as group theory is brought to bear via a series of approximations and idealisations ... In effect, the physics is manipulated in order to allow it to enter into a relationship with the appropriate mathematics, where what is appropriate depends on the underlying analogy. (French 2000, p. 114)

---

7. For more discussion of this case see Morrison (2000, ch. 3).

Although there are surely further details to be worked out about the motivation for these idealizations, the point remains that we can follow French to reconstruct the reasoning which led to isospin and ground the role played by group theory in this case.

That said, there remain many cases where a discovery was the result of deploying a mathematical analogy based on no independent scientific motivation. This poses a problem mainly for those who insist that scientific discoveries are the result of a rule-governed form of reasoning whose justification grounds the cogency of our scientific picture of the world. My impression is that hardly anybody working in the philosophy of science thinks that it is possible to reconstruct many scientific discoveries along these lines. The correctness of our scientific representations should be grounded in their success in generating predictions that are experimentally tested. If this is right, then we can concede Steiner's premise II and use our rejection of his premise I to block his argument for anthropocentrism. There may be a special way mathematics has contributed to the discovery of our current best science. At the same time, I have tried to explain why I disagree with Steiner on the significance of these contributions.

8.5 NATURAL KINDS

In section 4.5, I argued that the abstract varying contributions of mathematics could provide epistemic benefits. Put simply, if two representations share some of their mathematics and both are accurate, the successful track record of the first representation will make it easier to confirm the second. Even when one of the representations is accurate and the other is inaccurate, the mathematical link between them can give the scientist a framework within which to probe for the source of the errors. In both the direct and indirect cases, then, the ability to vary the interpretation of the mathematics makes a contribution to the eventual success of science. It is not immediately clear how significant this claim is for Steiner's argument. The examples in chapter 4 bear out French's point that the application of a given area of mathematics can depend on idealizations. In the complex potential case, covering both irrotational fluids and electrostatics, these idealizations involved picking out just those systems that would obey the Cauchy-Riemann equations, under some interpretation of the ϕ and ψ functions. The desire to deploy these well-understood areas of mathematics or to extend mathematical techniques in this or that direction makes a lot of sense for a scientist who does not know where else to turn in reformulating a failed representation. It is unreasonable for her to expect any particular gambit along these lines to work. But the overall strategy involves many attempts, and the scientist realizes that if she can find a place where one mathematical extension can be found, then she will be in a good position to draw on all of the benefits that this allows.

A further element that Steiner draws on in his discussion is related to this last point. Steiner explains how Goodman argued that certain generalizations are not even candidates for confirmation via successful prediction because the predicates are "unprojectible" (Steiner 1998, p. 57). In Goodman's classic example, we have the usual generalization "All emeralds are green" along with the defective "All emeralds are grue." "Grue" here is defined as a predicate possessed by an object if it is green at time t prior to 2100 or if it is blue at time t in 2100 or afterward. All our observations are instances of both "This emerald is green" and "This emerald is grue" because they are carried out prior to 2100. But we do not take "All emeralds are grue" to be supported by these observations because "grue" is not a viable predicate for such generalizations. Goodman himself tried to ground this feature in terms of our past usage with some predicates and not others, so that only the "entrenched" predicates are viable. Steiner does not endorse this solution, but he seems to think that the naturalist must ground the predicates they employ in associations to natural kinds or some other naturalistically acceptable surrogate. This is why he complains that discoveries prompted by mathematical analogies are not candidates for confirmation for the naturalist. They would be like generalizations involving "grue," unless the analogy could be grounded in naturalistically acceptable descriptive terms.

This point can be distinguished from the argument for I that we discussed earlier. There the claim was that mathematical concepts are species-specific and so deploying them to discover new representations was already anthropocentric. The Goodman-style argument suggests that even if the mathematical concepts are objective, it remains problematic to support these new discoveries through experimentation. "Grue," as defined, is just as objective as green, at least in the sense that an object is green or grue independently of the mental states of humans. This suggests that taking the new representations to be supported, however they were discovered, involves the assumption that the predicates in question are projectible. And this assumption has not been discharged by the naturalist. To my mind, this is a stronger consideration in favor of Steiner's I, where we now take the "systems" to be a strategy of formulating and confirming representations via observation.[8] Indeed, if this argument can be vindicated then the entire apparatus of this book will be undermined. This is because I have repeatedly relied on the confirmation of mathematical scientific representations via ordinary experimental testing. At the same time, I have not explained the respects in which the predicates associated with these representations are projectible in Goodman's sense.

My response to this argument of Steiner's is to deny that the naturalist must meet this Goodman-style challenge. Recall the basic picture of the content of a scientific representation: a given mathematical structure is interpreted by assigning some mathematical properties to properties of the target system. At this stage the agent formulating the scientific representation should have some prior referential access to these scientific properties. This is the only way open for them to assign a content to the representation. The second step is to indicate the sort of structural correspondence that must obtain for the representation to be accurate. Here the possibilities are quite intricate as we move into more highly idealized representations. At the end of this process, we may represent the system to be a certain way and yet lack the ability to say what this comes to independently of the mathematical structure we have deployed. Furthermore, we may remain uncertain about which aspects of the representation really do wind up accurately characterizing the target

---

8. See also Steiner (1998, pp. 161–162).

system. We have no recourse then but to proceed to the stage of testing. As the results of these tests come in, we should raise or lower our confidence that we have gotten things right. It is only at this stage that we can conclude that a given aspect of our representation corresponds to a genuine feature of the target system, for example, a "natural kind." If Steiner is to be believed, we must first decide which predicates correspond to natural kinds and only then take seriously experimental results with respect to these predicates. But this recipe fails to track scientific practice and would indeed undermine the development of scientific representations of the sort we have seen. More often than not, we come to appreciate the accuracy of our representations only as we work with them and test them. There is no simple test for which representations scientists should take seriously as candidates for testing. At the same time, we should not take the absence of this test to indicate a tacit rejection of naturalism of the sort imagined by Steiner.

9

Indispensability

9.1 DESCRIPTIVE CONTRIBUTIONS AND PURE MATHEMATICS

So far we have discussed several different ways that mathematics can make a contribution to successful scientific representations. These contributions have been presented in terms of their content, so it is fair to summarize our classification scheme as being focused on descriptive success. In this and the next three chapters, my aim is to determine what implications this contribution to descriptive success has for the interpretation of pure mathematics. There is an influential argument, known as the indispensability argument, which claims that these contributions to descriptive success support platonism about mathematics. This is the view that the subject matter of mathematics is a domain of abstract objects that are causally isolated from the physical world. Though I have more or less assumed platonism since chapter 2, I raise a number of objections to this argument and ultimately conclude that it can support only a very weak conclusion. One major criticism I develop is that most discussions of indispensability have focused on the global question of whether science as a whole needs mathematics. This has obscured the sorts of contributions mathematics makes to this or that representation and these contributions are what I have been at pains to emphasize in my discussion so far. Once we turn to the details of these contributions, it becomes far less clear why they should have any implications for our interpretation of pure mathematics. After considering the most influential presentations of the argument in Quine, Putnam, and Colyvan, I argue that they are not successful.

9.2 QUINE AND PUTNAM

According to Quine, our scientific beliefs are continually confronted with the observations we make. When the beliefs lead to an incorrect prediction, a number of options are available to adjust our beliefs to account for the failure. Furthermore, when the beliefs lead to correct predictions, the credit for this success should be shared widely among the beliefs that allowed for the prediction. In this sense, then, Quine is a confirmational holist. We do not confirm individual beliefs, but only large bodies of beliefs because only these collections of beliefs allow successful prediction.

Indispensability

This process of belief adjustment goes hand in hand with another process that Quine calls "regimentation." When we regiment our beliefs, we aim to represent them in a simple logical language. For Quine, it was clear that this logical language would be the language of first-order logic. The main reason for choosing this sparse language is that Quine thought it was adequate to regiment the entire language of science. It also possesses a certain kind of transparency. One of the main things that Quine is after with his process of regimentation is the clarification of our ontological commitments. Our ontological commitments are the kinds of entities whose existence follows from the truth of our beliefs. If it is possible to regiment our beliefs into the language of first-order logic, then these ontological commitments are easy to determine. We need only check which sentences of the form $\exists x\ Fx$ are entailed by our beliefs.

Although we have distinguished the process of belief choice from the process of regimentation, in a number of places Quine insists that these processes are really only aspects of the single process of doing science. As he puts it in "On What There Is,"

> Our acceptance of an ontology is, I think, similar in principle to our acceptance of a scientific theory, say a system of physics: we adopt, at least insofar as we are reasonable, the simplest conceptual scheme into which the disordered fragments of raw experience can be fitted and arranged. ... To whatever extent the adoption of any system of scientific theory may be said to be a matter of language, the same—but no more—may be said of the adoption of an ontology. (Quine 2004, p. 190)

So, the goals of simplicity and transparency that guide regimentation are the same as the overall goal of science of assembling a collection of beliefs that allow us to predict and control our experiences. Among other things, Quine's approach to ontology raised a strong challenge to the pretensions of traditional linguistic analysis. It is not as if the meanings of our beliefs place rigid constraints on their proper analysis. Instead, the acceptable ways of regimenting our beliefs into the language of first-order logic, consistent with the overall aims of science, help us clarify what these beliefs could mean and what, in the end, we should believe exists. That is, "a fenced ontology is just not implicit in ordinary language" (Quine 2004, p. 236).

This approach to science and ontology led Quine to offer an indispensability argument for platonism. The argument was that, all things considered, the scientific goal of making our beliefs as clear as possible, consistent with our best attempts to predict and control our experiences, dictates that we include sentences like "$\exists x(x$ is a set)." So in "The Scope and Language of Science," Quine insists that "certain things we want to say in science may compel us to admit into the range of values of the variables of quantification not only physical objects but also classes and relations of them; also numbers, functions, and other objects of pure mathematics" (Quine 2004, p. 207). Fortunately, these other objects can be replaced by set-theoretic constructions. Or as Quine presents the issue elsewhere, we may bring in sets or classes when we say that "there are over a quarter-million species of beetles" (Quine 2004, p. 240). The best regimentation of this statement concerns sets of beetles, each of whom are of the same species. When we take the set of these sets, we find that this set has a certain cardinality. So,

> Limited to physical objects though our interests be, an appeal to classes can thus be instrumental to pursuing those interests. I look upon mathematics in general in the same way, in its relation to natural science. But to view classes, numbers, and the rest in this instrumental way is not to deny having reified them; it is only to explain why. (Quine 2004, p. 241)

Quine's argument, then, does not depend on what is absolutely indispensable or on what our ordinary beliefs tacitly commit us to. It is rather that scientific standards of investigation and regimentation point to a collection of sentences whose truth requires the existence of abstract objects.

There is a further assumption that Quine makes here that has been very influential in the debates about indispensability and platonism. This assumption is that the indispensable role of the assumption of mathematical entities for science is an empirical justification of our belief in the existence of these entities. This is because the only considerations Quine allows in favor of the acceptance of a regimented sentence is its contribution to the success of science, and this success is ultimately judged by the empirical standards of which experiences we can predict and control. But I argue that we should also leave open the possibility that these commitments to the existence of mathematical entities obtain on the basis of some nonempirical justification. In this alternative view, this nonempirical support would give the scientist a reason to believe in mathematical entities prior to his work in science. It is possible that the indispensable role of mathematics is tied to the ways nonempirically justified mathematics contributes to the success of science. In fact, once this possibility is raised, we can see that the indispensable role of mathematics in science does not dictate that our mathematical beliefs are supported empirically. I return to this issue in chapter 10 after clarifying how some different indispensability arguments work.

The nature of Quine's argument is worth keeping in mind when we consider Putnam's influential characterization of indispensability arguments:

> quantification over mathematical entities is indispensable for science, both formal and physical; therefore we should accept such quantification; but this commits us to accepting the existence of the mathematical entities in question. This type of argument stems, of course, from Quine, who has for years stressed both the indispensability of quantification over mathematical entities and the intellectual dishonesty of denying the existence of what one daily presupposes. (Putnam 1979b, p. 347)

The main example that Putnam discusses leading up to this summary is Newton's law of gravitation. Here Putnam starts by assuming a "'realistic' philosophy of physics" according to which "one of our important purposes in doing physics is to try to state 'true or nearly true' (the phrase is Newton's) laws, and not merely to build bridges or predict experiences" (Putnam 1979b, p. 338). Then, assuming that Newton's law of gravitation is one of these "true or nearly true" laws, Putnam investigates what this law is saying and how statements with similar expressive power alter our quantification over abstract entities. Focusing on claims of the form "the distance from a to b is $r_1 \pm r_2$," Putnam concludes that they "cannot be explained without using the notion of a function from points to real numbers, or at least to rational numbers" (Putnam 1979b, p. 341). Here he appeals to the results of measurement theory.[1] These results show how it is possible to find axioms expressed in terms of physical properties and relations that are sufficient to ensure a domain of magnitudes whose structure is adequately captured by a function that matches each magnitude to a real number. The specific function results from picking a standard unit that is to be mapped to the real number 1. But the family of functions that are adequate to capture this domain of magnitudes are related by a simple transformation.

An important contrast between Quine's argument and Putnam's argument should be clear at this point. As we have presented it, Quine's argument turns on the claim that scientific standards push us toward a regimentation that involves quantification over abstract objects like sets. Putnam instead emphasizes the need to capture the original meaning of a scientific claim like the law of gravitation. This assumes that the law starts with a more or less clear meaning which our analysis must respect. This further constraint on the process of regimentation seems to make it easier to achieve the platonist conclusion that abstract objects like sets exist. For it seems to block the sort of wholesale reconstruction of the language of science that might be needed to get rid of quantification over abstract objects. Though Quine clearly would opt for this reconstruction if it could be argued that it would lead to a better science, Putnam presumably would object that this sort of reconstruction fails to do justice to the scientific claims we started with. Putnam's argument, then, makes certain assumptions about language and analysis, which Quine's argument does without. At the same time, Quine's argument is harder to defend. It is difficult to see how we can know that Quine's first-order language is really the best way to regiment our language. If our standards are scientific standards, then the fact that few scientists seem interested in Quine's way of regimenting the language of science should worry the Quinean. Quine presumably would appeal here to some kind of division of labor where the philosopher presents a first-order regimentation of the language of science for the purposes of clarifying the ontological commitments of science. There is no requirement that scientists actually use the language. But as we will see briefly, this division of labor makes Quine's argument quite vulnerable to Field's program of providing nominalist versions of our best scientific theories.

There is a further complication in assessing the links between Quine's argument and Putnam's argument.[2] The conclusion of Quine's argument is clearly that abstract objects exist. A remark that Putnam makes at the end of "Philosophy of Logic" strongly suggests that he did not think the conclusion of the argument was that platonism is supported by our best science. There he says that "the existence of what I might call 'equivalent constructions' in mathematics" (Putnam 1979b, p. 356) complicates the issues he had focused on, including the indispensability

---

1. See, e.g., Krantz et al. (1971).

2. For additional discussion and references, see Liggins (2008).

argument. But given that "the realm of mathematical fact admits of many 'equivalent descriptions'" (Putnam 1979b, pp. 356–357) it seems that Putnam is holding out the possibility that a description in terms of abstract objects is avoidable. And when we turn to an earlier paper of Putnam's called "Mathematics without Foundations," we find him endorsing just such an interpretation. To start, "the chief characteristic of mathematical propositions is the very wide variety of equivalent formulations that they possess" (Putnam 1979a, p. 45). One of these descriptions is in terms of sets, but an alternative "mathematics as modal logic" (Putnam 1979a, p. 47) is also deemed viable. This does not require the existence of any abstract objects, but only the truth of certain claims which involve modal operators. Contrary to what Putnam suggests in the middle of "Philosophy of Logic," there may be no need to include abstract objects in our ontology.

Given his insistence on first-order logic as the language of regimentation, it is easy to see how Quine would reject the possibility of getting by with modal operators instead of abstract objects. At the same time, it is hard to determine how he can convincingly argue that the language of modal logic is ruled out of bounds. It may be that the gains in expressive power that result from using modal operators lead to an overall increase in simplicity and clarity. This is the point of view pursued by Hellman in his book *Mathematics without Numbers* (Hellman 1989).[3] There Hellman explains how we can regiment the language of science using a second-order language that includes modal operators. The basic strategy is to take the ordinary axiomatic presentation of a mathematical theory like the Peano axioms for natural numbers. Then we consider the claim that it is possible for a structure to exist that would make these axioms true. Finally, any theorem of that area of mathematics is rendered as a necessary hypothetical to the effect that if a structure satisfies the axioms, then the theorem holds in that structure. The approach traces back to Putnam's "Mathematics without Foundations" paper and so must be consistent with the premises of the indispensability argument as Putnam conceives it. If this is right, then the conclusion of Putnam's argument can only be that a realist approach to the truth-value of the mathematical statements that are used in science is indispensable. That is, we must accept that mathematical statements have objective truth-values and that our aim is to discover what they are. This is the sense in which "realism" is used in Putnam's description of his approach to scientific theories more generally.

Putting together the pieces of Putnam's discussion, we can offer the following argument for realism about mathematics:

1_P. We ought rationally to believe in the truth of any claim that plays an indispensable$_P$ role in our best scientific theories.
2_P. Mathematical claims play an indispensable$_P$ role in science.
3_P. Hence, we ought rationally to believe in the truth of some mathematical claims.[4]

---

3. See also Burgess and Rosen (1997) for a survey of these sorts of strategies.

4. For ease of comparison, I present these indispensability arguments in the form recently deployed by Baker. See section 10.1 for Baker's argument.

Indispensability

The sense of indispensability$_P$ at work here concerns what must be assumed to be true for a claim to have its original meaning. The indispensable$_P$ role of mathematical claims in our theories is in formulating scientific laws like Newton's law of gravitation or presenting more mundane claims like Quine's claim about the number of species of beetles. But the fact, as Putnam sees it, that the mathematical domain admits of equivalent descriptions shows that the truth of the mathematical claims necessary to formulate our best science need not commit us to the existence of abstract objects. By contrast, we can ascribe to Quine the following argument for platonism about mathematics:

1_Q. We ought rationally to believe in the existence of any entity that plays an indispensable$_Q$ role in our best scientific theories.

2_Q. Mathematical entities play an indispensable$_Q$ role in science.

3_Q. Hence, we ought rationally to believe in the existence of some mathematical entities.

As with Putnam, the notion of indispensability$_Q$ is tied up with the formulation of our scientific claims. But we also saw that Quine was not willing to appeal to any well-defined meaning of our scientific claims prior to the process of regimentation. So the advocate of Quine's argument must be open to a consideration of the overall benefits of a regimentation that avoids abstract, mathematical entities. This shows that Quine's notion of indispensability$_Q$ concerns what is required by the best regimentation of the language of science. These alternative regimentations can be criticized only on the basis of scientific criteria like simplicity and empirical adequacy.

Beyond Hellman's modal-structuralist proposal, Lewis's mereological interpretation of mathematics poses a significant challenge to the Quinean argument for platonism. In *Parts of Classes* (Lewis 1991) and Lewis (1993), Lewis presents an interpretation of mathematics in terms of the part/whole relation that involves only concrete objects. He shows how the mathematics used in science can be interpreted in these mereological terms if we are willing to make strong enough assumptions about the mereological structure of the actual world. As a result, premise 2_Q of the Quinean argument is rejected for we need not accept the existence of any abstract, mathematical entities to formulate the claims of our best science. It is true that the additional claims about the mereological structure of the world are not motivated by traditional scientific considerations. For example, they go far beyond what our physics tells us about the structure of the Earth or the solar system. But the Quinean is not really in a position to insist that physics has a special place in the system of sciences in determining what we are to believe about the concrete world. Instead, the totality of our scientific beliefs determines which regimentations are viable. It seems just as clear and simple to embed the structures found in mathematics in the physical world as it is to extrude them into an abstract, nonphysical realm. Indeed, in some sense Lewis's mereological regimentation of mathematics seems better than an appeal to abstract sets. This is because it is at least conceivable that ordinary scientific methods could eventually clarify some further details of these posited concrete structures, for example, where they are located and how large they are.

While Quine's argument is vulnerable to Lewis's proposal, there is little problem for Putnam's argument. For even though Lewis's mereological interpretation of mathematics greatly affects the ontological commitments associated with mathematics, it does not affect the objective truth-values of the mathematical claims. So, like Hellman's modal-structuralism, Lewis's mereological approach is a realist interpretation of the truth-values of the statements of mathematics. It is not fair to object to Putnam's argument that these interpretations of mathematical claims deviate from the ordinary meaning we associate with mathematical claims. For Putnam clearly takes it to be a special feature of mathematical claims that they are subject to these reinterpretations and that the resulting perspective on the mathematical domain is "equivalent" to the more straightforward platonist one. It is true, of course, that a platonist could reject Putnam's claims about the availability of equivalent descriptions of the mathematical domain. That is, she may insist that only the platonist interpretation of our ordinary mathematical claims is acceptable. This might be based on the claim that we can see that a "literal" reading of mathematical statements entails platonism or on some more elaborate criticism of nonplatonist interpretations like Hellman's and Lewis's. But if the platonist offers this criticism, then it is hard to see how she is engaging with Putnam's indispensability$_P$ argument. This argument depended on considering which mathematical claims must be accepted to formulate the statements of science. It is a separate issue to determine the best reading of these mathematical claims. Putnam, of course, says that there are several readings and this is what is distinctive about mathematical claims. The platonist we are imagining responds that only her reading is acceptable. This is not an argument that has anything to do with how mathematics contributes to science. It is a question about pure mathematics independently of how it is deployed in science. So it must be resolved by an argument that is not an indispensability argument.

9.3 AGAINST THE PLATONIST CONCLUSION

A shared feature of Quine's and Putnam's indispensability arguments is that they operate at the level of what science as a whole requires from mathematics. From this lofty perspective they are unable to distinguish the different sorts of contributions we considered in part I. Considering these contributions affects Quine's and Putnam's arguments quite differently. I argue in this section that Quine's argument is unsound based on the falsity of 2_Q and the shift to the consideration of contributions in this or that case cannot overcome this limitation. At the same time, the epistemic contributions from mathematics are sufficient to defend Putnam's premises from its most significant challenge. This derives from Field's program of developing nominalistic versions of our scientific theories. If this program is successful, then mathematical claims need not be true to aid in the formulation of the claims of our scientific theories as they are deployed in scientific practice. That is, premise 2_P of Putnam's argument is false.

In the last section I pointed out that it is far from clear that the regimentations offered by Hellman and Lewis are flawed when compared to Quine's regimentation in terms of abstract objects. The question then becomes how the consideration of the contributions articulated in part I might affect this comparison. Although there were many different cases discussed, they were grouped into five different kinds of contributions. First, we had cases where the mathematics tracked causal interactions. Then there were cases where the mathematics contributed to the abstract nature of the representation, where this abstraction involved either the removal of causal content or the variation in interpretation of the mathematics. In chapter 5 we saw how mathematics can contribute to the scale of the representation, and in chapter 6 I argued that there was a defensible notion of a constitutive representation that must be believed for the associated derivative representations to be confirmed. Throughout I emphasized that a structural account of the content of these representations was adequate. This is the view that a mathematical scientific representation asserts that there is a certain structural relation between the mathematical structure and certain features of the target system. Based on this approach to the content of these representations, we were able to explain how these mathematical contributions were broadly epistemic. These discussions went far beyond the more elementary point that mathematics was needed to formulate the representations in the first place. For example, I argued that abstract acausal representations were easier to confirm, when they were accurate, than the associated causal representations.

The hard work of part I supports the conclusion that a structural conception of the content of these representations is sufficient to account for the contributions that mathematics makes to the success of science. If this is right, then the key question in evaluating Hellman's, Lewis's, and Quine's proposed regimentations is how well each proposal can handle this structural conception of content. It seems clear that if a proposed regimentation of this sort is able to assign the right truth-values to the claims of mathematics used in science, then it will also be able to assign the right contents to these representations. The content of these representations is analyzed in terms of the existence of a structural relation whose features depend on the mathematical structure involved and in some cases additional aspects of the relation itself that must be specified in mathematical terms. But if Hellman and Lewis can recover the right truth-values of the mathematical claims, then there is every reason to think they can pick out the right structural relation and the right mathematical structure, and so deliver representations with the same content that are available to the platonist. This is perhaps easiest to see in the case of Lewis's mereological interpretation. Given that Lewis makes the right kind of assumptions about the physical world, there will be copies somewhere in the physical world of any abstract mathematical structure available to the platonist. A given representation can be assigned a content, then, in terms of the existence of a structural relation between this concrete structure and the target system being represented. There is no problem with Lewis specifying this structural relation and its features. He, like the platonist, has the whole apparatus of mathematics available.

This point about content suggests that even though we have more or less assumed platonism since chapter 2, this assumption can be discharged in favor of realism about the truth-values of mathematical claims. It is harder to visualize the situation if we think of it in terms of a relation between concrete structures and target systems or possible structures and target systems, as with Hellman. But this point is irrelevant from the perspective of the Quinean project of regimentation. To object to these nonplatonist (though realist) alternatives, one must say how they violate the scientific standards of simplicity and clarity. It is not clear how this can be done, and if it can be done, the assumptions needed go beyond Quine's indispensability$_Q$ argument as we have presented it.

The most promising strategy for the platonist is to consider a contribution that goes beyond the sort discussed in part I. The contributions I have discussed seem to me to be especially important for an understanding of how mathematics contributes to the success of science. But I do not intend my survey to be exhaustive, and there may well be important aspects of how mathematics helps in science that I have overlooked. In the next chapter we turn to explanatory contributions and consider how they might rescue Quine's indispensability$_Q$ argument. At this point, though, we can see what form these other contributions must take if they are to vindicate the claim that premise 2_Q is true. They must show how the structural approach to content is flawed. If these other contributions can be explained by appeal to structural relations between mathematical structures and target systems, then Hellman's and Lewis's proposals seem back on par with the platonist.

I have just argued that the three interpretations of pure mathematics that are realist in truth-value are able to adopt our account of the broadly epistemic contributions mathematics makes to the success of science. But there are other antirealist interpretations of mathematics, and it is not obvious how they fare in this task. The most influential proposal along these lines is Field's claim that mathematical scientific theories need not be true to be good.[5] The reason for this is that each scientific theory T formulated using mathematics has a nonmathematical analogue N. Field's test for the acceptability of N is that the addition of mathematics M to N yields only a conservative extension of N. That is, for any claim p expressed only in the nonmathematical vocabulary of N, if p is a consequence of $N + M = T$, then p is a consequence of N. The original theory T may have certain pragmatic advantages over its nominalistic analogue N, but for Field these pertain only to the ease with which we can determine what the consequences of the theory really are. This difference in convenience is not sufficient to defend a preference for T over N. Largely following Quine's views on regimentation, Field argues that N is in fact superior to T in several scientific respects. The main advantage is that N eliminates the arbitrary choices that are inevitable when we represent a physical magnitude like distance using mathematical entities like the real numbers. As we saw with Putnam's discussion of this process, to get a determinate function from distances to real numbers we must choose a standard unit of distance. But Field argues that these arbitrary choices obscure the physical facts underlying the theory and that scientific criteria dictate a preference for an "intrinsic" version of the theory that avoids such choices.

Building on Putnam's discussion of measurement theory, Field argues that Newtonian gravitational theory can be presented in a nonmathematical form by

---

5. See Field (1980, 1989).

assuming a background ontology of space-time points and regions along with an array of physical properties and relations. Axioms are presented for these properties and relations that ensure the existence of a function of the sort countenanced by Putnam between the target systems and mathematical structures envisaged by the platonist. This allows Field to prove a series of representation theorems. Each representation theorem claims that the target system can be accurately represented by an associated mathematical structure. These theorems lead Field to conclude that he has shown how his axioms for space-time and physical properties have all the features necessary for the nominalistic analogues N of our scientific theories T. Among other things, we see that premise 2_P of Putnam's indispensability$_P$ argument is false. The mathematical claims M need not be true to make their contribution to the formulation of our ordinary scientific claims. For with N we have all that the scientific realist could want, while at the same time we have dropped the need for any mathematical claims to be true. So for Field, the scientist may appeal to N when fixing his ontological commitments and reject Putnam's conclusion that realism about mathematics is necessary.

Many objections have been made to Field's nominalistic theories (MacBride 1999), but the most obvious one comes directly from our discussion of the epistemic contributions in part I. We have seen how mathematics does much more than just allow us to derive consequences from nonmathematical claims. The epistemic aspects of what mathematics contributes far exceed this sort of deductive power. For example, we can formulate a representation that is largely acausal and is thus consistent with a wide variety of conflicting causal representations of the target system. As we saw in chapter 3, this is especially useful when we lack evidence about the correct way to represent these causal details of the target system. This sort of benefit is not achievable by Field's nominalistic theories. To get his representation theorems to go through, he is forced to make very specific assumptions about space-time and how the physical properties he invokes are arrayed. These assumptions may go beyond the evidence we have, and precisely in these circumstances a more abstract, mathematical representation is most valuable. Similar points can be made about the contributions enumerated in chapters 4 and 5. In each case, the mathematics helps us confirm the representations we have. These representations are presented in terms of a structural relation between a target system and a mathematical structure. Often we are quite ignorant of many of the features of the target system. But an appeal to a mathematical structure that can be interpreted in different ways or to a representation at a certain mathematically specified scale can get the process of scientific testing off the ground. This is not a merely heuristic or pragmatic point about what is convenient for the scientist. These contributions turn on how it is possible to confirm the representations we have available.

On the account I have developed, then, mathematics is so useful in science precisely because we are unsure which nonmathematical description of our various target systems we should adopt. This is consistent with the eventual elimination of mathematics along the lines Field envisages. But this elimination could only happen at the "end of science," and it is clear that we are not yet at that stage. For science as it is practiced now, mathematical claims play a central epistemic role. This shows how clarifying the different epistemic contributions that mathematics makes can aid in the defense of Putnam's indispensability$_P$ argument. Unfortunately, neither Putnam nor his later defenders have taken this route.[6]

9.4 COLYVAN

Colyvan's *Indispensability of Mathematics* (Colyvan 2001) is the most careful and extensive presentation of the indispensability argument for platonism to date. As we will see, his preferred version of the argument is not easily identified with either Quine's argument or Putnam's argument. Colyvan seems to agree with Quine, against Putnam, that the argument is an argument for platonism. Thus, he summarizes his argument as

1. We ought to have ontological commitment to all and only those entities that are indispensable to our best scientific theories.
2. Mathematical entities are indispensable to our best scientific theories. Therefore:
3. We ought to have ontological commitment to mathematical entities. (Colyvan 2001, p. 11)

But his notion of indispensability is different than Quine's. As we saw, Quine is concerned with a process of regimentation that is not interested in what scientists mean by their claims or even what is pragmatically useful for scientists to say in their day-to-day work. When Colyvan comes to clarify what he means by "indispensability," he includes much more than Quine. He says that an entity is dispensable from a theory just in case "there exists a modification of the theory in question resulting in a second theory with exactly the same observational consequences as the first, in which the entity in question is neither mentioned nor predicted" and "the second theory is preferable to the first" (Colyvan 2001, p. 77). So an entity like a real number is indispensable$_C$ when the theory T in which it appears has no preferable competitor T' that drops that entity from its ontological commitments. Colyvan goes on to characterize his notion of preferable using the whole host of criteria scientists appeal to in their theory choice. These include "Simplicity/Parsimony," "Unificatory/Explanatory Power," "Boldness/Fruitfulness," and "Formal Elegance" (Colyvan 2001, pp. 78–79). So an entity is indispensable$_C$ to a theory when each of its competitors either does worse by these criteria or else includes the entity. For ease of comparison, we can put Colyvan's argument in a form that is similar to how we have presented Putnam's and Quine's arguments:

1$_C$. We ought rationally to believe in the existence of any entity that plays an indispensable$_C$ role in our best scientific theories.
2$_C$. Mathematical entities play an indispensable$_C$ role in science.
3$_C$. Hence, we ought rationally to believe in the existence of some mathematical entities.

---

6. Considerations of space have led me to skip Resnik's pragmatic indispensability argument (Resnik 1997). At the same time, I believe the problems I raise here also undermine Resnik's platonist conclusion.

Indispensability

The difference between Quine's indispensability$_Q$ and Colyvan's indispensability$_C$ is that Colyvan explicitly articulates several different desirable features for our scientific theories, whereas Quine seems focused mainly on simplicity and clarity. The inclusion of "explanatory power" on Colyvan's list turns out to be crucial.

In the last section we reviewed the objections to Quine's argument based on Hellman's and Lewis's interpretations of mathematics and argued that the Quinean had little to say against them. Somewhat surprisingly, Colyvan concedes at several places in his book that the conclusion of his indispensability argument is consistent with these interpretations. At the beginning of his presentation he notes that he will use the term *Platonism* to include any view that entails the view that mathematical entities exist even if the entities are not abstract (Colyvan 2001, p. 4). For this reason, Colyvan's argument for his form of Platonism is hard to distinguish from Putnam's realism about objective truth-values. At the end of the book, when Colyvan is summing up the implications of accepting the conclusion of his indispensability argument he notes that his conclusion

> simply asserts that there *are* mathematical objects. They might be constituted by more mundane items such as universals and/or relations (…), patterns or structures (…) or the part/whole relation (as David Lewis (1991) claims). Perhaps they are constituted by *more* exotic items such as possible structures (as Hilary Putnam (1967) and Geoffrey Hellman (1989) claim). In short, any (realist) account of mathematical objects is all right by the indispensability argument. (Colyvan 2001, p. 142)

I believe that there are two issues involved here that Colyvan is running together. First, there is the important question of whether the existence of mathematical entities entails the existence of abstract objects. Second, there is the question of how the indispensable role of a mathematical claim like "There are numbers" is to be reconciled with our test for ontological commitment. Colyvan seems to think that accepting a claim like "There are numbers" in our best scientific theories requires belief in the existence of mathematical entities, while at the same time leaving open the question of the nature of these entities. But as we have seen, if the mathematical domain is subject to "equivalent descriptions" of the sort Putnam envisaged, then we can accept the indispensable role of a claim like "There are numbers" and yet deny that special mathematical entities exist. This is sufficient to undermine Quine's indispensability$_Q$ argument. Nothing that Colyvan says in his account of indispensability$_C$ affects this criticism. The criteria that scientists offer that lead us to include a claim like "There are numbers" in our best theories do not dictate how this purely mathematical claim is to be interpreted. In particular, they do not rule out either Lewis's or Hellman's interpretations of pure mathematics. So at the end of day, we can conclude that this claim must be given an objective truth-value of true, but not that any mathematical entities exist.

As our last quotation shows, Colyvan recognizes the inability of his argument to rule out the interpretations offered by Lewis and Hellman. He is more successful in his objections to Field. Colyvan quite rightly draws attention to several examples where including mathematics allows scientific theories that do better by the criteria he enumerates. For example, the inclusion of complex numbers allows a greater methodological unification in the techniques for solving differential equations (Colyvan 2001, pp. 81–83). Based on this example and others, Colyvan argues that Field's nominalistic theories are worse by ordinary scientific standards. I am, of course, in agreement with Colyvan on this point and expect that the additional examples deployed in part I of this book will only reinforce these objections against Field. At the same time, I have been keen to emphasize epistemic benefits that seem to go beyond what Colyvan has made explicit. This becomes clear when we note that the weaker degree of confirmation of Field's theories that I have noted is not due to any of the four factors Colyvan lists. They are based, instead, on the need for Field's theories to include claims about the physical world that the mathematical scientific theory can remain neutral about. This is an important factor in scientific theory choice and leads to many of the epistemic contributions from mathematics I have enumerated.

We can reconstruct Colyvan's considered view on his argument, then, as an argument in favor of realism in truth-value based on the wider notion of indispensability$_C$ which goes beyond both Quine's notion of what results from the best regimentation and Putnam's notion of what is needed to formulate the claims of our scientific theories. Colyvan's argument seems better than Putnam's based on the greater focus on the criteria that scientists deploy in choosing their theories. I also take Colyvan's notion of indispensability$_C$ to subsume the sorts of epistemic contributions I emphasized in part I. This gives us our most compelling indispensability argument for realism about mathematical claims:

1. We ought rationally to believe in the truth of any claim that plays an indispensable$_C$ role in our best scientific theories.
2. Mathematical claims play an indispensable$_C$ role in science.
3. Hence, we ought rationally to believe in the truth of some mathematical claims.

The indispensability$_C$ of claims can be characterized in a way that is completely parallel to Colyvan's explanation of the indispensability$_C$ of entities. That is, a claim is indispensable$_C$ to a theory when all competitors that remove that claim do worse by ordinary standards of scientific theory choice.

Two issues remain to be resolved before I can endorse this argument. First, the way mathematical claims play their indispensable$_C$ role is far from clear, so the basis on which one is supposed to assent to 2 is shaky. I believe that an understanding of why 2 is true involves significant assumptions that complicate an evaluation of this indispensability argument. Second, Colyvan has recently emphasized explanatory contributions from mathematics to science. A discussion of explanation involves additional complications and may strengthen the conclusion of the argument from mere realism to platonism. Both of these issues are pursued in the next chapter.

10

Explanation

10.1 EXPLANATORY CONTRIBUTIONS

Colyvan and Baker have recently initiated a new kind of indispensability argument that turns on the explanatory contributions from mathematics. This has moved the debate forward in a number of respects, the most important of which is the consideration of several actual examples of scientific explanations that involve mathematics in an especially central way. By analogy with the inference to the best explanation (IBE) argument for the existence of unobservables, Colyvan and Baker urge us to accept the existence of abstract objects. Unfortunately, except for a few remarks about unifying power, neither Colyvan nor Baker has given an account of how mathematics contributes to scientific explanation. Although there are any number of possibilities, our discussion is structured around the earlier descriptive contributions identified in chapters 3–6. If we can link any of these contributions, which we have argued are in the end broadly epistemic, to explanatory power, then it might seem like the new explanatory indispensability argument can be vindicated. I argue, however, that these hopes prove illusory. For mathematics to make its explanatory contributions, the agent must have already confirmed her mathematical beliefs to a large degree. This undermines not only the conclusion that abstract objects exist but even the weaker view that explanatory contributions can be our primary form of evidence for the truth of our mathematical beliefs.

Since the publication of his book, Colyvan has emphasized the special way the existence of mathematical entities contributes to superior scientific explanations of physical phenomena. Along with similar work by Baker, this has led to the articulation of a new explanatory indispensability argument for platonism.[1] Although Colyvan had mentioned this prominently in his book (Colyvan 2001, p. 7), the connection between explanatory power and the existence of mathematical entities was not put at the center of his exposition. Now, though, there is a special focus on explanatory considerations motivated by parallel arguments for scientific realism. In both cases, it seems that an application of IBE warrants the conclusion that a new kind of entity exists. In the case of traditional debates about scientific realism, these entities are unobservables, like atoms. Now, in the case of platonism, the

---

1. See especially Colyvan (2002), Lyon and Colyvan (2008), Colyvan (2010), Baker (2005) and Baker (2009). An important recent paper is Psillos (2010).

claim is that our best explanations involve an appeal to mathematical entities in a way that is relevantly similar. So the scientific realist who accepts IBE should also accept the existence of mathematical entities. If this style of argument could be vindicated, then Colyvan would be able to achieve something that was missing from his book. This was to move from the acceptance of a claim like "There are numbers" in our best theories to the conclusion that mathematical entities exist. As I have presented things, he would be able to move from a realism in truth-value to a platonist conclusion.

Critics of the explanatory indispensability argument (such as Melia) have argued that there really are no mathematical explanations of physical phenomena.[2] After reviewing this debate, I conclude that both sides have severe limitations. First, after clarifying what the existence of such mathematical explanations requires, I argue that these explanations do in fact exist. Second, I object that Colyvan and Baker have not presented an adequate account of the ways mathematics contributes explanatory power. But third, against Melia, I claim that this does not mean that mathematics plays a merely "indexing" or "representational" role. This is based on the now-familiar point that we often lack any knowledge of what the mathematics is representing. Finally, I endorse a complaint made by Melia that IBE fails to justify mathematical truths when they appear in a scientific explanation. But I argue that this difference between unobservables and mathematical entities is related to the special character of mathematical truths in a way that is different than Melia.

The first step is to establish the existence of mathematical explanations of physical phenomena. The issue can be clarified in at least two ways. First, we may wonder if any explanations make use of a mathematical claim. A mathematical claim here is just any statement about a mathematical domain. I propose a simple replacement test to answer this sort of question. Starting with an explanation that involves a mathematical concept or theorem, the test involves considering what happens if this concept or theorem is removed. If the resulting argument is no longer an explanation or else has less explanatory power, then I conclude that the mathematical concept or theorem is making an explanatory contribution to the original explanation. But this is not the only sense in which mathematical explanations of physical phenomena may exist. A second way to proceed is to ask if a mathematical claim allows a scientific explanation that has greater explanatory power than would be available otherwise. The replacement test is not sufficient to answer this sort of question. What is needed is what I call a comparison test. This test considers all available explanations of the phenomena at issue. If the explanation that uses the mathematical claim has the greatest explanatory power, then we have a mathematical explanation in this second sense. The comparison test is clearly the test that is relevant to an application of IBE. It is only the best explanation that is a candidate for the use of this inferential principle.

In many explanations offered in contemporary science we find mathematical claims. As we often lose the explanation if the mathematical claim is removed, the

---

2. See Melia (2000), Melia (2002) and Daly and Langford (2009). I have also benefited from Saatsi (2011).

replacement test is passed. So we have mathematical explanations of physical phenomena in this weak sense. But given that there is typically no way of formulating a competing nonmathematical explanation with as much explanatory power, the comparison test will be passed as well. This suggests that there are many candidates for the application of IBE to justify claims of pure mathematics.

The comparison test fits well with Colyvan's indispensability$_C$ if we restrict the relevant scientific criteria to explanatory power. As we clarified indispensability$_C$ for claims at the end of chapter 9, a claim is indispensable$_C$ to a theory when all competitors that remove that claim do worse by the ordinary scientific standards of theory choice. If we focus only on a single explanation and its explanatory power, then it is natural to say that a claim is indispensable$_C$ to that explanation when all competing explanations that lack that claim have a lower degree of explanatory power. But all such cases also pass our comparison test.

Both the replacement and comparison tests are passed by the main example in Baker's paper "Are There Genuine Mathematical Explanations of Physical Phenomena?" (Baker 2005). Baker describes a case where scientists wished to explain why the life cycle of three species of periodic cicada were all prime. He then reconstructs the explanation offered by the scientists in terms of an argument whose conclusion is the explanandum.

1. Having a life cycle which minimizes intersection with other (nearby/lower) periods is evolutionarily advantageous. [biological "law"]
2. Prime periods minimize intersections (compared to nonprime periods). [number theoretic theorem]
3. Hence organisms with periodic life cycles are likely to evolve periods that are prime. ["mixed" biological/mathematical law] (Baker 2005, p. 233)

Inspection of the argument reveals that premise 2 is needed to connect the concept of *prime* in the conclusion to the biological feature of minimizing intersection with predators and similar competing species. If we simply removed the mathematical premise, we would no longer have an explanation. The only delicate issue is whether there is some alternative explanation that is just as strong and makes the link between the biological premise and the explanandum. That is, the comparison test is not obviously met. Adopting a broadly naturalistic approach, Baker seems to argue that the practice of scientists suggests that they think this is the best explanation of the primeness of the life cycles. As these scientists are the experts in this domain, strong evidence is needed to show that there is a superior alternative explanation.[3]

Lyon and Colyvan appear to endorse this approach in their discussion of the mathematical explanation of the hexagonal character of the bee's honeycomb. What needs to be explained is "why hive-bee honeycomb has a hexagonal structure" (Lyon and Colyvan 2008, p. 228). As with the periodic cicada case, the explanation involves a biological claim and a mathematical theorem. The theorem links a part of the biological claim to the mathematical concept that appears in the explanandum. The biological claim is "that those bees which minimise the amount of wax they use

---

3. Saatsi (2011) attacks Baker on this point.

to build their combs tend to be selected over bees that waste energy by building combs with excessive amounts of wax" (Lyon and Colyvan 2008, p. 228). This point is then conjoined with the mathematical theorem that says "a hexagonal grid represents the best way to divide a surface into regions of equal area with the least total perimeter" (Lyon and Colyvan 2008, pp. 228–229). Lyon and Colyvan note that this explanation "is arguably our best explanation for this phenomenon" (Lyon and Colyvan 2008, p. 229). They do not dwell on its mathematical character, but it is fair to see them working with our replacement and comparison tests. Removing the mathematical theorem removes the explanatory power of the argument. Furthermore, there does not seem to be a competitor with as much explanatory power.

Both examples are drawn from evolutionary biology, but it is clear that nothing turns on this. As a third example of this general sort, we can recall the Königsberg bridges case (section 3.3). Euler explained why it was impossible to cross all the bridges exactly once in a circuit that returns to its starting point. The explanation involved the observation that the bridges of Königsberg formed a graph of a certain kind, namely, one with at least one vertex with an odd valence. Euler added his mathematical theorem that there is no return path through a graph that crosses each edge exactly once when at least one vertex has an odd valence. This is why there is no such return path across the bridges of Königsberg. Again, the explanation passes our replacement test. Removing the mathematical theorem severs the link between the claim that the bridges form a certain graph and the conclusion that no path of a certain kind exists. A consideration of the available alternatives suggests that it has a reasonable claim on being our best explanation of this phenomenon. That is, the comparison test is met as well.

So far I have only argued that mathematical explanations of physical phenomena exist. Now we must see how the existence of such explanations is supposed to inform an explanatory indispensability argument for platonism. As Baker has recently stated it, in line with earlier discussions by Colyvan, the argument is

1_E. We ought rationally to believe in the existence of any entity that plays an indispensable explanatory role in our best scientific theories.

2_E. Mathematical objects play an indispensable explanatory role in science.

3_E. Hence, we ought rationally to believe in the existence of mathematical objects. (Baker 2009, p. 3)[4]

The first thing that should strike one about this argument is how unconnected 2_E is from the existence of mathematical explanations, at least as I have clarified this. This is because the replacement test turned only on what happened to an explanation if a whole claim was removed or transformed by the removal of a concept. Similarly, the comparison test involved comparing an explanation with this claim to explanations that do without it. It is not immediately clear what the contributions of whole mathematical claims has to do with the existence of mathematical objects. Indeed, this is one respect in which the debate about the explanatory indispensability argument has come to resemble the debates about the original indispensability argument, which we reviewed in chapter 9.

---

4. I relabeled the premises for ease of exposition.

The difficulties for the platonist are shown by Baker's attitude toward our replacement and comparison tests. He distinguishes between "explanations involving mathematics" and "ipso facto mathematical explanations" (Baker 2005, p. 234) and restricts the indispensability argument to the latter. Unfortunately he never provides a test that distinguishes these two cases except to argue that in genuine mathematical explanations the mathematical entities are relevantly similar to unobservables like electrons (Baker 2005, p. 236). Baker could be charitably read as restricting his ipso facto mathematical explanations to those that meet our comparison test. In that case, it is still incumbent on Baker to clarify how these mathematical claims lead to explanations with the greatest explanatory power.

Daly and Langford respond to Baker by invoking "Melia's indexing argument":

> Even granted that the formulation of these explanations involves indispensable mention of mathematical entities, it does not follow that these entities play a part in the explanations. Melia's indexing argument shows that the part which mathematical entities play may only be to pick out features of space-time which provide the whole of the geometric explanations. (Daly and Langford 2009, p. 645)

Here they refer to Melia's strategy of taking reference to a mathematical entity like a real number to be merely the means to pick out or "index" a physical relation like a distance relation (Daly and Langford 2009, p. 642). As a result, Daly and Langford, along with Melia, would presumably reject premise 2_E of the explanatory indispensability argument. We are not committed to the mathematical entities because what really contributes to the explanation are the physical entities or relations that make up the physical system (Melia 2000, p. 474; Melia 2002, p. 76). Melia's position is in many ways a further development of Field's program. The main difference is that Melia does not think it is possible to formulate nominalistic theories of the sort Field insisted on. Still, even in the absence of these theories, Melia insists that our mathematical scientific theories offer explanations only in virtue of some underlying physical entities and relationships. It is true that we can only represent these entities and relationships by appeal to mathematical entities, but we do not thereby commit ourselves to the existence of these mathematical entities.

I would suggest that the availability of Lewis's and Hellman's interpretations of pure mathematics shows that we need not resort to this "indexing" strategy to block an explanatory indispensability argument for platonism. Even when a claim meets the comparison test it, is hard to see why that claim must be given a platonist interpretation. For this reason it is worth casting the argument in terms of a realism in truth-value.

1_{ER}. We ought rationally to believe in the truth of any claim that plays an indispensable$_C$ explanatory role in our best scientific theories.
2_{ER}. Mathematical claims play an indispensable$_C$ explanatory role in science.
3_{ER}. Hence, we ought rationally to believe in the truth of some mathematical claims.

The comparison test gives content to the notion of indispensability$_C$. The existence of mathematical explanations of physical phenomena (in this sense) now directly supports 2_{ER}. The platonist can then try to supplement 3_{ER} with additional arguments for the conclusion that the belief in the truth of the relevant mathematical claims rationally supports the belief in the existence of some mathematical objects.

Suppose, then, that we shift to a debate about my revised explanatory indispensability argument. It appears that even though the critics of the original argument would grant 2_{ER}, they would now question 1_{ER}. That is, they seek to sever belief in truth from explanatory contributions, at least when the claims in question are mathematical. This is clearly because they have a model of explanation according to which a claim can contribute to an explanation and not be true. This is the indexing approach, which insists that the mathematical claims in these explanations merely serve to pick out some nonmathematical features of the system in question. What is needed, then, is an alternative characterization of the source of the explanatory contributions that mathematics makes in these cases. Unfortunately, this has not yet been provided by advocates of the explanatory indispensability argument. They sometimes allude to unification, but do not clarify the link between unification and justification.[5]

I argue that we can distinguish at least three ways in which mathematics can contribute to the explanatory power of a scientific explanation. They correspond to the main epistemic contributions we have enumerated in part I. These are by (i) tracking causes, (ii) isolating recurring features of a phenomenon, and (iii) connecting different phenomena using mathematical analogies. The different contributions need to be treated separately because the interpretive implications of one sort of contribution need not carry over to explanations that depend on another sort of contribution. To consider a pure case of i, recall the dynamic representation of the cars in section 3.2. Analogously, consider a particular discrete system of physical objects operating under a small number of forces, such as gravitational and electromagnetic. Suppose further that the initial and boundary conditions of the system are such that the system has a unique way of evolving in some time interval, say $t = 0$ to $t = 10$. Then we can explain the state of the system at $t = 10$ by describing its state at $t = 0$ along with the relevant laws governing the forces and the initial and boundary conditions. Whenever we obtain this sort of explanation, it turns out that it is highly mathematical. It is a mathematical explanation of a physical phenomenon in both of the senses articulated earlier. To see why, notice that removing the mathematical aspects of the explanation deprives the argument of its explanatory power. Furthermore, we do not have a competing explanation available that can be given in nonmathematical terms.

Interestingly enough, the cicada, honeycomb, and bridge explanations noted earlier are not cases in which causes are tracked in this fashion. It might seem, in fact, that this example is an ideal case for the indexing approach defended by Melia and Daly and Langford. I want to argue, though, that this strategy faces an insurmountable problem with these tracking causes explanations. On the assumption

---

5. See, for example, Colyvan (2002, p. 72) and Baker (2009, p. 11).

that we really have a mathematical explanation where the mathematical claims are indispensable$_C$ to the explanation, it follows that we lack the means to characterize the causes in question in nonmathematical terms. The reason for this is not hard to find. We lack a knowledge of the detailed causal interactions that are responsible for the state of the system we are trying to explain. It is true that we can track these causal interactions using the mathematics, but we do not know which physical features of the objects are responsible for the evolution of the system. In the end, this is what frustrates the search for a nonmathematical analogue of these mathematical explanations. If we knew all the genuine features of the objects in question, maybe we could replace these mathematical explanations with superior nonmathematical analogues as the indexing strategy suggests. But our inability to do this shows that we lack some knowledge of the features of the system and pinpoints the special contribution that mathematics is making in these causal cases. This is also my diagnosis of the reasons behind the failure of Field's program, that is, we lack knowledge of the axioms of Field's nominalistic theories.[6] These epistemic limitations also block the indexing approach. We cannot restrict our commitments to the features of a physical system that are indexed by the mathematics without saying what these features are like. If we fail to know much about these features, then merely picking them out does not constitute an adequate explanation. The burden is then on the advocate of the indexing approach to clarify some sense of "explanatory power" according to which these indexing explanations have greater explanatory power than their competitors that exploit the truth of these mathematical claims.

The existence of contributions ii and iii shows that mathematics need not always contribute by tracking causes. The cicada, honeycomb, and bridge explanations are indeed of type ii where a recurring feature of a phenomenon is isolated and shown to be relevant to the explanandum. For the cicadas, the recurring feature is the prime length of the life cycles of the three species. To explain this, we mathematically relate primeness to a distinct biological feature of the species we have reason to believe is present. With the honeycomb, the recurring feature is the hexagonal structure of the honeycomb. Again, this is connected to the minimal use of wax via the mathematical theorem. Finally, we can view the bridges case as a type ii case if we think of the many attempted circuits of the bridges. Each of these bears the property of crossing at least one bridge twice. This feature is then linked to the valence of the vertices of the associated graph using Euler's theorem.

Based on these three examples, it should be clear how different the mathematical contribution is to the explanation when it isolates recurring features as compared to when the mathematics tracks causes. The explanations where recurring features are isolated have a good claim on being instances of noncausal explanations.[7] They seem to have been emphasized by Colyvan and Baker precisely because they fail to fit the indexing approach. Indeed, it is hard to see how the indexing approach is even

---

6. Colyvan (2010) provides another argument linking the failure of Melia's indexing strategy to the shortcomings of Field's program.

7. The same point holds for Lyon and Colyvan's more elaborate phase space explanation in Lyon and Colyvan (2008). I hope to engage with this case elsewhere.

supposed to get off the ground when the mathematics is isolating recurring features. As with the previous case, we typically lack the knowledge sufficient to characterize these features in nonmathematical terms. But even if we could, a physical characterization of those features risks introducing irrelevant aspects of the systems in question. For example, in the bridges case, it is irrelevant to the explanandum that the bridges are in fact made of stone. Similarly, the specific genetic coding responsible for the actions of the cicada or the bees are not relevant to the sort of explanation we have offered. For type ii cases, then, we can account for why the mathematical explanation is the best, that is, why it has the greatest explanatory power. It selects the relevant features of the systems that are responsible for the explanandum and leaves out features which are irrelevant.

Type iii cases have received less discussion in the context of indispensability arguments. I merely indicate their existence here and reserve a discussion of their interpretive significance for chapter 11. A core family of cases involves the abstract varying contributions surveyed in chapter 4. Here we link different phenomena using mathematical analogies by deploying the same mathematical equations to represent both phenomena. For example, we can use Laplace's equation to represent both irrotational fluids and electrostatic systems (section 4.2). The explanatory value of this link could of course be questioned. But as we will see in chapter 11, Batterman has written extensively about another sort of type iii case and argued that it does involve a special kind of "asymptotic explanation." This is aligned with some of the material on scaling from chapter 5. Our main example will involve the explanation of three features of the rainbow. The best explanation of some of these features involves mathematically relating the wave representation of light to the ray representation of light. The specific mathematical analogy is that the latter representation results from the former when a certain limit is taken. This is clearly not a type i case because we are not tracking causes when we make this mathematical transformation. Similarly, there is no recurring feature corresponding to the limit operation. So, this part of the mathematics is not making a type ii contribution. Of course, some patterns recur across all the rainbows, so in this respect we have a distinct type ii contribution from the mathematics of the ray representation. Batterman has argued, and I agree with him, that there are special interpretative issues connected with these explanations that do not arise in the ordinary type ii cases. We will see how the limit operation makes a distinct kind of contribution to our explanation of the rainbow and what interpretive significance this has. There is little support here for indispensability arguments for platonism, but we will see how mathematics can play a central role in an IBE argument for novel conclusions about the proper interpretation of the mathematics involved.

10.2 INFERENCE TO THE BEST MATHEMATICAL EXPLANATION

In the last section I defended 1_{ER} against one sort of objection based on the indexing strategy and supported 2_{ER} by providing a more satisfying characterization of some of the ways mathematics can make explanatory contributions. Surely other sorts of contributions can be clarified, leading to yet more examples of mathematical explanations of scientific phenomena.[8] It might seem then that I accept the revised indispensability argument and should conclude that scientific practice gives us good reason to believe in the truth of some mathematical claims. There is one problem with this position, though. Colyvan and Baker follow Quine's assumption that the indispensable role of mathematics in science indicates that mathematics is justified empirically. The effects of this assumption come through clearly in Colyvan's discussion of defenders of the Eleatic principle. They restrict inference to the best explanation to causal explanations. Colyvan notes, "They have all accepted the view that we have ontological commitment to the entities of our best scientific explanations. That is, the defenders of the Eleatic Principle we've met so far accept *inference* to the best explanation" (Colyvan 2001, p. 53; my emphasis). But someone could accept inference to the best explanation and still place reasonable preconditions on its application to mathematical entities. For example, it may be that mathematical claims can only contribute to explanations if the mathematical claims are independently supported by purely mathematical means. I want to argue in this section that our justification for a given mathematical claim cannot depend primarily on its contribution to a scientific explanation. If this is right, then there must be a prior nonempirical source of justification for the mathematical claims that make explanatory contributions. On this view, explanatory contributions can only provide additional boosts in justification for a belief that was already substantially justified. But if prior justification is required for explanatory contributions, then the explanatory indispensability argument is completely undermined. Mathematics is, indeed, indispensable to science, but only because its claims receive substantial support independently of its application in science. So accepting 2_{ER} presupposes that we have accepted 3_{ER}. That is, the argument begs the question at issue between those who believe in the truth of these mathematical claims and those who do not.[9]

There are two routes to this conclusion, only one of which I think is worth pursuing in this context. First, one might reject any use of IBE. On this view, associated with van Fraassen's pragmatic conception of explanation, explanations are arguments that we offer only after we have elected to accept a scientific theory. Explanatory power, then, is not an indicator of truth, and it should not be used to decide which claims or theories to accept. Though, popular, this strategy is completely uninteresting in the context of explanatory indispensability arguments. All sides start from the position of some form of scientific realism that accepts at least some instances of IBE. In the classic cases of atoms and electrons, for example, the ability of the theories that posit atoms or electrons to explain this or that phenomenon is used to support our belief in the existence of these unobservable entities. So, I opt for the second route according to which IBE is appropriate for some claims involving unobservables, but is not available to provide the basis on which to believe in the

---

8. See, for example, Lange (2009).

9. The worry that the indispensability argument is circular is also raised by Wilholt. See especially Wilholt (2004, p. 36) and Pincock (2005, p. 331).

mathematical claims at issue in this debate. My challenge is then to articulate the relevant difference between claims concerning unobservables and these mathematical claims.

There are two problems that undermine the application of IBE to mathematical claims. After raising these problems, I explain why they fail to arise in the case of unobservables. The basic difference is that the truth of many mathematical claims goes far beyond what is needed for the explanation to be successful. The first problem is what I call the problem of weaker alternatives. When we have a mathematical explanation of a physical phenomenon, we can typically find a weaker mathematical claim that can do the explanatory work of the stronger mathematical claim. As a result, it is not clear which claim should be justified by the use of IBE. A more serious problem is what I call the sensitivity problem. The truth of many mathematical claims depends on aspects of a mathematical structure that are irrelevant to the success of a given explanation. As a result, the explanatory contributions of these mathematical claims do not bear on the truth of these claims in the right kind of way. In developing the sensitivity problem I articulate a principle that restricts IBE to a special sort of epistemic situation. This restriction can be further motivated by reminding ourselves of the failures from chapter 7. There it became clear that the proper physical interpretation of a mathematical scientific representation must reflect the scope of the representation. If this scope is wide enough, then we are warranted in accepting the explanation and the existence of a new kind of entity. What I argue here is that this sort of demand is not met for the mathematical claims that appear in these explanations.

To approach the problem of weaker alternatives, we need to consider an agent who does not already believe the relevant mathematics. In the cicada case, for example, the agent does not yet accept that prime periods minimize intersection (as compared to nonprime periods). Analogously, with the bees, the scientist is in doubt about whether the hexagonal tiling minimizes the perimeter among all the tiling patterns using polygons. Finally, the agent does not know if the circuits are impossible if a vertex has an odd valence. Suppose further that the agent is in doubt about the following weaker mathematical claims: prime periods *of less than 100 years* minimize intersection (as compared to nonprime periods), the hexagonal tiling minimizes the perimeter among all the tiling patterns using *regular* polygons, and the circuits *through graphs with fewer than 10 vertices* are impossible if a vertex has an odd valence. Each of these claims is weaker than its analogue in the sense that the original claim implies the new version of the claim but not vice versa. So any agent who believes the original claim should also believe the new version of the claim. But of course it is not the case that any evidence for the weaker claim is also evidence for the stronger claim. In fact, reasonable rules for the use of IBE suggest that, other things being equal, the explanation that employs the weaker claim is superior to the explanation employing the stronger claim. This is the challenge from weaker alternatives to the explanatory indispensability argument. There are any number of weaker mathematical claims that contribute to explanations of the phenomena at issue. These alternative explanations seem to have as much explanatory power as the explanations that employ the stronger mathematical claims. So, our comparison test does not decide between them. The use of IBE appears to license only the belief in the weaker claims and not the belief in the stronger claim. Furthermore, once we see how many weaker claims are available, it is not clear on what basis the scientist is supposed to choose one over another.

The cicada case and the bridges case involve explananda where, in fact, the periods are less than 100 years and there are fewer than 10 vertices. The original explanations can be weakened using our new mathematical claims, and no other changes are necessary to preserve an explanation. In the bees case things are a bit more subtle. If we adopt the weaker mathematical claim that concerns only regular polygons, we must adjust the biological part of the explanation to claim that the bees are constrained to produce tiling patterns using regular polygons. This could be questioned on the basis of biological considerations. At the same time, there are also biological reasons in favor of selection for a species that proceeded using a regular tiling. For example, this might be easier to "program" into the genetic code of the bees than an intricate irregular polygon pattern. More generally, it seems that there will be a trade-off between the mathematical and nonmathematical assumptions that go into a mathematical explanation. Sometimes a weaker mathematical claim can be simply substituted in for a stronger mathematical claim. In other cases a weaker mathematical claim must go along with a different nonmathematical claim to recover an explanation. The consideration of these sorts of corresponding changes will yield any number of candidate explanations for our physical phenomena.

The reply I imagine at this point is that it is arbitrary to restrict the scope of these mathematical claims. I agree that the weaker alternatives do seem arbitrarily chosen from the perspective of someone who already accepts the axioms that entail the fully general claims. But if we put ourselves in the position of someone who genuinely does not know what to believe, then the caution implicit in the preference for a weaker alternative is easily defended. There are any number of general mathematical claims that worked for many test cases, but then unexpectedly failed. For example, Euler's conjecture that no natural numbers satisfied $v^5 + w^5 + x^5 + y^5 = z^5$ lasted for nearly 200 years, until 1966 when a counterexample was found.[10] Analogously, the move from the regular polygon case to the irregular polygon case took hundreds of years to make. Though the theorem was known by Pappus for regular polygons, the irregular polygon version of the theorem was proven only recently by Hales.

A related reply is to insist that the weaker mathematical claims lead to explanations with less explanatory power than the explanations that use the stronger mathematical claims. This could be motivated by the view that the stronger mathematical claims allow for a more unified scientific theory. Note, though, that the weaker mathematical claims are able to cover all the actual instances of the phenomena at issue, for example, cicada, bees, and bridge systems. I do not believe that the ability to explain nonactual instances of these phenomena should heighten the explanatory power of these explanations. At a minimum, this puts the burden on Colyvan and Baker to further motivate this sort of approach to explanatory power.[11]

---

10. See Nahin (2006, p. 250).

11. See Weslake (2010) for an approach to explanation that may serve this purpose.

We can consider a second kind of case that sheds light on my second, sensitivity, problem. Suppose through the checking of cases or other minimal means that our scientist is able to convince herself of the truth of the weaker mathematical claims that I have noted. That is, she compares the periods of intersection for the primes and nonprimes from 1 to 100 and finds that prime periods minimize intersection. Similarly, she carries out Pappus's proof for regular polygons by first determining the three regular polygon tilings that are possible (triangle, square, and hexagon) and then calculating that the hexagonal tiling minimizes the perimeter.[12] Finally, she checks graphs with up to 10 vertices and concludes that Euler's theorem restricted to these graphs is correct. At this point I think we can make sense of how the mathematics contributes its explanatory power. Based on her prior belief in the mathematical claim, the scientist can use her understanding of the relevant mathematical structure to account for the explananda. This is because the mathematical claim provides the link between the feature and its reoccurrence in the systems in question. At the same time, the scientist's limited mathematical knowledge does not give her any reason to believe in any strong axioms for the domain in question. Extrapolating in mathematics is dangerous, and the explanatory contribution that a weak mathematical claim provides does nothing to support any further extrapolation.

My claim, then, is that the ways mathematics helps with scientific explanation are not sensitive enough to sort out the various options that agents must choose between when they need to decide which stronger mathematical claims to believe. The differences between these options are so fine grained that it would be unreasonable to base such a choice on their application in explanation. To make the problem as vivid as possible, I want to consider the choice faced by an agent who does not yet believe in any of the usual strong axioms for the natural numbers. For that person, it is an open epistemic possibility that there are only finitely many natural numbers. Thus, the choice is between believing that there are numbers 1 through n for some n or that the natural numbers continue indefinitely. In saying that both options are epistemically possible, I mean only that as far as the agent is concerned, either claim could be true. This point about epistemic possibility is consistent with one of these options being metaphysically necessary. In fact, I believe that it is metaphysically necessary that there are infinitely many natural numbers. But my point is that this belief is not justifiable based on the cicada explanation. The totality of the natural numbers does not manifest itself in this explanation. This explanatory contribution does not tell against any of the relevant epistemic possibilities.

These considerations suggest the following restriction on the application of IBE:

Sensitivity: A claim that appears in an explanation can receive support via IBE only when the explanatory contribution tells against some relevant alternative epistemic possibilities.

---

12. See Hales (2000, p. 448) or Adam (2003, pp. 231–234).

Explanation

This is not a causal or constitutive condition of the sort imagined by those who restrict IBE to concrete claims. For example, Melia allows explanatory arguments for quarks because

> By postulating few different fundamental kinds of objects and a few fundamental ways of arranging quarks, we can account for the existence of a wide range of apparently different kinds of objects simply, elegantly and economically. Moreover, this is a genuine ontological account—the complex objects *owe their existence* to these fundamental objects and their modes of recombination. (Melia 2000, p. 474)[13]

But I am not saying that IBE is applicable only when the claims are about the causes or parts of the phenomenon being explained. Tracking causes is one way to tell against alternative possibilities, but it is not the only way. For example, Woodward has noted the noncausal explanation that links the stability of the planetary orbits to the dimensions of space-time:

> it has been argued that the stability of planetary orbits depends (mathematically) on the dimensionality of the space-time in which they are situated: such orbits are stable in four-dimensional space-time but would be unstable in a five-dimensional space-time ... it seems implausible to interpret such derivations as telling us what will happen under interventions on the dimensionality of space-time. (Woodward 2003, p. 220)

In my view, someone who already believed the associated mathematical theorem could explain why the orbits are stable. This explanation depends on the claim that there are four dimensions to space-time. The dependence is not causal or constitutive on any of the usual ways of clarifying what this comes to. Nevertheless, an application of IBE to this case would give an agent a reason to believe that there are four dimensions to space-time. This is because the success of the explanation is sensitive to the dimensionality of space-time and undermines the dimensions besides four.[14]

How does our requirement that an application of IBE respect sensitivity fare when we consider its canonical application in support of the belief in atoms and electrons? As a number of historical studies have shown, the appearance of the claim that atoms exist in some explanation was not sufficient to convince the scientific community to believe in atoms. Scientists were clearly waiting for the right kind of explanation. For atoms this seems to have come in connection with Brownian motion. The way the existence of atoms manifested itself in this case was much more intimate than in previous explanations that involved atoms. It would be a

---

13. See also (Field 1980, p. 43) for Field's preference for intrinsic explanations.

14. See also the discussion of inference to the best explanation in chapter 11 for other nonmathematical claims that pass the sensitivity test.

delicate historical matter to pin down this difference in a definitive way. According to Mayo's detailed reconstruction, the crucial derivation linked the core features of atoms making up a fluid to the displacement of a Brownian particle immersed in that fluid (Mayo 1996, ch. 7).[15] The features of this representation placed very strong constraints on what the atoms making up the fluid had to be like because only atoms of a certain sort would give rise to this displacement function. When the Brownian particle's trajectories were matched with the predictions of the representation, an application of IBE was warranted because this evidence undermined the relevant epistemic alternatives. What scientists were waiting for was a phenomenon that depended on the central features of atoms in a transparent fashion. The crucial difference between the successful atoms case and the unsuccessful cicada case is that all the central properties of atoms are brought to bear in the explanation of Brownian motion while, as I have argued, very few features of the domain of natural numbers contribute to the cicada explanation. For atoms the properties with which scientists concerned themselves were their weights and the number of atoms in a given amount of material. The calculations relating to Brownian motion involved both features. By comparison, the earlier use of atoms in Dalton's law of proportions merely posited the existence of atoms in a given proportion as an attempt to explain some chemical phenomena. This involved the assumption that atoms had weights and that a given number of atoms could be found in a given amount of material, but it left these central features completely unspecified.[16] Similarly, the cicada explanation posits numbers and appeals to a distinction between primes and nonprimes. But it does nothing to circumscribe the domain of natural numbers by indicating how many natural numbers there are or which operations are defined on them. Another historical case concerns the existence of electrons, thought of as particles with a minimal negative charge. Prior to Thomson's and Millikan's experiments, the value of this charge and the charge to mass ratio were not relevant to explanations that involved the posited electrons. But once an explanation was developed that took advantage of these particular values, our restricted form of IBE found a convincing application and the claim that electrons exist could be accepted.

The advocate of the explanatory indispensability argument needs something like this to provide support for their premise 2_{ER} if they are to avoid begging the question in favor of 3_{ER}. I am not optimistic that this sort of case will be found. The reason is quite simple: the mathematical domains in question are highly structured, and there is unlikely to be an explanation exploiting this specific structure in a way that is sensitive enough to tell in favor of the truth of claims that are substantial enough to single out this structure. Instead, on the account of mathematical explanation I have developed, scientists first justify the relevant mathematics by mathematical means and then use this mathematics to explain scientific phenomena. So we

---

15. See also Chandrasekhar (1943, §2), noted at Mayo (1996, p. 229).

16. Here I also draw on Maddy's discussion in Maddy (1998), although she does not take this lesson from her discussion.

see that mathematical explanations are central to the success of science, but these explanations do not provide us with the evidence we have for the truth of these mathematical claims.

10.3 BELIEF AND UNDERSTANDING

The main claim of the last section was that mathematical claims must be justified prior to their use in explanation if they are to make their explanatory contribution. More generally, I would argue that a scientist must assign a high degree of confirmation to a mathematical claim if that claim is to make any of the epistemic contributions to the success of science that I have emphasized. The reason for this goes back to our characterization of the content of these mathematical scientific representations. Recall that the content involved the existence of a structural relation of this or that sort between a specified mathematical structure and the target system. So, for an agent to *understand* this sort of representation, he or she must *believe* that the claims describing this structure are true. The process of prediction and testing operates only on those representations that a scientist can understand. If this understanding really presupposes that the agent already believes that the mathematical claims are true, then this process of testing cannot be the main source of justification for these beliefs.

One way to resist this conclusion is to opt for what I call fictionalism. This is an alternative way to characterize the content of our scientific representations. Crucially, it insists that there is no genuine subject matter for mathematical claims to be about. So, any claims that imply the existence of any mathematical entities are false. This is a different position from Lewis's or Hellman's, according to which the objective truth-values of mathematical claims do not require the existence of any mathematical entities. Instead, for the fictionalist, the only sense in which some mathematical claims come out true is as a result of stipulations concerning what is "true in the story" of mathematics. The fictionalist, then, aims to make sense of mathematical practice using some analogy with some antiplatonist theories of literary fiction.

The fictionalist objects to my claim that an understanding of a scientific representation requires a belief in some mathematical statements by insisting that an agent must merely understand these mathematical statements to understand the scientific representation. In virtue of understanding these statements, they come to see what is "true in the story" of that domain of mathematics. This is supposed to be sufficient to underwrite an understanding of how a target system would have to be for that representation to be an accurate reflection of the features of the system. If this sort of fictionalist position could be defended, then it would spell the end to all of the indispensability arguments reviewed here, for the fictionalist explains how mathematics can make a contribution to science without bringing in the actual truth of any mathematical claims. We turn to a review of this position and its problems in chapter 12. I argue that the fictionalist is not able to deliver an acceptable account of the content of our scientific representations.

A second alternative approach is to insist that we need merely believe the mathematical claims in question, but we are free to do so without any further justification, at least on a temporary basis. Considering this alternative returns us to the issues pursued in chapter 6. There I argued that constitutive representations must be believed for their derivative representations to be confirmed by the ordinary process of testing. A crucial step in this argument was that prior to believing the constitutive representations, it was possible for an agent to realize that this sort of presupposition relation obtained between the constitutive representation and the associated derivative representations. For example, prior to coming to believe Newton's laws of motion, an agent can understand that these laws must be believed for an "empirical law" like the law of universal gravitation to ever be empirically confirmed. This is what gives the agent the practical reason to tentatively adopt the laws of motion in the service of the overall scientific goal of arriving at true beliefs about this domain.

Now we are in a better position to see where the mathematics involved in these constitutive representations figures in and how this provides a response to the confirmational holist. I have just claimed that agents must believe some mathematical claims if they are to understand our mathematical scientific representations. These representations include the constitutive representations. So, for the mathematics involved in these constitutive representations, an agent must justify his or her belief in these mathematical claims for the whole process of testing to get off the ground. Once these mathematical claims are believed, then the agent can appreciate the need to believe some constitutive representations. This justifies the practical decision to tentatively adopt some set of constitutive representations and begin the process of testing the derivative representations.

At this point the holist will of course reply that there is no need for agents to justify their mathematical beliefs before starting this process. It seems equally possible that agents can tentatively adopt the mathematical beliefs to start based on no justification at all. Then, on the basis of their belief in these claims, they will be in a position to appreciate the links between a set of constitutive representations and the confirmation of some derivative representations. After the decision to believe some mathematical claims, the whole story proceeds just as I have presented it. The holist adds that the success of the entire package of beliefs is sufficient to ultimately justify the beliefs in the mathematical claims. But this just shows that there is no need to justify a belief in a mathematical claim prior to its use in science.

My objection to this proposal is based on an appeal to how science is practiced. I maintain that the holist picture of science is in conflict with how science is actually done. This argument comes in two parts. In the first part I pinpoint the conflict based on the ways scientists deal with failed predictions or explanations. Then I argue that the holist's account of this reaction is not adequate. To start, let us bracket out the disputed distinction between constitutive and derivative representations and assume that a scientific claim is simply tested by conjoining it with other claims. The holist admits that these claims include mathematical claims, but insists that we need not have any prior justification for our acceptance of these claims. We then engage in the usual process of prediction or explanation and see how our whole package of beliefs fares. The problem I want to raise develops when this package fails to yield accurate predictions or the best explanation. Given that many (or maybe even all) of the mathematical beliefs were adopted without any prior justification, the holist then allows that it is reasonable to reject these mathematical beliefs. But this allowance conflicts with how scientists actually proceed.[17] Except in those cases where a flaw can be discerned in the mathematics by ordinary mathematical means, the scientist will blame the nonmathematical beliefs for the failure of their package of beliefs. The mathematics is treated like a "fixed point" in a way that goes beyond what the holist can make sense of. This conflict shows that scientists treat the mathematics in their package of beliefs as already largely justified by ordinary mathematical means.

Here, then, is a way the prior justification of mathematical claims manifests itself in scientific practice. There is an asymmetry in the way these claims are treated when compared to their nonmathematical companions in the package of beliefs. What can the holist say to account for this asymmetry? It is common to talk about the central place of mathematical beliefs in our "web of belief." But when the metaphorical character of this point is stripped away, all that is left is the descriptive claim that scientists are reluctant to give up their mathematical beliefs. Some holists make the further claim that this serves some broader "maxim of minimal mutilation." Unfortunately, no holist has made this point precise in a convincing way. Although mathematics as a whole does reach widely throughout our package of beliefs, that feature is not relevant to the sort of choice we are considering. Only a small family of mathematical claims is relevant to a given failed prediction or explanation, and it is not clear why adjusting these mathematical claims would affect things more widely than some adjustment in our nonmathematical claims. So, it seems that the holist is not able to resolve the conflict between scientific practice and what their account of science would predict.

The same problem with the holist proposal becomes clear if we allow the distinction between constitutive and derivative representations. For the many cases in which a given package of constitutive representations is judged to fail, there is no case where the mathematics itself was rejected. But when constitutive representations fail, the agent imagined by the holist should be in doubt about whether she should blame her mathematical beliefs or her other beliefs summarized by the constitutive representations. As the holist imagines that the only source of justification for mathematical claims is their role in a successful package of beliefs, there is no other avenue for determining when to blame the mathematical beliefs. This sort of standoff fails to fit with the actual practice of science. Even with the failed constitutive representations of catastrophe theory discussed in chapter 7, the mathematics remains intact and important for other sorts of applications.[18] The holist has no explanation for this central feature of the history of science.

If these objections are right, then there must be a way for scientists to confirm their mathematical beliefs prior to their use in science. In chapter 14 I explore some possibilities for grounding this sort of justification, but eventually conclude that

---

17. See Busch (2010) for more discussion of this point. I do not believe Busch would endorse the argument I am developing here.

18. These include some features of the rainbow discussed in chapter 11.

we currently lack a compelling explanation of how it is possible. This is perhaps a disappointing result and explains the hope that some kind of indispensability argument could do better. I have argued, though, that these hopes are misplaced. Once we recognize the intricate nature of the structures studied in mathematics, the sensitivity problem looms large. But given the central role of mathematics in determining the contents of our scientific representations and the resulting epistemic contributions, there is little hope of doing science without already having a reason to believe the mathematics.

11

The Rainbow

11.1 ASYMPTOTIC EXPLANATION

In the philosophy of science there is another debate about mathematical explanation in science that has been initiated by Batterman's work on asymptotic explanation. Here the question is what the explanatory contributions from mathematics implies for the proper interpretation of our scientific representations. Batterman has argued that these explanatory contributions, especially in connection with the sort of scaling techniques reviewed in chapter 5, show that some representations cannot be reduced to the more fundamental theories of modern physics. This suggests that these theories are unable to fix our beliefs about the systems being represented, and so there is a role for nonfundamental theories in correctly representing how the world is. Although Batterman is often focused on the relationship between quantum mechanics and classical mechanics, we discuss only an example from within the domain of classical mechanics. This case involves the best explanation of three aspects of the rainbow. I argue that Batterman is basically correct to maintain that this case involves explanations that make use of concepts originating with the ray theory and that are unavailable from within the fundamental wave theory. However, I pin down what this tells us about the rainbow in what I believe is a clearer way than what Batterman has achieved to date. This will show, among other things, what our mathematical contributions achieve in terms of fixing the best interpretations of our most successful representations. These interpretations will result from an application of inference to the best explanation (IBE). The cases we discuss in this chapter involve the third type of mathematical explanation noted in the last chapter: connecting different phenomena using mathematical analogies. If the arguments of this chapter are successful, we will see how mathematical explanations can contribute to the proper interpretation of our representations. This is possible even though the explanatory indispensability arguments were deemed a failure.

I clarify the role of asymptotic explanation using a case that is relatively free of interpretive controversy. This is the rainbow. As Batterman has put it, "the problem of explaining the basic features of rainbows is less encumbered with philosophical puzzles than is the interpretation of quantum mechanics" (Batterman 1997, p. 395). This is due, at least in part, to our comfort with the most fundamental theory involved in this explanation. This is classical electrodynamics and the associated conception of light as an electromagnetic wave. But Batterman has argued that our best representation of the rainbow also deploys the ray representation, which treats light in terms of geometric lines. The ray representation can be seen to result from the wave representation through the operation of taking a limit. On a first pass, we can say that the rays of the ray theory result from the waves of the wave theory when the wavelength is taken to 0.[1] Following Batterman, we will see to what extent the ray representation is able to contribute to our best explanation of certain features of the rainbow. This is initially puzzling because it is hard to see how the clearly false claim that light travels along rays can help explain aspects of the rainbow. A further important part of the rainbow case is that some features of the rainbow are explained by combining elements from the ray representation and the wave representation. This involves what can be called "intermediate asymptotics" and is relevant to our best explanation of the existence and spacing of the part of the rainbow known as supernumerary bows. An appeal must be made to what happens "on the way" to the limit, even though the limit operation does not correspond to any physical process or aspect of the rainbow as it really is. The mathematics of the representation allows considerable explanatory insight based on an appeal to "structures" whose special status seems to depend on both the ray and wave representations.

Batterman has summarized the situation by saying, "It seems reasonable to consider these asymptotically emergent structures to constitute the ontology of an explanatory 'theory,' the characterization of which depends essentially on asymptotic analysis and the interpretation of the results. This full characterization must make essential reference to features of both the ray and wave theories" (Batterman 2002b, p. 96). But both sympathizers and critics have not agreed on what Batterman takes the interpretative significance of this "theory" to be. To help clarify the situation, I make use of the distinction between representation and theory that Batterman does not employ.[2] On this approach, the theories we accept tell us what a system is made up of and how these parts interact. I argue that even though we use representations that were originally derived from the ray theory, their role in our best explanations need not lead to an acceptance of the ray theory. Instead, the only theory we accept is the wave theory. But this is consistent with an important role for the ray representation and its interpretation in shaping our beliefs about the rainbow. I believe this conclusion is completely in the spirit of Batterman's many remarks on the interpretive implications of asymptotic reasoning. So I do not intend my discussion here as a criticism of his views. At the most, what I argue is that Batterman's views are not as clear as they should be, and this has hampered the appreciation of the significance of cases like the rainbow.

A further benefit of this discussion is that it will highlight how important the prior justification of the mathematics is to these sorts of explanations. At several stages it will become clear that only because we already have a reason to accept a mathematical claim can it contribute to a successful scientific explanation. Without this confidence, we could not end up with the well-motivated physical interpretation

---

1. We return to the question of whether this is the best way to think about the relationship between these two representations.

2. See section 2.1.

The Rainbow

of the mathematics. We will see how central this sort of interpretation is to a correct understanding of this type of explanation.

An important conclusion of this chapter is that it is helpful to distinguish two different explanatory benefits associated with asymptotic reasoning. The first benefit is that such reasoning allows us to ignore details that are not relevant to the phenomenon being explained. This sort of contribution obtains for many ways of deploying mathematics in scientific reasoning and has nothing specific to do with the taking of limits. We have already seen cases of this sort with our explanations of the cicada, bees, and bridges of Königsberg. The second benefit can be more directly tied to asymptotic reasoning and can also be used to help motivate Batterman's emphasis on the existence of the singularities that sometimes arise when limits are taken. This benefit is that the representations that result from such limits must be given a different physical interpretation than the original representations. This marks an important difference from the case of wave dispersion discussed in chapter 5. There the representation that resulted from taking the limit did not require a new interpretation. Now, as we will see, the singular character of the limit can be linked to the need to offer an interpretation in terms of different physical concepts. The challenge then becomes to say how different this physical interpretation must be and what consequences this has for our overall conception of the phenomenon being represented.

I argue that in the rainbow case, at least, the strong interpretive conclusions that Batterman has suggested should be weakened. At the end of the day, there is no role for the ray theory in shaping our beliefs about the rainbow. Instead, we use the properly interpreted ray representation to help us interpret other representations that arise in the course of asymptotic analysis. As I put it, we use the results of one idealization to inform the proper interpretation of another idealization. This shows how concepts that were originally deployed in the context of the ray theory are relevant to our best explanations. But it does not show that we need to accept any aspects of the traditional ray theory when we consider the rainbow. Throughout we will see how the account of idealization developed earlier in this book is crucial to avoiding interpretive mystery. Again, I emphasize that I take my exposition of the details of the case to be primarily a clarification of what Batterman has offered, and not a criticism. But in the concluding section I argue that Batterman's suggestion that some of our explanations are "theory laden" (Batterman 2005, p. 159) with respect to the ray theory is misleading.

11.2 ANGLE AND COLOR

Rainbows can be initially characterized in terms of the following observable features. They appear when an observer looks towards rain with the sun behind her. Rainbows involve a relatively bright band centered on a 42° angle above the direction of sunlight. An initially puzzling feature is that this angle remains the same as the observer moves toward or away from the rainbow. The band itself is divided into colors with red, orange, and yellow appearing at the top followed down through the spectrum to blue and violet at the bottom (see figure 11.1). Finally, in the

Figure 11.1: A Rainbow

blue-violet region some rainbows exhibit dark bands. The spacings of light and dark in this region of the rainbow are the supernumerary bows, whereas the rainbow itself can be characterized as the primary bow.[3] Given these many features of a rainbow and the reoccurrence of the rainbow across various conditions, there are many different explanatory questions that a scientist can ask. I will focus on three: (a) Why does the rainbow appear at 42° relative to the direction of sunlight, when it does appear? (b) Why does the color pattern of the rainbow appear as red on top through violet at the bottom? (c) Why do dark lines appear at the bottom of the rainbow in the blue-violet region with this or that pattern of spacing? These explanations involve different relationships between our fundamental wave theory and the ray theory.

A purely ray theoretic explanation seems available for a.[4] It involves the idea that the rainbow results from light rays passing through individual raindrops. Crucially, we can account for the 42° angle if we assume that the raindrops are spherical. Based on a mathematical argument to be given shortly, the spherical shape of the raindrop has the effect of focusing the sunlight so that much of the light that hits each drop is directed backward and downward toward the observer. The perceived light at 42° thus corresponds to light that has undergone a minimal total deflection of 138°. This "rainbow angle" θ_R obtains for each spherical raindrop, independently of its size, and so the accumulated effect is the band of light perceived by the observer.

The main physical assumptions underlying this explanation of a are Snel's law concerning the refraction of light as it passes from one medium to another and the claim that the angle of reflection equals the angle of incidence when light is reflected.

---

3. Considerations of space preclude me from discussing the secondary bow, which sometimes appears above the primary bow.

4. My exposition is based mainly on Nahin (2004). See also Adam (2003) and Nussenzveig (1977).

The Rainbow

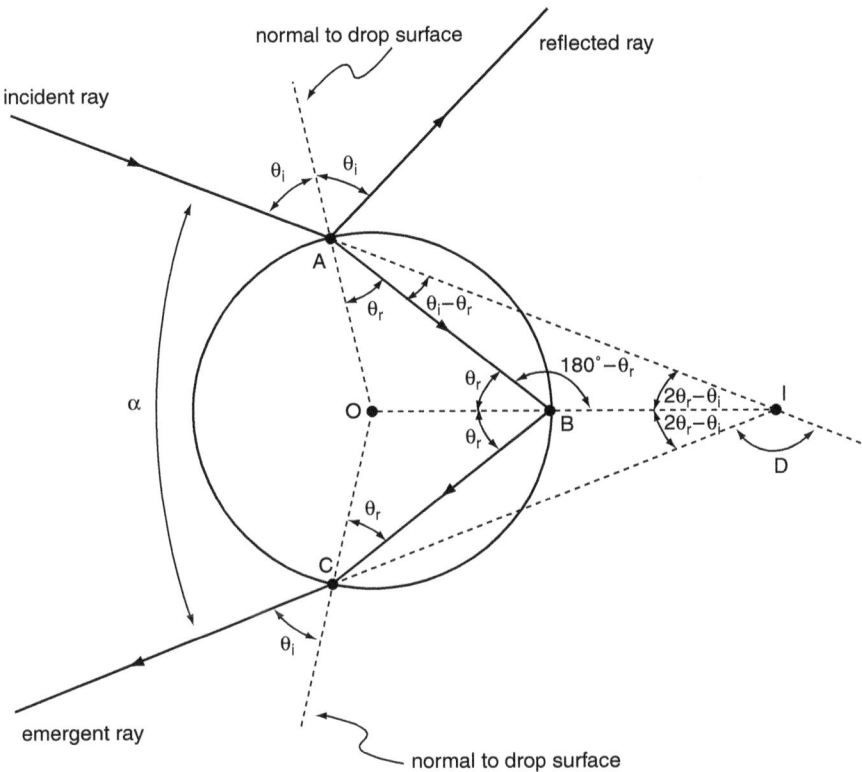

Figure 11.2: Ray Path

Source: Nahin (2004, p. 181)

What we want to calculate is how the deflection of a ray of light from the sun changes as we consider rays that strike the drop at different angles. As figure 11.2 shows, a path that involves an initial refraction, followed by a reflection on the back of the drop, followed by a second refraction, can lead to a ray that will be deflected backward and downward toward the observer. The angles in question here are the initial angle of incidence θ_i and the initial angle of refraction θ_r. Snel's law and the geometry of the circle allow us to determine the total angle of deflection D using just θ_i:

$$D = 180° + 2\theta_i - 4\sin^{-1}\left(\frac{1}{n}\sin\theta_i\right) \tag{11.1}$$

The basic idea of our explanation of α is that 11.1 tells us that much of the sunlight that hits the raindrop at angles $0° < \theta_i < 90°$ will have D at or close to $D = 138°$. This corresponds to the observed angle of $\alpha = 42° = 180° - D$ that we are trying to account for. This proposal can be made more precise by noting that D obtains a minimum, and so α obtains a maximum, as we move from $\theta_i = 0°$ to $\theta_i = 90°$ when $D = 138°$. Furthermore, this minimum corresponds to a large amount of light because many θ_i lead to this value for D or something close to it. $42°$ is the maximum value for α, so according to this explanation no sunlight from these rays appears above the primary rainbow.

So far no mention has been made of the colors of the rainbow. That means that our explanation of a is clearly not an explanation of b. But a small addition to our account of where the rainbow appears is sufficient to rectify this gap. In the foregoing argument, we treated only the case where $n = \frac{4}{3}$. Treating n as a constant is sufficient to explain where the rainbow appears, but a more accurate perspective on n is that it is a variable that depends not only on the media involved but also on the color of light undergoing the refraction. At one end of the visible spectrum, the value of n for red light is roughly 1.3318, whereas for violet light at the other end of the spectrum, n is approximately 1.3435. This slight change in n can be used to account for the angular spread between the colors of the rainbow and their order. For example, our new value for n for red light yields a value of D to be $137.75°$, and with the new value of n for violet D becomes $139.42°$. As a result, α is greater for red than for violet and the red band appears above the violet band in the sky (Adam 2003, p. 89).

Though this is clearly *an* explanation for b that can be easily combined with our explanation of a, it is not clear that this is the *best* explanation of b. Furthermore, when we press on this explanation for b and see an explanation that has a better claim on being the best explanation for b, we will encounter the worry that our explanation of a is problematic as well. To see the limitations of our explanation of b notice that we have treated the link between color and n as a series of brute facts. Here an appeal to the wave theory of light promises to do better. Once we make the link between the decreasing wavelength of light and the change in color from red through the visible spectrum to violet, we can account for the change in n. According to the wave theory of light, what is going on here is the more general phenomenon of light dispersion, where the behavior of light can be affected by the wavelength of the light wave. A beam of white light is then represented as a superposition of waves of various wavelengths that are separated as each component wave is refracted at a slightly different angle. Going further, we can use this wave representation to account for the physical claims that the ray representation takes for granted. These claims include Snel's law and the law of reflection.

On this view the best explanation of b takes light to be made up of electromagnetic waves. This is the best way to make sense of the colors of light and how the colors come to arrange themselves in the characteristic pattern displayed by the rainbow. Now, though, we have a reason to doubt the explanation we have given for a because it presented light as traveling along geometric rays. Without some relationship to our wave representation of light, this account of a seems to float free of anything we should take seriously. This point raises an instance of the central question of this chapter: how can we combine the resources of the clearly incorrect ray representation with the correct wave representation to provide our best explanations of features of the rainbow? The answer I suggest follows immediately from our earlier discussions of idealization: the ray representation we have deployed need not require the assumption that light is a ray. It may assume only that in certain circumstances *some features* of light can be accurately represented using the ray representation. But what are those circumstances, and how do they relate to these features? A first attempt to answer this question might draw attention to one of the mathematical relationships between the two representations. This is that the ray representation results from the wave representation when the wavelength is taken to 0. Another way to put this makes use of the wave number k, which is equal to 2π divided by the wavelength. We then can put this limit in terms of the wave number going to infinity. This leads to a representation in which the wave-like behavior of light is completely suppressed. We can then track the path of the light using geometric lines and deploy simple rules like the law of reflection. The lines result from connecting the crests of the light waves. As the wavelength is decreased, these crests get closer and closer and at the limit produce a line.

The problem with this first attempt is that this sort of relationship between the wave representation and the ray representation would also block any ray-theoretic representation of the connection between color and the index of refraction n. For if we interpret the scope of our ray representation as going along with the $k \to \infty$ limit, then we have erased the very variations in the light waves that are responsible for color and the changes in n. It is hardly coherent to say that if the wave number is close to infinity, then we can offer our explanations of a and b. To start with, the wave number is not close to infinity in any absolute sense. More important, the differences between the wave numbers of red and violet light are manifestly part of the explanation for the colors of the rainbow. So we cannot defend the view that it is correct to assume that the wave number is infinity even if this would render the links between the ray and wave representations particularly transparent.

The lesson we should draw from the failure of the first attempt is that we need to pay attention to both the mathematical links between the two representations and their proper physical interpretation. Some mathematical links preclude any viable interpretation. This fits with the way singular perturbation theory was used to develop the boundary layer theory representation in chapter 5. There we saw that the width of the boundary layer posed an interpretive challenge based on its central place in the representation. My proposal here is that the circumstances under which our ray representation can be used to provide explanations of a and b should be specified in terms of a dimensionless parameter known as the size parameter β. β is the product of the wave number k with the radius a of the raindrop. It is a dimensionless quantity because its value is independent of the units used to represent k and a. On this view, the ray representation results from the wave representation when the product $\beta = ka$ is taken to infinity.[5] This ray representation can be reliably used if there is a size of raindrop above which the wave-theoretic aspects of light are not relevant to the path of the light through the drop. If the raindrops have a circumference of roughly 1 mm, then the product in question here is about 5000 (Nussenzveig 1977, p. 124). This makes it possible to treat the paths of light as straight lines and draw on the link between the wave number and the index of refraction. Even though we have assumed that $\beta \to \infty$, this assumption is consistent with appealing to the values of k that are responsible for the dispersion of the light. Analogously, we do

---

5. When he is being careful Batterman will also invoke this parameter, and not simply the wavelength. See, for example, Batterman (2005, p. 154) and Batterman (2010, p. 21, n. 23).

not represent the wave crests as forming a continuous straight line, but only claim that the distance between crests is so small with respect to the radius of the drops that it is not relevant to the path of the wave.

11.3 EXPLANATORY POWER

The conclusion of the previous section is that we can account for our best explanations of a and b by linking the wave representation to the ray representation using a limit specified in terms of the size parameter β. There are two ways this two-legged approach provides superior explanations to a one-leg approach that tried to get by with just the wave representation by itself. The most obvious advantage is that we see that many detailed wave representations of a specific rainbow will be related to a single ray representation. This is because the limiting relationship drops many details that are not relevant to the features of the rainbow we are trying to account for. Our best explanations of a and b aim to account for features that all rainbows have in common, and we come to understand how these features arise independently of many features that would be included in a viable wave representation of a particular rainbow. Here the parameter used to link the wave representation to the ray representation can be seen to have special significance. The successful explanation, which results from taking $\beta \to \infty$, shows that the size of this parameter is crucial to these features. When this parameter fails to be large, we can then also explain why these features fail to materialize. For example, if the raindrops are too small, then a rainbow will not be observed, or its colors will be distorted.

The two-legged explanation gains part of its explanatory power from its function of unifying our description of the different rainbows at the right level of abstraction. This contribution to explanatory power can be tied directly to the mathematical operation that links the wave representation to the ray representation. But it would be a misunderstanding of this sort of contribution to think that a unified description at the right level of abstraction must result, or even usually results, from applying these sorts of limits. Another means to obtain just this sort of explanatory contribution from mathematics is to attempt to describe a phenomenon directly at the right level of abstraction. This is what we saw with the cicada, bees, and bridges cases and that type of explanation more generally. So this sort of explanatory contribution from mathematics need not involve limits for the simple reason that we may not articulate two different representations of the sort deployed in the rainbow case.

Still, there is a second sort of mathematical contribution to the explanatory power of our explanations of a and b that can be more directly tied to the presence of our limit operation. This is the way the mathematical character of the limit operation guides the physical interpretation of the ray representation. If we consider the ray representation independently of its relationship to the wave representation, then we may be tempted to assign physical significance to its parts in line with the traditional ray theory of light. But when we view the ray representation as the product of our limit operation, we see that this traditional interpretation is problematic. To take a feature of the rainbow that will assume greater significance shortly, the relative intensity of the light that makes up the parts of the rainbow is not accurately captured by our ray representation. We see this when we consider what happens to the wave representation's way of handling intensity when the limit is applied. The intensity is calculated by squaring the amplitude of a light wave. But there is no correlate in the ray representation to the intensity of a light wave in the wave representation. This problem is not really avoided by the change to the size parameter β. This shows how taking the limit operation can have positive as well as negative effects. The positive effect is that we erase the representation of irrelevant details, while a negative effect is that we erase the means to represent the intensity of a ray of light using the link to the wave representation.

Limit operations force us to withdraw the physical interpretation from some parts of the representation that results. In this respect they are like many other techniques of idealization and simplification. The need to withdraw an interpretation should be obvious for the parameter β. If we take it to infinity to derive our ray representation, then we cannot make use of its actual finite values when we interpret the ray representation. But sometimes other aspects of the representation will be affected in a more subtle way so that some care is needed to ascertain which parts of the resulting representation retain their original significance. This is especially important to keep in mind when the limiting operation produces singularities, that is, quantities that diverge or become otherwise undefined.

How, though, does this sort of interpretive shift contribute to explanatory power? My suggestion is that the interpretive flexibility necessary to handle these shifts helps us explain *other* aspects of the rainbow that would otherwise remain inexplicable. This comes out clearly in our best explanation of c, the existence and spacing of the supernumerary bows. As I summarize in the next section, the best explanation of c is obtained by combining the wave representation and the ray representation in a more nuanced way than what was required for our explanation of a and b. Roughly, the best explanation of c will result from using information from the ray representation to guide the interpretation of a new, third representation. This guidance seems to draw on the novel interpretations that we assign to the ray representation over and above what is grounded in the wave representation. So, it seems that the ray representation makes a genuine contribution to the explanation. The central issue, though, is what interpretive significance this has. After considering the explanation of c, we turn to Batterman's and Belot's contrasting views on its significance for our beliefs about the rainbow.

11.4 SUPERNUMERARY BOWS

The best explanation of c the existence and spacing of the supernumerary bows (see figure 11.3) involves both the wave theory and the ray theory in a more intricate way than what we saw in the case of a and b.[6] The supernumerary bows result from the constructive and destructive interference of the light waves that travel through a raindrop. Although this satisfactory qualitative explanation was arrived at in the

---

6. My main sources are Nussenzveig (1992) and Adam (2002).

Figure 11.3: Rainbow with Supernumerary Bows

nineteenth century, as late as 1957 the scientist van de Hulst could complain that a "quantitative theory of the rainbow for the broad range of size parameters occurring in nature was still lacking" (Nussenzveig 1992, p. 29).[7] This was despite the fact that a representation was formulated in 1908 by Mie which described the situation adequately in terms of electromagnetic theory. The problem with the Mie representation is that it involves an infinite series of terms that converges very slowly. The development of computers allowed a reliable estimate of the values of this sum for different parameters, but it was widely felt that this sort of brute force calculation was inconsistent with genuine scientific understanding. As Nussenzveig puts it, "a computer can only calculate numerical solutions: it offers no insight into the physics of the rainbow" (Nussenzveig 1977, p. 125).[8] An adequate explanation was achieved only when Nussenzveig and his collaborators applied sophisticated mathematical techniques that transformed the slowly converging series of the Mie representation into a more transparent representation. The resulting "complex angular momentum theory," or CAM, provides the best explanation of the existence and spacing of the supernumerary bows.[9] My purpose here, then, is to consider what interpretive significance this explanation has. In particular, given the acceptance of some form of IBE, what should we infer about rainbows given this explanation? My ultimate conclusion is that the mathematical techniques of CAM are central to the superior explanatory power of the explanation because the techniques depend on a particular physical interpretation of the mathematics. CAM is thus best seen as radically

---

7. See van de Hulst (1957, p. 249).

8. Given at Batterman (1997, p. 407).

9. Although CAM is sometimes referred to as "CAM theory," I drop the word *theory* to avoid confusing CAM with the restricted sense of *theory* defined earlier.

different from a more straightforward numerical calculation. The "insight into the physics of the rainbow" which results has implications for what we should believe about the rainbow. At the same time, it does not provide us with a distinct theory that should shape our beliefs about the constituents and interactions giving rise to the phenomenon. The only genuine theory we are left with is the electromagnetic theory.

The explanation offered by CAM is obtained from the Mie representation in three steps. The Mie representation uses electromagnetic theory to describe what happens when monochromatic light, represented as a plane electromagnetic wave, is scattered by a spherical, homogenous drop of water.[10] The central parameter of the problem is the size parameter β encountered earlier. A further familiar parameter is n, the index of refraction. For the air to water case, the value typically employed is 4/3. The problematic part of the Mie approach is the representation of the *scattering amplitudes* at a given angle θ. The amplitude is given by two terms $S_j(\beta, \theta)$ $(j = 1, 2)$ corresponding to two perpendicular directions of polarization. Each $S_j(\beta, \theta)$ term is expressed as an infinite sum of terms for $l = 1$ to $l = \infty$ that involve what are called partial waves $S_l(\beta)$ as well as p_l terms, which are defined using the Legendre function. The partial waves $S_l(\beta)$ are identified with a complicated fraction involving Bessel and Hankel functions (Nussenzveig 1992, p. 39).[11] A reliable numerical calculation of the S_j functions requires an evaluation of around $\beta + 4\beta^{\frac{1}{3}} + 2$ terms (Nussenzveig 1992, p. 40). For an ordinary rainbow, β may exceed 4500, so more than 4500 terms may need to be considered before the calculation is terminated. Accurate predications concerning the supernumerary bows result, but the explanatory power of this calculation is limited.

The first step away from the Mie representation is to transform the infinite sums for each $S_j(\beta, \theta)$ into what is known as a Debye expansion. This step can be motivated by considering the account of the rainbow offered by the ray representation. The light that we see results from an initial refraction, followed by an internal reflection and then by a second refraction. But some light from the initial refraction will be reflected, and some light at the internal reflection point will be refracted out of the raindrop. We can track these interactions by considering the reflection coefficients R and the transmission coefficients T. This allows us to rewrite a partial wave term $S_l(\beta)$ as a sum of terms with index $p = 0$ to $p = \infty$. p here indicates how many internal reflections are being considered:

$$S_l(\beta) = S_{l,0}(\beta) + S_{l,1}(\beta) + S_{l,2}(\beta) + \Delta S_{l,P}(\beta) \tag{11.2}$$

$S_{l,0}(\beta)$ tracks how much light from this term will be directly reflected, while $S_{l,1}(\beta)$ captures the light that is directly transmitted through the drop. Crucially, $S_{l,2}(\beta)$ tracks the light that is reflected internally once. By appeal to the ray representation, we know that this is the light that makes up the rainbow. The remaining terms are

---

10. A steady-state approach is deployed here in line with some of the techniques discussed in chapter 3.

11. These are some of the so-called special functions. See Batterman (2007) for some discussion.

grouped together as $\Delta S_{l,P}(\beta)$. They are not relevant for the rainbow, but could be examined in more detail for other phenomena where they become important.

Substituting the new representation of the partial waves into the sum provided by the Mie representation yields the Debye expansion of the total scattering amplitudes (Nussenzveig 1992, p. 92):

$$S_j(\beta,\theta) = S_{j,0}(\beta,\theta) + S_{j,1}(\beta,\theta) + S_{j,2}(\beta,\theta) + \Delta S_{j,P}(\beta,\theta) \qquad (11.3)$$

The values for the transmission and reflection coefficients for water and air show that nearly all of the light which hits the drop is found in the first three terms of this Debye expansion.[12]

This is some progress, but much of the complexity of the original Mie representation is now simply packed into the evaluation of the first three terms of the Debye expansion. The way forward is to apply our second step, namely, to make an appeal to the Poisson sum formula. This formula allows us to identify a sum of infinitely many terms with a sum of terms involving an integral (Nussenzveig 1992, p. 45):

$$\sum_{l=0}^{\infty} \phi\left(l+\frac{1}{2},x\right) = \sum_{m=-\infty}^{\infty} (-)^m \int_0^{\infty} \phi(\lambda,x)\exp(2im\pi\lambda)d\lambda \qquad (11.4)$$

Crucially, we require that the ϕ on the right-hand side be a complex-valued function that agrees with the real-valued function we started with on the left-hand side when $\lambda = l + \frac{1}{2}$ ($l = 0, 1, 2, \ldots$). Applying the Poisson sum formula to the first three terms of our Debye expansion might seem to make the problem even worse, as we are introducing a complex-valued function and are now required to evaluate an integral involving that function. It turns out, though, that it can be much easier to evaluate an integral involving this sort of complex-valued function. The main reason for this is that there is considerable freedom in changing the path along which we integrate our function. This freedom allows us to choose a path where the dominant contributions to the integral become manifest.

The use of the Poisson sum formula and the consequent use of complex analysis to evaluate the terms of the Debye expansion makes clear what the word *complex* refers to in the complex angular momentum theory. But the interpretation of λ as some kind of angular momentum remains opaque and until it is clarified the whole procedure risks collapsing into some form of numerical calculation. The link is made by a claim known as the localization principle. It associates each value of l in the Mie representation sum with a ray with the impact parameter $b_l = (l+\frac{1}{2})/k$.[13] Each impact parameter is the distance from an axis that passes through the center of the drop (figure 11.4). So, when $b_l < a$ the light will hit the drop. By analogy with similar "scattering" problems in particle physics, we can treat the way the impact parameter leads to a given deflection angle θ on the model of a particle like an

---

12. Nussenzveig gives the amount as 98.5% (Nussenzveig 1992, p. 95).

13. Nussenzveig (1992, p. 8). The principle's use in optics is summarized at van de Hulst (1957, pp. 208–209).

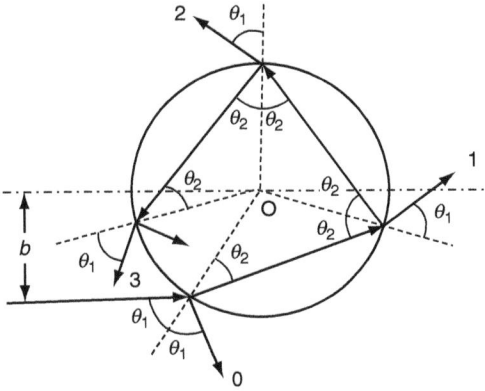

Figure 11.4: Impact Parameter

Source: Nussenzveig (1992, p. 16)

electron being shot through a spherical force field. This makes it useful to think of λ in terms of the angular momenta of the components of the light that hit the drop.

The third step of our explanation involves evaluating the three terms of the Debye expansion that have been transformed by the Poisson sum. For ease of exposition I consider only the third term $S_{j,2}(\beta, \theta)$ corresponding to one internal reflection, but the same techniques can be used to evaluate the other terms. Here we have an integral involving the complex function $S(\lambda, \beta)$ which extends the real function $S(l, \beta)$. The basic idea of evaluating these integrals is that we need only consider the contributions from critical points. These critical points come in two kinds. First, there are *saddle points* where the first derivative of S with respect to λ is 0.[14] Second, there are *poles* where S lacks a derivative of some order. In the simplest kind of case, as with the function $\frac{1}{z}$ at the point $z = 0$, the function is undefined. To evaluate an integral with a complex function it is often sufficient to know the saddle points and the poles. The trick is to consider a path of integration that departs from the real axis but goes through the saddle points, as in figure 11.5.[15] Drawing on these facts, we can then see how the value of the integral changes as we vary θ by considering how the saddle points change. Unsurprisingly, an important change occurs when θ is close to θ_R, the rainbow angle. When θ is π radians or 180°, there is a single saddle point at $\lambda = 0$. As θ decreases, the saddle point moves to the right along the real axis. There is then a critical value θ_L at which a second saddle point appears even further down the real axis at $\lambda = \beta$. Finally as θ gets closer to θ_R, these two saddle points approach closer to each other. At $\theta = \theta_R$ the two saddle points merge.[16] If θ is decreased further, then the saddle points diverge from each other and move into the complex plane.

---

14. This is a saddle point, and not a minimum or maximum, because the second derivative changes sign around this point.

15. Saddle points are indicated by circles; poles correspond to crosses.

16. Here I omit some complications this merging causes. See Nussenzveig (1992, p. 106).

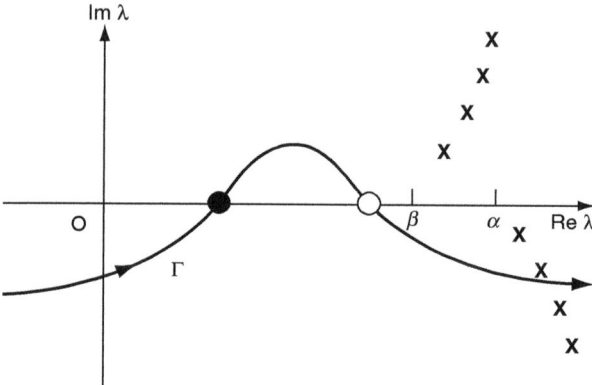

Figure 11.5: Critical Points
Source: Nussenzveig (1992, p. 104)

The physical interpretation of these saddle points is crucial to an evaluation of the CAM explanation of the supernumerary bows. The basic idea behind this interpretation is that the saddle points correspond to geometric rays. The existence of a single saddle point in the $\pi > \theta > \theta_L$ region indicates that only a single geometric ray appears at these deflection angles. But when $\theta_L > \theta > \theta_R$ the two saddle points correspond to two geometric rays combining to increase the amount of light scattered in this direction. This focusing effect is most dramatic when $\theta = \theta_R$ when the saddle points coincide. As Nussenzveig puts it, "a rainbow corresponds to the collision between two saddle points" (Nussenzveig 1992, p. 102). Finally, for $\theta_R > \theta$ the absence of real saddle points shows that there are no corresponding geometric rays.

The justification of this interpretation of the saddle points is that we assume a situation where the dominant contributions to the light arise from electromagnetic waves which approach the behavior of rays of light. In particular, we have assumed throughout that we are operating in a context where the wavelength of light is much smaller than the radius of the drop, the only other relevant length parameter. So in terms of our size parameter, $\beta \gg 1$. Our mathematical theory tells us that the dominant contributions to our integral will come from saddle points. So, as with the localization principle, there is an interpretive conjecture that these points correspond to rays that would appear more and more sharply if the ratio between the drop radius and the wavelength increased further. This conjecture is supported by some preliminary mathematical analysis of the Mie representation where we find that the angle of deflection of the dominant term corresponds to the angle of deflection predicted by the ray theory (Nussenzveig 1992, pp. 9–11). At the same time, it is important to emphasize the experimental nature of this association. It is not dictated by the Mie representation because the Mie representation does not include light rays in its scope. Similarly, the link between rays and these saddle points is not based on anything that the ray theory of light could tell us. This is mainly because we reject the ray theory. But it is also clear that the ray theory has nothing to say about the saddle points that appear in our CAM treatment.

The Rainbow

A similar point can be made about the interpretation of the poles of $S(\lambda, \beta)$, which are also used in the evaluation of our integral. The contributions the poles make, known as residues, correspond to light that has traveled along the surface of the drop for some distance. Again, the link between the features of S and their physical interpretation is initially conjectural. On the side of the physics, we recognize that the ray theory cannot be capturing all of the light relevant to the rainbow. In particular, the way light is diffracted by a sphere is ignored. So we conjecture that some light beyond that corresponding to geometric rays is visible in the rainbow and that

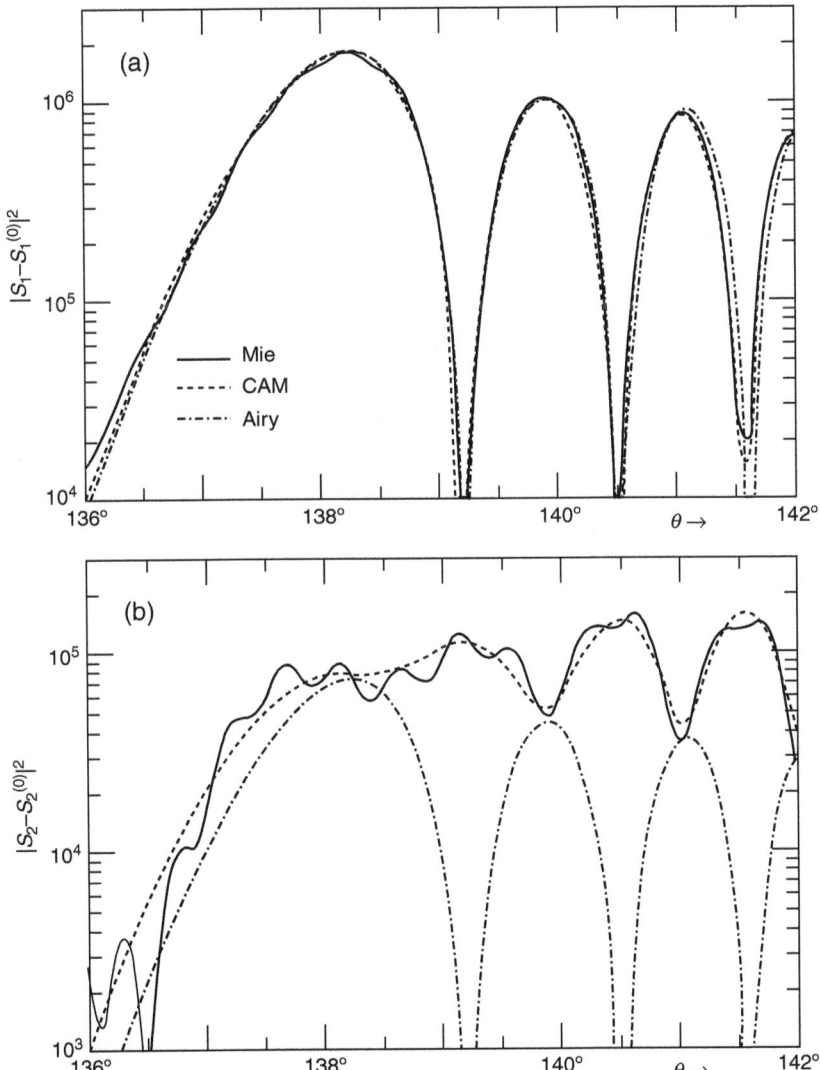

Figure 11.6: Comparing Predictions

Source: Nussenzveig (1992, p. 113)

this light may result from surface waves that are captured by the surface of the drop. On the mathematics side, we know that we are able to evaluate our integral using the poles of our function based on the contribution they make to the amplitude in a given direction θ. This leads us to match the poles with the contributions from surface waves. As with the saddle points, some preliminary mathematical analysis can reinforce this conjecture.

Because the Mie representation can be rendered tractable by a computer, it is possible to compare its predictions with the results of the CAM analysis. We find that CAM outperforms its predecessors and is able to more or less match the Mie representation for a wide range of the size parameter β. For example, figure 11.6 shows how CAM compares to the Mie representation and the earlier Airy representation when $\beta = 1500$. These lines represent intensities, which are obtained by squaring the amplitudes. The Airy representation does much better for one direction of polarization than for the other. The dotted line of the CAM approach tracks the Mie representation quite closely, both for the rainbow around $138°$ and the two additional peaks corresponding to two supernumerary bows. By contrast, the Airy representation incorrectly puts peaks for supernumeraries where there are in fact dips in the light intensity. From a standpoint of predictive accuracy, then, CAM is comparable to the Mie representation.

11.5 INTERPRETATION AND SCOPE

Predictive success is not sufficient for explanatory power, so it is worth asking why CAM is deemed such a good explanation of the existence and spacing of the supernumerary bows. Two factors seem crucial. The first is the way each step in the derivation is susceptible to a physical interpretation. As a result, it is possible to adopt the physical interpretation and come to understand why the rainbow has the features it does. For example, the reasons and details of how the size parameter β affects the supernumeraries becomes much clearer. Using the terms of the Debye expansion, we can see that light reflected off the drop and light internally reflected are relevant to the rainbow. Furthermore, the localization principle and the association between saddle points and rays lets us appreciate how the light behaves in some respects as the ray theory would predict. But the additional role of the poles and their link to surface waves points to the importance of diffraction and the wave character of light even when the size parameter β is large.

A second factor that clearly contributes to the explanatory power of CAM is its wide range of applicability. This range has a number of aspects. First, restricting our focus to a particular rainbow with some fixed β, CAM delivers a "uniform" representation that is valid for the whole range of θ. By contrast, there are "transitional" representations that hold only for limited ranges of θ. Although better than nothing, such transitional representations often result from more or less ad hoc adjustments that do not indicate where the representation will fail or how it can be extended. By contrast, a uniform representation will account for the success of more limited transitional representations and perhaps even correct their mistaken interpretations.[17] A related aspect of the range of applicability concerns the acceptable variation of the parameter β. The Airy representation can be used only when β is very large, for instance, 5000. CAM can account for this limitation and is valid in a much larger range, that is, $\beta > 50$ (Nussenzveig 1992, p. 112).

There is an additional aspect of the range of applicability that fits with the popular link between explanation and the unification of otherwise disparate phenomena. Other meteorological phenomena, such as the secondary rainbow and the glory, can be successfully treated using CAM. The relationship between the rainbow and the glory, in particular, was a subject of much investigation, and only with the CAM approach was an adequate explanation of the glory found.

Finally, there are aspects of CAM that allow it to be extended to other sorts of physical phenomena. This is possible because the mathematics can be interpreted in terms of a completely new range of physical properties. This was one of the contributions from mathematics we emphasized in chapter 4. There are many cases of "scattering," such as in particle physics or even acoustics where phenomena that are analogous to the rainbow appear. Thus it is possible to speak of nuclear rainbows and glories and to treat them using the basic framework provided by CAM. This is a different sort of unification of phenomena than the way CAM unifies the meteorological rainbow and glory, and philosophers seem reluctant to assign it equal importance. Nevertheless, it is an important part of scientific research to search for these commonalities and the insight they provide.

We are now in a position to ascertain what the best explanations of a, b, and c mandate as far our beliefs about rainbows. My suggestion is that the explanations are successful not just because some mathematical transformations are successfully applied to the Mie representation. As we have seen, the explanations involve a series of conjectures about how the representations resulting from these transformations should be interpreted. To the extent that these explanations are the best ones available, it seems reasonable to take these interpretations seriously. This is just a form of IBE noted earlier. However, in its usual role in the philosophy of science, the IBE licenses beliefs in a new kind of entity, like electrons. This is the wrong way to think of how IBE shapes our beliefs in the rainbow case. The explanations do not posit a new class of entities. Instead, the conjectures involve linking this or that part of the mathematics with features of the rainbow that are deemed to be important for the aspect of the phenomenon being explained. This happened in two different ways. In the explanation of a and b, we considered what occurred if a certain limit was taken. Once this limit was conceived as involving the size parameter β, it was possible to see that the explanation did not turn on the assumption that light is a ray as the traditional ray theory would have it. So this aspect of the traditional ray theory is completely idle in the explanation. At the same time, we came to see how a and b were due to those aspects of light that could be correctly described by the ray theory.

---

17. For example, the earlier flawed Fock theory relied on "transverse diffusion" (Nussenzveig 1992, p. 60) as opposed to surface waves. As Nussenzveig emphasizes, the success of CAM over the Fock theory vindicates van de Hulst's earlier proposal that surface waves are important in diffraction and undermines Fock's proposal (Nussenzveig 1992, pp. 34, 85–86). See van de Hulst (1957, §17.52), drawing on work by Beckmann and Franz, noted at p. 227.

This explanation adds to whatever descriptions of the rainbow are provided by the fundamental electromagnetic theory. This is the sort of "physical insight" these limit techniques can provide.

Things worked out somewhat differently in the explanation of c. There we do not consider what happens if β is taken to infinity, only what happens when β gets sufficiently large. How large is large enough is part of the explanation, and as we have seen CAM proves empirically successful when β gets larger than 50. CAM explains which features of the rainbow play an important role in this context. But now we draw on aspects of both the wave theory and the ray theory. It is not possible to explain the existence and spacing of the supernumeraries using just concepts deployed in the ray theory. This is because interference and diffraction are central to what is being explained. At the same time, it is not possible to explain this phenomenon by appealing only to what we find in the wave theory. Important links were made between the critical points and aspects of the rainbow. These links were given in terms of rays and surface waves. They involved aspects of the light scattering that are not available from the perspective of the wave theory alone. Instead, a scientist must ascend from the wave theory to the ray representation before she is able to get the "physical insight" into the supernumeraries that CAM provides. This does not mean that she must believe that the ray theory is correct. Instead, she must use the results of one idealization to inform the proper interpretation of another. The techniques deployed in the explanation of a and b and their proper interpretation let us explain c.

It is helpful at this point to see how these explanations fit the restricted form of IBE articulated in chapter 10. One of the most impressive aspects of our explanation of c is its wide range of applicability. This fits with our requirement, inspired by the failures of chapter 7, that an explanation must have a wide scope if we are to apply IBE to it. Furthermore, we see in this example how the new aspects of light that we come to believe in manifest themselves in the rainbow. The most important new aspect is the role of surface waves, which were associated with the poles of the S function. The contributions from these surface waves are needed to achieve predictive accuracy and are not tacked on in an ad hoc or incomplete fashion. In this case there is a clear choice between two relevant epistemic alternatives. Either surface waves are important for the pattern and spacing of supernumerary bows, or surface waves are not important for this phenomenon. The best explanation of c vindicates a representation whose physical interpretation invokes surface waves, so the former alternative receives a boost in confirmation.

A further point that arose in chapter 10 was the need for the prior justification of the mathematics deployed in a scientific explanation. It should now be clear how the role of the Poisson sum 11.4 requires that it be justified by mathematical means prior to its use in this explanation. If we did not have this equation at our disposal, the whole CAM approach would become shaky and dubious. It would no longer be reasonable to draw the strong interpretive conclusions from this explanation of c that scientists like Nussenzveig do in fact draw. More to the point, we can see how the truth of this equation cannot be supported by the success of this explanation. The phenomena in question bear only tenuously on the truth of this equation. It would be foolish to use the success of this explanation as evidence for the truth of the

Poisson sum. In chapter 14, I suggest an alternative route that I believe is sufficient to justify our belief in the truth of this sort of equation.

11.6 BATTERMAN AND BELOT

In assigning explanatory power to asymptotic techniques and emphasizing the importance of interpretations derived from nonfundamental theories, I am of course echoing Batterman's many discussions of these issues. In his first discussion of the rainbow, Batterman (1997) focuses on Stokes's analysis of the Airy representation. This representation provided at least a partial explanation of c for the case when β was very large. Stokes's asymptotic analysis of the Airy function at the heart of the Airy representation allowed this explanation to succeed. Stokes's techniques are analogous in many respects to the way CAM was applied to the later Mie representation. The improved understanding provided by Stokes' analysis leads Batterman to conclude that "these 'methods' of approximation are in effect more than just an instrument for solving an equation. Rather, they, themselves, make explicit relevant structural features of the physical situation which are so deeply encoded in the exact equation as to be hidden from view" (Batterman 1997, p. 396). This is certainly one aspect of CAM as well. In this respect we are in agreement with Batterman that these mathematical techniques can contribute explanatory power.

The Stokes case involves the Airy representation, and we saw that it has been superseded by CAM. However, Batterman (2002b, ch. 6) provides an additional discussion of the rainbow focused on the explanation of c. His discussion considers the way the mathematical theory known as catastrophe theory can be used to obtain many of the results that we have seen follow from CAM.[18] Although the mathematical techniques are quite different in appearance, they both involve an analysis of the critical points of functions that result from the Mie representation when a parameter gets large. The catastrophe theory analysis concerns how the singularities that emerge in the ray theory behave as the relevant parameters are changed. So, the proper interpretation of the mathematics here involves the ray theory just as much as the CAM techniques do. As Batterman puts it, "one cannot interpret these purely mathematical results in wave theoretic terms alone. An understanding of what the mathematical asymptotics is telling us requires reference to structures that make sense only in the ray theory" (Batterman 2002b, p. 96). Unfortunately, Batterman does not say exactly what beliefs about the rainbow should be adjusted based on this analysis. However, everything he does say in his book is consistent with the proposal we arrived at in the last section—that the conjectures linking the mathematics to aspects of the ray representation are vindicated by the successful explanation of c. This gives us "insight" into what features of the light are responsible for c, and these features are characterized in terms borrowed from the ray representation.

---

18. Catastrophe theory was discussed in chapter 7. See Nussenzveig (1992, pp. 115–116) and Adam (2002, §7) for a comparison of the results of the two approaches.

There is a much more deflationary way to read what is going on here, and we can see Belot (2005) as taking this route in reaction to Batterman's book.[19] According to Belot,

> The mathematics of the less fundamental theory is definable in terms of that of the more fundamental theory; so the requisite mathematical results can be proved by someone whose repertoire of interpreted physical theories includes only the latter; and it is far from obvious that the physical interpretation of such results requires that the mathematics of the less fundamental theory be given a physical interpretation. (Belot 2005, p. 151)

This may just mean that the success of the mathematical techniques deployed by CAM or catastrophe theory need not lead us to accept any theory besides the fundamental theory. If this is all that Belot intends to say, then he is correct. The insight associated with our explanation of c does not involve belief in any *theory* beyond the wave theory. But saying only this underestimates what scientists themselves say about the importance of our best explanation of c. As we have seen, it shows us which contributions are most important to the supernumeraries, and the importance of these contributions is not accessible from the perspective of the wave theory. Belot's attitude, then, threatens to undermine the crucial distinction between a brute-force numerical computation and a well-motivated physical analysis of the situation. The former lacks explanatory power and interpretive significance for our understanding of the rainbow, whereas the latter increases our scientific understanding by isolating what is responsible for this or that aspect of the rainbow. These important factors are isolated using the interpretation of the ray representation arrived at in our explanations of a and b. So, although we do not ever come to adopt a new and competing theory of the rainbow, our insight into the rainbow must be expressed in the vocabulary of representations associated with a nonfundamental theory, in this case, the ray theory.

This is essentially the line Batterman takes in his reply to Belot. But he also claims that Belot's "pure mathematician" needs the less fundamental theory to motivate the steps taken to develop the ray representation in a stronger way than I have suggested. These developments go beyond simply taking a parameter like the wavelength to 0. In particular, Batterman suggests that the initial and boundary conditions of the ray representation make sense only if we interpret them using the ray theory of light:

> those initial and boundary conditions are not devoid of physical content. They are "theory laden." And, the theory required to characterize them as appropriate for the rainbow problem in the first place is the theory of geometrical optics. The so-called "pure" mathematical theory of partial differential equations is not only motivated by physical interpretation, but even more, one

---

19. See also Redhead (2004).

cannot begin to suggest the appropriate boundary conditions in a given problem without appeal to a physical interpretation. In this case, and in others, such suggestions come from an idealized limiting older (or emeritus) theory. (Batterman 2005, p. 159)

If using a theory this way involves beliefs about the subject matter, then it seems to me that Batterman has gone too far. It is true that the ray theory originally motivated the ray representation, which played a role in the explanation of a and b. Furthermore, I have accepted that this very representation plays a role in the interpretation of the steps leading to our best explanation of c. But we have not seen a need to adopt any beliefs about light that the traditional ray theory incorrectly offered. For example, the ray theory maintains that light travels along rays. All we have come to accept as a result of our explanations is that in certain contexts there are aspects of light that are accurately captured by the ray representation. The explanatory power of the ray representation turned on the way the parts of it that we take seriously can be grounded in the fundamental wave theory. So the ray theory as it was traditionally presented obscures what is really going on, and this is why we reject it. This is consistent with using the ray representation as an idealized representation that conveys accurate information about this or that aspect of the rainbow. As we have seen, it is consistent with the best explanation of c drawing on certain features of this idealization to guide a further idealization. I believe that this position fits with Batterman's considered view on the issue, as he says things like "asymptotic explanation essentially involves reference to idealized structures such as rays and families of rays, but does not require that we take such structures to exist" (Batterman 2005, p. 162). Again, I hope to have presented a clearer view of what these explanations tell us about rainbows.

Perhaps what has led Batterman into these stronger statements is his focus on theories and the traditional assumption that if we deploy a representation that derives from a theory, this theory must play a role in shaping our beliefs about the systems being represented. The central role of the ray representation and its interpretation indicate for Batterman that the ray theory must play a role in our beliefs about the rainbow. Belot can also be seen to be making this assumption, but he starts from the fact that we reject the ray theory. So for Belot, the rejection of the ray theory entails that we cannot really employ any ray representation in shaping our beliefs about the rainbow. If we reject the traditional assumption linking theory and representation (as I have argued we should), then we can make sense of the widespread scientific practice of using representations that are associated with outdated theories. These representations help reveal what is going on in a given situation and are crucial to the explanations we have. But these explanations are consistent with accepting only the fundamental theories, even though we use concepts from other representations to help us to decide what to believe. Mathematical transformations link these theories to our representations in unanticipated ways, and philosophers must focus on these transformations if they are to come to terms with the commitments of practicing scientists.

11.7 LOOKING AHEAD

To sum up chapters 9, 10, and 11, in chapter 9 I surveyed several indispensability arguments and concluded that an argument for realism in truth-value for some mathematical claims had the best chance of success. But in chapter 10 I argued that even this sort of indispensability argument fails to confront the possibility that our mathematical claims must be justified nonempirically before they can make their contribution to science. One major gap in my discussion is the fictionalist approach to mathematics and scientific representation. I propose to close this gap in chapter 12 by arguing that the fictionalist has yet to offer an acceptable account of the content of our scientific representations.

Chapter 11 has involved an extended case study where mathematics allows explanations of three different features of the rainbow. This case illustrates many of the lessons from earlier in these chapters but was primarily aimed at clarifying how these explanations guide the physical interpretation of the representations we accepted. At the heart of the proposal I developed is the idea that some mathematical techniques, like taking limits, impose their own restrictions on the sorts of interpretations we can wind up taking seriously. This is also a central theme in the work of Mark Wilson, and in chapter 13 we explore to what extent his conception of scientific representation can be reconciled with the proposals developed so far.

12

Fictionalism

12.1 MOTIVATIONS

I summarized one of the lessons of chapter 10 by saying that to understand a given mathematical scientific representation, an agent must believe some mathematical claims. My main aim in this chapter is to argue against an alternative approach which insists that understanding a mathematical scientific representation requires only that an agent understand some mathematical claims. In this view, science does not require any mathematical beliefs, only the contemplation of mathematical claims. This approach faces a challenge that also came up in the context of Field's program. This is to distinguish between claims like $2 + 3 = 5$ and claims like $2 + 3 = 4$. Mathematicians and scientists work with the former kind of claim, but generally criticize uses of the latter kind of claim. If we cannot distinguish between these claims based on their truth, then what is the basis for the different treatment? A popular response to this predicament leads directly to fictionalism about mathematics. This is to say, even though both of these claims are false to the extent that they entail the existence of mathematical entities, the former kind of claim becomes true if we preface it with the qualifier "In the story of mathematics..." So, "In the story of mathematics, $2 + 3 = 5$" is true, whereas "In the story of mathematics, $2 + 3 = 4$" is false. The problem is solved, then, by making a link between literary fiction and mathematics. For example, with the Sherlock Holmes stories, "In the story, Sherlock Holmes lives in London" is true and "In the story, Sherlock Holmes lives in Paris" is false. We can make sense of this discrepancy without believing that Sherlock Holmes lives in London. Analogously, the thought goes, we can make sense of the use of mathematics in science without appealing to the truth of any mathematical claims.

There is a more recent development within the philosophy of science that I would also like to combat in this chapter. This trend insists that we can make sense of scientific representation without believing any claims about what we call scientific models. The models we have discussed in this book are usually the abstract mathematical structures we structurally relate to a target system. But in a few cases we have also seen that models can be concrete entities, as with the scale models of ships (section 5.3). A fictionalist about models insists that we can make sense of how models contribute to the success of science without believing any claims about abstract models. On this approach, understanding a scientific representation merely requires that we understand some claims about abstract models. Again, as with the mathematical case, there is the problem of distinguishing between acceptable and unacceptable claims about a given model. The fictionalist about models argues that this distinction is merely between true "In the story of the model..." claims and false "In the story of the model..." claims. Building on a similar account of literary fiction, the view is that this difference can be made out independently of the truth of any of these claims "outside" the story of the model.

Some similarities between fictionalism about mathematics and fictionalism about models are suggested by the way I have defined these two positions. Both try to get by without any beliefs of a certain sort and rely instead merely on understanding. Furthermore, the claims at issue are sorted into two kinds based on an account of the relevant story and which claims come out true when we consider adding the relevant "In the story of X..." operator. These similarities make both versions of fictionalism vulnerable to the same sort of general objection. This objection is based on the difficulty in clarifying how the claims that are true in the story relate to beliefs about the real world. I argue that in both cases our limited epistemic situation makes the solution of this "export" challenge quite daunting. In the mathematics case, fictionalism about mathematics is ruled out. For models, fictionalism winds up requiring the very sort of direct specification of the content of our models that fictionalists seem intent on avoiding. In both cases, then, an export challenge is sufficient to remove much of the motivation for the fictionalist position.

Before proceeding to these objections, I carefully review some options for making sense of literary fiction. The salient issues here are how we might learn about the world from engaging with literary fiction and what we learn about the world when we successfully engage with literary fiction. Although a thorough investigation of these issues would take us too far from our main topic of mathematics and science, getting clear on how literary fiction helps us is essential if we are to see why mathematics helps science in such a different way.

12.2 LITERARY FICTION

When philosophers think about literary fiction, one of the central concerns is how we are to make sense of our talk of fictional characters like Sherlock Holmes and Emma Bovary. In her excellent survey of the main options for doing this, Stacie Friend argues that there are two main strategies (Friend 2007).[1] The realist about fictional characters posits abstract objects that can be used to explain what we are talking about when we say "Sherlock Holmes lived in London." The antirealist about these characters tries to get by without the existence of any new entities. Only the antirealist positions interest us because only these "fictionalist" approaches to literary fiction can serve as useful templates for mathematics and scientific models in a way that departs from my realist position. So, for the sake of argument, I assume that some antirealist approach to literary fiction is adequate.

---

1. Noted by Frigg (2010b).

Friend helpfully contrasts two antirealist approaches. The most straightforward is the sort of operator proposal noted in section 12.1. In this view a claim like "Sherlock Holmes lived in London" is not true, but becomes true when it is "implicitly prefixed with a story operator" (Friend 2007, p. 143). The features of this story operator are determined by the claims which make up the story, along with some additional rules concerning which claims should be imported into the story. Even though it is not explicitly claimed that Sherlock Holmes has parents, it is appropriate for the reader to believe that "In the Sherlock Holmes stories, Sherlock Holmes has parents" is true. These additional rules are determined in part by the conventions of the genre of literature of these stories and in part by what one takes the author to be intending to communicate in the stories. For example, a science fiction story may allow fewer kinds of claims to be imported into the story, and an author writing in the Middle Ages probably did not intend to import scientific developments of which he was unaware.

One might worry about the indeterminacy of the rules for spelling out which claims come out true in the story, but more serious problems undermine the operator proposal and motivate an alternative pretense proposal. As Friend notes, there are many claims that do not include a story operator, but which we nevertheless wish to assert. These include claims about how the fictional character relates to other fictional characters and how the character relates to things in the real world. For example, some readers wish to assert that they admire Sherlock Holmes, and literary critics defend the view that Sherlock Holmes is more brilliant than Hercule Poirot. There is no story operator in question for one who admires Sherlock Holmes because they are not asserting that in the story they admire Sherlock Holmes. Similarly, there is no single story operator that we could prefix to a comparative claim about two fictional characters from different stories. As a result, it seems that the operator proposal is not a viable position.

Friend argues that a pretense theory is a better option for the antirealist. According to the version of the pretense theory developed by Walton, a work of fiction is offered by the author as a prop in a game of make-believe (Friend 2007, p. 144). The reader who takes up the invitation to engage in the game then adopts the pretense that the story describes actual individuals engaged in real actions. Assertions about fictional characters are then continuations of this game. Some assertions, such as that Sherlock Holmes lived in London, are appropriately prompted by what the author has written in the stories. They thus find their justification by appeal to the rules of the "official game" begun by the author. Other claims, such as the relational claims that caused problems for the operator proposal, can be seen to involve further "unofficial games" that are parasitic on the official game initiated by the author (Friend 2007, p. 145). For example, when I tell someone that I admire Sherlock Holmes, I extend the pretense of the original game so that in the extended game it is appropriate to pretend that I can stand in relations to Sherlock Holmes. In a similar fashion, a critic can extend the games initiated by the authors of two stories so that comparative claims between two fictional characters become available. But as with the original game, only the rules of these new games make a given assertion appropriate or inappropriate.

Although there are many further details of the pretense proposal left to be worked out, we have seen enough to return to our main questions: what do we learn about the world from literary fiction, and how do we learn it? The pretense proposal suggests the following answers. We do not learn that any claims about the world are true, but we can learn how to do things we might not have been able to do before. The skills we learn concern how to play the various games of make believe which are initiated by the works of literature we engage in. Appreciating a story requires that we play a complicated game initiated by the author, and literary criticism requires us to invent new, unofficial games that extend the original game in creative ways. Engaging with literature develops a host of creative skills, then, which we can apply in our lives beyond literature. I can increase my ability to initiate games of make-believe and use this creativity to understand others and even myself. This is because the ways we understand a work of literature and relate it to other works of literature involve the same sort of intense use of imagination that is required when we try to understand the actions of others and perhaps even our own actions.[2]

I do not wish to downplay the significance of these sorts of skills. Even in science the ability to understand others and think in creative ways is of undoubted importance. But it seems unwise to try to make too much of a link between this sort of learning from fiction and what happens with mathematics and scientific models. What we are after is an account of how engaging with literary fiction can provide us with a reason to believe claims about the world outside the fiction. For the detour through fiction to make a genuine contribution to these beliefs, we need to see how fiction is helping us obtain reasons for these beliefs, which would be otherwise hard to obtain. When the question is put in these terms and we consider the Sherlock Holmes stories, I think it is fair to conclude that engaging with these stories does not give us reason to believe any new claims about the world outside the fiction. We may learn that cocaine was available in Victorian England and that some people used cocaine in line with Holmes's "seven percent solution." But the reason we come to believe this sort of claim is derivative on our prior belief that the author believes this claim to be true and our confidence in the author's reliability. There is no special virtue in learning the truth of this claim via engagement with fiction as opposed to, say, consulting an encyclopedia entry on the history of cocaine abuse. If this is all learning from fiction comes to, then it is just learning via testimony. The detour through fiction only complicates the task of figuring out what the speaker intends and how reliable they really are.

A fair complaint at this stage is that my preoccupation with the Sherlock Holmes stories has biased the discussion. What we need to do is consider genres of fiction besides detective stories. We would expect to learn more about the world from engagement with others kinds of fiction, especially those that decades of criticism have established as the great or canonical works of this or that genre. I take two novels from different genres of fiction and see what we can learn from them about

---

2. Here I am indebted to the work of Elisabeth Camp. She links the appreciation of literature to the cultivation of what she calls perspectives. See, for example, Camp (2009).

the world. First, there is Dickens's novel *A Tale of Two Cities* (1859), which falls into the genre of historical fiction. Second, I discuss the classic instance of the "realistic novel," Flaubert's *Madame Bovary: A Tale of Provincial Life* (1857). A reasonable working assumption is that historical fictions allow us to learn something about a specific historical episode that we would otherwise find difficult to appreciate. Flaubert himself and many of his contemporaries seemed to have thought that his realism led his readers to learn some general truths about personal relationships that are supported by the novel in a special way.

A Tale of Two Cities traces the lives of many characters in Paris and London both before and after the French Revolution. Characters who are introduced independently eventually come to interact in unexpected ways. Along the way Dickens provides an in-depth description of the living conditions of different classes of people. One theme of the story is the unfair way poor people are treated in Paris before the French Revolution and in London throughout the entire story. Dickens suggests that these conditions are part of the reason for the French Revolution and the ensuing Terror that engulfs the central characters of the novel. The continual juxtaposition of Paris and London leads one to consider what the future of London will be given the rampant injustice in England. On this interpretation the novel is offered as a warning to Dickens's mid-nineteenth-century English readers: support the reforms necessary to improve the state of the poor in England or else face a horror equivalent in many respects to what occurred in France. To convince readers of the need for these changes, we can see the novel as providing evidence for the following claim, thought of as offered in 1859: (*A*) if no changes are made to the condition of the poor in London, then a revolution and terror analogous to the French Revolution will have a good chance of occurring in England. *A* is a substantial claim with real predictive power which Dickens may aim to convince his readers of. And, the proposal continues, historical fiction has a better chance of supporting this sort of claim than a more direct nonfictional argument.

Let us consider how this proposal could be articulated by a supporter of the pretense view of fiction. Dickens offers his novel as a prop in a game of make-believe. His readers take up the game and generate a wide range of claims they rightly take to be licensed by the rules of the game. These relate initially to the inclusion of many details about Paris and London as they actually were in the historical time period relevant to the story. These "import" practices derive largely from the genre of historical fiction with its special rules concerning the truths about events associated with the history. But to arrive at a claim like *A* that a reader will actually believe independently of the pretense, there must also be a stage of interpretation where the reader thinks through what Dickens intends one to think about the real world and whether these claims are supported by the fiction. That is, which claims should one export from the fiction, and what should one think of them? Noticing certain patterns in the way the story develops might alert the reader to a special category of claims that Dickens, at least, intends to be exported. Furthermore, the details of the story associated with these claims could give the reader a reason to believe that the exported claims are true.

By considering a paradigm case of historical fiction, we have come to an answer to our two initial questions: readers of historical fiction can come to learn the truth of some historical claims, and they support these claims by the evidence amassed in the story itself. I have two criticisms of this proposal. First, it is not clear how rational it is for a reader to shape their belief in A by engaging with *A Tale of Two Cities* in this fashion. The reader is fully aware that Dickens has structured the story in such a way that his own point of view will seem eminently reasonable. A balanced consideration of the evidence in favor of and against A is needed before a rational agent should assent to A. But on the assumption that Dickens believes A and one of the reasons he wrote the novel was to convince the reader of A, the reader should be very suspicious of the evidence that appears to favor A. At best, then, engaging with the literary fiction could provide the reader with an interest in A and a motivation to investigate the matter further. But it would be unreasonable to take the evidence provided in the story as sufficient to assent to A.

My second criticism of this proposal is to emphasize that the role of a historical fiction in licensing these sorts of claims about the real world is not that central to the value of the work of fiction. Even when we come to isolate these claims like A and find them unsupported by the evidence provided by the story and any other historical evidence we can come up with, we do not fault *A Tale of Two Cities* or its author. The novel remains a classic, and we continue to admire its author for the amazing way the novel is executed. This suggests that the significance of these exportable claims is fairly low when we consider the virtues of a great historical novel. A criticism of the novel based on the implausibility of its exportable claims seems misguided and to ignore the more important features of the novel.

The low priority that we assign to isolating and evaluating the exportable claims from a work of literary fiction accounts for the relative lack of discussion of how the rules for the pretense generate these exportable claims. Most of the debates about the pretense view seem to concern the rules for importing claims into the fiction or how the rules of the official game licensed by the novel relate to the additional unofficial games proposed by readers and critics. By contrast, the main discussion of how to export claims seems to be focused on the special problem of "imaginative resistance." As Gendler puts it, there are some claims, such as false moral claims, that produce a "pop-out" reaction where a reader is unable or unwilling to engage in the pretense that they are true (Gendler 2006, §5). The clash between our moral beliefs and the deviant proposals found in the novel blocks our participation in the game. But this phenomenon is a far cry from the supposed practice of exporting a well-supported claim from a fiction. Pop-out cases show how reluctant we are, at least for some topics, to even take the first step that would be necessary to learn the truth of this claim from fiction.

There are other genres of fiction, though, where the failure of these exportable claims would arguably undermine the value of the fiction itself. The case I have in mind is a realistic novel like *Madame Bovary*. There is considerable evidence that Flaubert took himself to be offering a correct representation of a type of person and how that sort of person could bring themselves to commit immoral acts like adultery. For example, in an 1853 letter to Louise Colet, Flaubert wrote

> Literature will increasingly adopt the appearance [allures] of science; it will be above all expository [exposante], which does not mean didactic. We must

present pictures, show nature as it is, but complete pictures, we must paint the underside as well as the surface.³

The contrast between an expository work and a merely didactic work is not immediately clear, but Flaubert may mean that his novel shows how certain acts arise for this type of person. A merely didactic work would instead simply state that this person commits this act without making the link to their type very clear. In this view of *Madame Bovary*, the fictional character of Emma Bovary comes to have special significance. For this genre, the task of the reader is not just to take up the game of make-believe initiated by the author. She must go on to isolate those features of Emma Bovary that Flaubert takes to be indicative of a certain type of person and then draw out the links between this personality and their actions. If this is such a central part of the novel, and the novel is judged a classic, then it seems fair to conclude that many readers and critics agree that Flaubert has presented an accurate picture.

It must be admitted that Flaubert is much more successful than someone like Dickens at presenting a plausible picture of his characters and their interactions. For example, the character of Sidney Carton in *A Tale of Two Cities* is motivated to make a significant sacrifice by the virtue of the main woman in the story, Lucie. The critical reader may find this action completely implausible as it clashes with the features ascribed to Carton leading up to his sacrifice. By contrast, there is a clear development between the psychological reflections of Emma Bovary early in the novel and her later adulterous actions. As a result the reader comes to appreciate not only why this character might act this way but also presumably why other people act this way. The novel is realistic, then, in the sense that it abandons the sentimental hopes that imbue a novel like *A Tale of Two Cities* and takes on the actions of real people and their less than ideal motivations. This seems to be one of the reasons Flaubert's novel was so shocking to so many readers at the time.

Granting all this, we can see how a reader could come to believe a psychological claim like: (B) A person with personality traits X in situation Y is likely to commit adultery. We come to believe a claim that is exemplified in the fiction, to adapt a term from Elgin.⁴ Emma Bovery exemplifies a more general truth that we intend to apply to the real world when, for example, we decide to whom we should propose. The rules that govern the game of make-believe surrounding this realistic novel license us to search for these sorts of exportable claims. Furthermore, there is supposed to be something about the way the claim is exemplified in the fiction that provides evidence to believe B. This is because the reader is meant to understand why having traits X in situation Y prompts actions of a certain kind. There is an inner plausibility or "ring of truth" to B that is absent from the analogous claims about Sidney Carton's sacrifice.

---

3. "La littérature prendra de plus en plus les allures de la science; elle sera surtout exposante, ce qui ne veut pas dire didactique. Il faut faire des tableaux, montrer la nature telle qu'elle est, mais des tableaux complets, peindre le dessous et le dessus" (Bruneau 1973, vol. 2, p. 298, noted in the preface to the Tantor Media audiobook).

4. See Elgin (2009) for a summary of her views on literary fiction.

For realistic fiction, then, it is reasonable for a reader to search for claims like B and take the author to intend some such claims to be true of the real world. These claims are not simply learned via testimony by the author of the fiction as the events portrayed in the fiction themselves give independent support to the claim. It is also arguable that the sorts of psychological claims illustrated by *Madame Bovary* stand a much better chance of being adequately supported by what occurs in the story than the historical claims that might be exported from *A Tale of Two Cities*. Although I am prepared to concede this point, I remain skeptical of the degree to which engaging with a fiction can provide genuine evidence for the truth of a psychological claim like B. On the supposition that B does seem plausible to us after we have read *Madame Bovary*, can we say that this is because some special evidence has been provided by the novel? It seems more reasonable to say that the events described in the novel have an inner plausibility because they match the theories of human behavior we accepted prior to engaging with the novel. If there was a disagreement between these theories and what occurred in the novel, then surely we would reject the events as implausible.

This is not meant to deny that engaging with a novel can dramatically affect the convictions of a reader. But at least when the claims in question concern psychology, history, or other comparable matters, the best way to proceed seems to me to be to consider the evidence we can gather by interacting with the real world, not a world described by a fiction. Considering a claim as it is exemplified in fiction can thus help a great deal in focusing our attention on it and making it the subject of further investigation. This additional investigation, though, should depart from the fiction and take up the ordinary, scientific methods of considering the actual evidence we have or else in searching for new evidence derived from new experiments. My conclusion, then, is that even though we can identify claims that are exported from some works of literary fiction, there is no special kind of evidence available to support these claims. Learning something from fiction, if by that we mean coming to believe something on the basis of evidence provided in the fiction itself, is something we just do not seem to be able to do.

12.3 MATHEMATICS

Fictionalism about mathematics grants that mathematical statements have ordinary contents of the sort embraced by the realist. However, the fictionalist also insists that we can use the statements in science and yet not believe these ordinary contents. This is because mathematical statements can be assigned a weaker representational content that is subject to genuine belief and assertion. The fictionalist about mathematics thus faces the same task as the antirealist about literary fiction. She must articulate a sense in which some mathematical statements are acceptable even though we reject their ordinary contents. The standard way to try to do this is to think of mathematics along the same lines as literary fiction. That is, a mathematical statement with its ordinary content, though false, is true in the story of mathematics. This accounts for our acceptance or rejection of a mathematical statement. But as I emphasize, there is an urgent task for the fictionalist for mathematics that was much less important for literary fiction. This is to explain what the acceptance of a mathematical statement as true in the story of mathematics entails for our remaining beliefs about the world. The fictionalist must isolate the representational contents from the fictional contents and explain why the weaker representational contents are the proper targets of our genuine beliefs.

Fictionalism about mathematics is a popular position and seems to owe its popularity to the failure of Field's program and the perennial appeal of anti-platonism. Among its most prominent advocates we find Yablo, Bueno, and Leng.[5] In addition to this, Balaguer and Rosen argue that fictionalism about mathematics is a viable position, even if, in the end, they do not endorse it.[6] Most of the discussion of fictionalism about mathematics is focused on articulating what is "true in the story" of mathematics. One attempt to do this is based on an appeal to what Balaguer calls a "full conception" of a mathematical domain like the natural numbers or real numbers: "a sentence is built into FCNN [our full conception of the natural numbers] just in case it is one of our (explicit or implicit) beliefs about the natural numbers or follows from these beliefs" (Balaguer 2001, p. 90). Implicit beliefs are fixed by an informal relation of entailment and "notions, conceptions and intuitions" (Balaguer 2001, p. 90) beyond explicit beliefs play a role in fixing the premises of these entailments. Balaguer claims that the fictionalist should appeal to these full conceptions when they explain what is true in the story of this or that part of mathematics:

> the obvious thing for fictionalists to say here is that "our story of arithmetic" is determined by what we (as a community) *have in mind* when we're doing arithmetic. In other words, it's determined by the intuitions, notions and conceptions, and so on that we have in connection with arithmetic. In other words, it's determined by our *full conception of the natural numbers*. (Balaguer 2001, p. 94)

To support this point, Balaguer considers the case where a new axiom is introduced to solve an open problem. He insists that the new axiom must be "intuitively obvious to all of us" (Balaguer 2001, p. 94), and so part of our full conception, if it is to solve the original problem and not simply change the subject. This is consistent with a full conception changing over time, but Balaguer seems to think these changes are rare and do involve changing the subject of investigation.[7]

Balaguer's account is reminiscent of the operator approach to literary fiction, which we reviewed in the last section. But one of the problems with an operator approach is that it is not able to make sense of either claims about fictional characters made from outside the fiction or comparative claims that consider characters from two different fictions. The same worries clearly arise for Balaguer's version of fictionalism. For example, G. H. Hardy denied that 1729 was interesting,

---

5. Yablo (2001, 2005), Bueno (2009), Leng (2007, 2009). Leng (2010) appeared as this book was being completed, but I hope to discuss this important book in future work.

6. Balaguer (1998, ch. 7), Balaguer (2001), Rosen (2001).

7. See also Yablo (2001, p. 89), but compare Yablo (2005, pp. 102–103).

whereas Ramanujuan insisted that 1729 was interesting. Similarly, I believe that the empty set is not identical to the natural number 0. None of these claims are part of our full conception of the natural numbers, so it is not clear how Balaguer can interpret them. They either do not involve any story operator or else involve more than one story operator whose interaction is not illuminated by what Balaguer says.

Problems of this sort motivate Yablo to adapt Walton's pretense theory to the ends of fictionalism about mathematics. Yablo goes further by aligning the fiction of the natural numbers with some aspects of the genre of realistic fiction such as *Madame Bovary*. Initially talk of the natural numbers is initiated as part of a game of make-believe. To engage with this official game, we adopt the pretense that there are such abstract objects. The point of this official game is to represent genuine facts that obtain in the world independently of the game. That is, the pretense of the numbers is adopted so that we can find out about things in the real world. Still, this official game is naturally extended by additional unofficial games. In these further games we take the numbers themselves as "props" and consider the number of primes or how numbers relate to things like sets. Finally, Yablo allows an additional sort of extension where considerations of "aptness" lead us to further refine our various games and their connections. As he summarizes his position,

> Our *ultimate* interest may still be in describing the natural world; our *secondary* interest may still be in describing and consolidating the games we use for that purpose. But in most of pure mathematics, world and game have been left far behind, and we confront the numbers, sets, and so on, in full solitary glory. (Yablo 2005, p. 100)

The link to Walton's account of literary fiction thus gives Yablo much more flexibility to account for mathematical practice.

For the sake of the current argument, I take for granted that Balaguer and Yablo have articulated notions of fictional content that are determinate enough to make sense of the practice of pure mathematics.[8] This leaves the problem of explaining what connection these claims have to our beliefs outside the fiction. This "export" challenge is to provide rules that will indicate, for a given context, which claims can be extracted from the fiction and taken literally as claims about the actual world. My argument is that this challenge cannot be met because any set of rules that are detailed enough to do the job will presuppose knowledge of the actual world that we do not have. This forces the fictionalist to undermine the epistemic contributions from mathematics to the success of science that I discussed at length in part I. To see the problem, consider Balaguer's and Yablo's views on how to meet the export challenge. There are confusing differences in terminology, but the basic strategy remains the same. In each case we have an attempt to specify what can count as part of the representational content in terms of a notion of "physical,"

---

8. This point will be criticized in chapter 13.

"concrete," or what "concerns" actual objects.[9] If this specification was successful, then we would have a way of moving from what I call the fictional content to the representational content. Balaguer uses a general test concerning physical objects based on their causal isolation from any other alleged kinds of object (Balaguer 1998, p. 110, p. 131). Yablo, at least some of the time, keys in on the reference to nonmathematical objects and insists on restricting his commitment to these (Yablo 2001, p. 97). The general problem with these strategies is that we are not told enough about these acceptable objects and their properties to meet the export challenge. To focus on the nearly trivial example at the heart of Yablo's discussion, consider s: "the number of sheep is three times the number of goats." Balaguer argues that there are no causal relations between the sheep and the goats, on one hand, and the numbers that represent their relative sizes like 6 and 2, on the other hand. He concludes that there must be a feature of the physical world in terms of causally connected things in virtue of which the acceptable part of s is true (Balaguer 1998, p. 133). Yablo claims that the "real content is that portion of the literal content [of s] that concerns the sheep and the goats" (Yablo 2001, p. 97). In our terms, the representational content is that portion of the fictional content that concerns only the sheep and the goats.

Rosen also considers a version of mathematical fictionalism. Unfortunately he does not make clear if he is opting for a story operator approach or a pretense approach when specifying the fictional contents.[10] But Rosen offers what seems to be the clearest recipe for extracting the representational content from the fictional content: "S is nominalistically adequate iff the concrete core of the actual world is an exact intrinsic duplicate of the concrete core of some world at which S is true" (Rosen 2001, p. 75). The concrete core of a world is defined as "the aggregate of all the concrete objects that exist in W" (Rosen 2001, p. 75). For our sheep and goats claim s, we are led to compare the concrete core of the actual world with the many concrete cores of the many worlds in which s is true. Although each of these worlds makes the whole fictional content of s true because it has both abstract and concrete objects, the existence of the abstract objects is deemed irrelevant to the truth of the representational content. All we need is a match at the level of the concrete aggregates between one of these possible worlds and the actual world.

Is the division between the concrete and nonconcrete determinate enough to render the representational contents determinate? As Yablo is well aware, there is no finitely long sentence whose vocabulary is restricted to the sheep, goats, and their clearly physical properties that captures how the sheep and the goats must be for s to be true of them.[11] So there seems to be no way to directly specify this feature of the world. But the alternative indirect indications offered by Balaguer, Rosen, and

---

9. Balaguer calls our representational content the nominalistic content, whereas Yablo opts for real content. Balaguer uses "platonistic" for our fictional content, and Yablo uses "literal".

10. He uses both "fictionalizing operators" (p. 75) and talk of "pretense" (p. 76).

11. To see why, notice that there is no upper bound on the size of the sheep population, so any sentence in first-order logic would need to be an infinite disjunction to cover all cases.

Yablo are insufficient to fix the representational content.[12] To see why this indirect approach is not adequate, consider s': "The number of sheep is three times the number of sheep minds." This claim might arise in the context of a theory of sheep where some are full-fledged sheep with minds and bodies and others are zombie sheep who behave strangely because they lack minds. The point of this example is that it is not clear if minds are physical, concrete or to be included in Yablo's real content. But the burden is on the fictionalist to make this matter clear or else their representational contents will not be clear. If their representational contents are not clear, then it is not legitimate for fictionalists to counsel us to accept the representational contents that correspond to their supposed fictional contents.

There are two ways out of this general problem that I can see. First, the fictionalist might insist that we look to the beliefs of the speaker to determine what is exported. That is, if the speaker believes that sheep minds are physical or concrete, then they will be included in the representational content just as sheep and goats were included for s. It is hard for me to see how this move will work, though, because I doubt that most speakers have any clear notion of what is physical or concrete. It is hard to specify this, and philosophical disputes remain open about how best to do it. I would not say that there is no distinction there to capture, but only that the typical speaker does not have beliefs about this distinction that can be used to fix the representational content.

A second way out would be for the fictionalist to claim that the actual facts fix the representational content, and not the speaker's beliefs. So if, in fact, sheep minds are physical or concrete, then s' would be handled like s and the speaker would wind up committed to the existence of sheep minds. But if, in fact, sheep minds are neither physical nor concrete, then the representational content of s' would not include sheep minds. There would be a determinate representational content there, but the speaker would not have any epistemic access to it. The problem with this proposal is that it ignores the link between evidence and ontological commitment. Notice that the representational content is used to assess ontological commitment. On the current proposal, the speaker will be committed to sheep minds if they exist and are physical and will not be committed to them if they do not exist or are not physical. But this commitment obtains or fails to obtain independently of the evidence that the speaker has. This makes the speaker fundamentally irrational, for they are taking on beliefs in the existence of entities independently of their evidence. The link between representational content and ontological commitment imposes special considerations on how we fix the representational content.

To show that this problem is not just idle speculation, recall a case from chapter 7: Laplace's correction of Newton's calculation of the speed of sound. Part of Laplace's argument made an appeal to caloric and the way the compression of air releases latent caloric and so raises the temperature of the air. So, we can ask our fictionalists: what conception of heat is part of the representational content of Laplace's fiction? The recipes provided by the fictionalists provide no help here. If, as Laplace believed,

---

12. Leng (2010, §6.2) notes that in Yablo (1998), Yablo agrees with this worry. I take such a concession to involve the rejection of fictionalism.

heat is a fluid, then it would seem to be as concrete, physical, and nonmathematical as anything a fictionalist would countenance. If, by contrast as we believe, heat is a kind of transfer of energy, then it seems highly mathematical and should appear in the representational content in a much more minimal way.

What this shows is that a given fictional content is subject to varying interpretations, and these different interpretations give different verdicts on the representational content. It is no help to be told to accept the representational content of some fictional content because the representational content turns out to be highly indeterminate. The problem is not that there is a genuine content picked out by these tests and we just fail to know what it is. Rather, our conception of the world will influence how we interpret a given fictional content, and so there is no general recipe that can be implemented by the mathematical fictionalist to solve her export problem.

Both options noted for the sheep case could be tried for the heat case. For example, we might say that the agent's beliefs about the physical or concrete character of heat are decisive. Here there may be more agents, perhaps even scientists like Laplace, who said enough about heat so that they can be reasonably interpreted as believing it is physical. Still, there is no good reason that the success of Laplace's equation should be thought of in this way. Many scientists at the time realized that the caloric interpretation was redundant to the derivation. This is a common feature of applied mathematics we discussed in chapter 7: its elements are subject to a variety of physical interpretations and typically also to the view that this part of the mathematics has no interpretation. This flexibility is undermined by the fictionalist, but this flexibility has been at the center of our reconstruction of the many epistemic benefits that mathematics provides to scientists.

Our other option is to consider what heat really is and only add it to the representational content if it is in fact physical. Again, this seems to sever any link between our evidence and what we believe exists. We cannot obtain the representational content from the fictional content by insisting that however the world actually is, the representational content is that part of the fictional content that comes out true. For fictionalists about mathematics, the point of engaging in the fiction is that it helps us eventually believe the right things about the real world. But nobody thinks we can wind up believing the right things by just insisting that we believe whatever happens to be true. We need some connection between the parts of the fiction that we export and our evidence about how the world is. In the case of *Madame Bovary*, we saw how difficult it is to find out what to export from the fiction. There is no special evidence about how the world is that we can gather by engaging with the fiction. The same point seems to apply to mathematical fictions. To decide which claims to export in the literary case, we consider what we believe about history or psychology prior to engaging with the fiction. This is why we reject Dickens's sentimental psychology and find Flaubert's pessimistic psychology more plausible. In the mathematics case, what seems to be going on is that Balaguer, Rosen, and Yablo take themselves to have a clear grasp on the domain of the physical, and they have already convinced themselves that only physical things exist. Given these commitments, it is understandable why they would develop a fictionalist approach to mathematics and propose this style of export procedure. But this picture of mathematics and its applications clashes with the many positive ways mathematics can contribute to science, which we surveyed in part I. I developed an account of how mathematics helps us come to know what the physical world is like. I conclude that the fictionalist interpretation of mathematics must be rejected. At a minimum, the fictionalist about mathematics owes us a positive account of how mathematics contributes to the success of science and some argument for the compatibility of these contributions with fictionalism.[13]

12.4 MODELS

Let us say that a fictionalist about scientific models is someone who maintains that we can make sense of scientific representation without any beliefs concerning abstract scientific models. The models that we have placed at the heart of our account of representation are either concrete objects like scale models or else abstract models like mathematical structures. So fictionalists will reject the need to believe anything about an abstract model when they characterize the content of a scientific representation. Instead, they try to make sense of scientific practice by assigning fictional contents to these representations. These fictional contents pertain only to what is "true in the story of the model." Drawing on antirealist accounts of literary fiction, the fictionalist about models then insists that what is true in the story of the model need not involve any actual truths about the model itself. So, they conclude, it should be possible to provide an account of representation which stops short of requiring scientists to believe anything about their models. The models, except perhaps concrete ones, are fictional entities that fail to exist just as Emma Bovary fails to exist. Nevertheless, we can learn about the world by engaging with these models, just as we learn about the world by engaging with literary fictions like *Madame Bovary*.

Fictionalism about models is a growing and somewhat amorphous group of positions. Some philosophers who make a link between models and fictions are not fictionalists in the sense just defined.[14] But among those who explicitly adopt antirealist accounts of fiction and relate them to a lack of belief in abstract models, we find writers like Frigg, Godfrey-Smith, Suárez, and Toon.[15] In my discussion here, I focus on Frigg. This is because he explicitly rejects a fictionalist position for mathematics (Frigg 2010b, p. 265), so the arguments from the last section do not apply to his position. Other fictionalists about models adopt fictionalism

---

13. Leng (2009, §4.2) notes a similar "problem of mixed claims," but does not offer a solution. See Leng (2010, §8.1) for additional discussion.

14. See, for example, Contessa (2010).

15. Frigg (2010a,b), Godfrey-Smith (2006, 2009), Suárez (2009a, 2010a) and Toon (2010). See also many of the essays in Suárez (2009b). Maddy (2008) expresses some sympathy for fictionalism about models as she says "it is hard to see why that abstract model must 'exist'" (p. 35). See Thomson-Jones (2010), Psillos (2010) and Psillos (2011) for additional discussion of these issues.

for mathematics, thus creating additional problems for their position. Frigg also has made a clear argument in favor of fictionalism about models and has drawn extensively on Walton's account of literary fiction. All these positive aspects of Frigg's position suggest that whatever vulnerabilities can be found for it should carry over to the other fictionalists about models.

As I have presented things in this book, a scientific representation gets its content in three steps. First, we must fix the abstract structure that we are calling the model. This is a purely mathematical entity. Then some parts of this abstract structure must be assigned physical properties or relations. At this second stage, the parts of the purely mathematical entity are assigned denotation or reference relations. Finally, a structural relation must be given which indicates how the relevant parts map onto the target systems of the representation. At the end of these three steps, the representation has obtained its representational content. This means that it is determinate how a target system has to be for that representation be an accurate representation of that target system. As we have seen, representational contents may be quite intricate, especially when they result from the deployment of mathematical techniques associated with simplification and idealization.

Frigg presents an important argument that aims to show that this kind of account is not viable. He starts by saying,

> The view of model systems that I advocate regards them as imagined physical systems, i.e. as hypothetical entities that, as a matter of fact, do not exist spatio-temporally but are nevertheless not purely mathematical or structural in that they would be physical things if they were real. (Frigg 2010b, p. 253)

He then points out that an abstract mathematical structure, by itself, is not a model because there is nothing about it that ties it to any purported target system. But "in order for it to be true that a target system possesses a particular structure, a more concrete description must be true of the system as well" (Frigg 2010b, p. 254). The problem is that this more concrete description is not a true description of the abstract structure. It is not a true description of the target system, either, in the case of idealization. So, for these descriptions to do their job of linking the abstract structure to their targets, they must be descriptions of some other system. Frigg argues that these systems are the models after all.

My objection to this argument is that there are things besides Frigg's descriptions that can do the job of linking abstract structures to target systems. As I have just summarized, a link is possible that does not involve the correct description of anything. The representation gets its content, and if the representation is accurate, then we can obtain correct descriptions of the target system. But there is no need to find the mathematical structure in the target system to make the representation about that target system.

Although Frigg's argument is not conclusive, it gives a good sense of what he wants his fictional models to do. They must satisfy the concrete descriptions that scientists provide when they introduce a model. It might seem like this requires some kind of realism about the models themselves and their features. But Frigg makes clear that he intends to draw on Walton's antirealist account of literary fiction so that the hypothetical descriptions are merely true in the story of the model. Fictional models thus mediate between abstract mathematical structures and target systems by allowing claims to be true in the model.

Frigg recognizes that this approach to models and representation faces many obstacles. One challenge, of course, is to isolate a determinate fictional content. Frigg separates out the purely mathematical structure from his fictional model. He faces the challenge of specifying which parts of representation are purely mathematical and which involve genuine features of objects in his fictional model. As he assumes a platonist interpretation of mathematics, he is able to provide an account of these purely mathematical structures. At the same time, it is far from clear which concrete features should be assigned to the entities in the model. In any ordinary presentation of a scientific model, there is no explicit line of demarcation that indicates what gets assigned to the mathematics and what winds up in the non-mathematical model. Though Frigg deploys Walton's account of ordinary fiction to help make this clearer, I fail to see how it comes to terms with the sort of indeterminacy worries I have just raised for mathematical fictionalism. I don't want my objection to rest on this point, though, so let us assume determinate fictional contents for Frigg's fictional models.

Even if we assume that the fictional content of the model is determinate, it still remains highly indeterminate which features to export, so the representational content of the model remains mysterious. In the proposal we have articulated, our mathematical beliefs provide a domain of purely mathematical structures whose features we can investigate. Using this knowledge, we can then come to assign and understand the representational contents of our scientific representations even when the models are abstract. Then, by deploying ordinary processes of experimental testing and refinement, we can assemble evidence that the representation is accurate in this or that respect. I claim that the precision and definiteness with which scientists test these representations and judge them to be accurate or inaccurate indicates that something equally precise and definite is required in meeting the export challenge. That is, the fictionalist must say how we are to obtain the representational content from the fictional content. The fictional content involves some features that are clearly irrelevant to testing. For example, a hypothetical description considered in chapter 5 would require a model in which it is true that the ocean is infinitely deep (section 5.4).[16] But clearly this part of the fictional content will not be exported to the representational content. How, though, do scientists sort out these different contents and determine using experimental testing that the representation is accurate?

As far as I can tell, the main proposals on the table from fictionalists about models rely on the similarities between the fictional entities and the objects that make up the target system. For example, Frigg has focused on the similarities in the properties possessed by the objects in the model and in the target: "when I say that the population of rabbits in a certain ecosystem behaves very much like the population in the

---

16. We might also ask the fictionalist what time it is in a model if the model is introduced by the description "Let time go to infinity."

predator-prey model, what I assert is that these populations possess certain relevant properties which are similar in relevant respects" (Frigg 2010b, p. 263).[17] But this similarity strategy is undermined in ways that are closely analogous to the problems faced by the mathematical fictionalist. It cannot be that the representational content is obtained from the fictional content by considering how the model and the target are actually similar, for we do not know how they are actually similar. If we use actual similarity to fix the representational content, then we face the analogous problem to the view that whatever is actually physical gets included in the representational content of the mathematics. There the problem was that we ended up committed to the existence of things for which we had no evidence. Now the problem is that the representational content of the model is fixed using things we fail to know, so we fail to know the basic features of the representational content. For example, if heat is a fluid, then this is included in the representational content, but the scientist using the model can only know this if she knows that heat is a fluid. This picture fails to fit with scientific practice where we determine the accuracy of the representation by comparing what we find in the system with the content of the representation.

A more plausible suggestion is that we move from the fictional content to the representational content by considering what an agent believes. Here we do not consider what the agent believes to be physical, only what the agent believes is similar to the features of the target system. With this approach, if the agent believes that the model is similar to the target in some respect, then this respect gets included in the representational content. Again we face the analogous problem to the mathematical case. In the mathematical case, it was doubtful whether agents had clear beliefs about what was physical. Now it is doubtful whether agents have clear beliefs about how the model is similar to the target system. It is not usually the case that beliefs about these similarities are in place before the model is tested. Instead, beliefs about these similarities are, if ever, in place only after the model has been tested. So, although similarity beliefs may appear as the result of scientific investigation, they cannot be used to fix the representational content at the earliest stages of modeling a target system.

It might seem like engaging with the model could itself provide evidence that what goes on in the model is similar to what is going on in the target system. The problems with this line of reasoning are the same as what we saw for the claim that we can support a general psychological claim by engaging with *Madame Bovary*. Considering the fiction might develop skills that would be otherwise absent. It could convey information to us via testimony based on the prior belief that the narrator was a reliable source on this or that topic. Or, perhaps most promising, the fiction could be "expository" because it links some initial conditions to an outcome by showing how they were likely to occur. In the case of literary fiction, though, I argued that this sort of exposition was supported in the end only by prior beliefs about the subject matter (e.g., psychology). The processes illustrated in the fiction seem plausible only because we have already accepted that processes like this occur outside the fiction.

---

17. This point is repeated in Frigg (2010a). See also Godfrey-Smith (2006, pp. 737–738) and Godfrey-Smith (2009, pp. 104–105).

We were unable to find any way to support the view that events portrayed in a fiction provide a special sort of evidence over and above what is better gathered by engaging with the real world via ordinary experiment. I would argue for the same conclusion when it comes to fictional models. Considering what is true in the model along fictional lines can develop skills and focus our attention on a scientific claim. But nothing that occurs in the model provides evidence for a claim that we might wish to export to the representational content.[18]

In a more recent paper, Frigg has tackled the export problem directly with the following proposal, for model X and target Y:

X t-represents Y iff:
(R1) X denotes Y.
(R2) X comes with a key K specifying how facts about X are to be translated into claims about Y. (Frigg 2010a, p. 276)

In our terms, the collection of translated facts is the representational content of X. He goes on to insist that "the detailed study of different keys is a research programme to be undertaken in the future" (Frigg 2010a, p. 278) and provides two examples of how keys can be devised. One is just the "identity" key, which translates all facts from the model to the target. As Frigg notes, this is not a plausible key for any actual scientific model. The second key is the "ideal limit" key, which appeals to specific features of the target system and translates facts about the model with reference to these features. In his example, Frigg explains how to apply the ideal limit key to a model of a spinning top without friction for a target system of a top where frictional forces operate. The strategy is to use the actual small value of friction for the table to impose a bound on the gap between the actual rotation of the top and the idealized rotation of the top in the model. The details are not worked out, but the proposal is to restrict the representational content so that it is approximately correct.

However, it is fair to ask Frigg at this point how these keys make use of the fictional contents he has been at pains to articulate. Far from showing that "it is the fictional scenario that provides the content to the model" (Frigg 2010a, p. 284), Frigg's discussion of keys shows that the contents delivered by the keys are central to scientific testing and refinement. On the proposal for content that I have developed, Frigg's keys appear when we specify the hoped-for structural relation between the abstract model and the target system. If such keys are allowed, then there is no need for anything beyond an abstract mathematical model. The facts about models are purely mathematical, but once the model is interpreted we obtain nonmathematical claims about the target system. So, beyond the rejection of fictional contents, my favored proposal for representational content has all the flexibility and

---

18. There is an additional problem connected with representations that we deliberately construct so they are not similar to any target systems. See, for example, the three-sex species model noted at Wimsatt (2007, pp. 128–130). On my view, the representational content is inaccurate, but it is far from clear how the fictionalist can move from their fictional content to the representational content using similarity.

context-sensitivity that Frigg would wish for. For example, there is no difficulty obtaining a representation with the representational content of the spinning top model discussed by Frigg. We have already discussed a closely analogous case: the representation of irrotational fluids that we obtain from the Navier-Stokes equations when the viscosity is set to 0 (section 4.2).

To summarize, then, two major tasks face the fictionalist about models. First, for a given model, the fictional content must be specified. This fleshes out what is true in the story of the model, and I have granted for the sake of argument that this is possible. Second, for a given model with a fictional content, the representational content needs to be isolated. This content is used to determine if the model is correct for a given target system. My claim is that there is no need to appeal to features of the fictional content when giving these accuracy conditions. If this is right, then the fictional content is idle to any account of the success of science. We can assign our representations a fictional content, but the direct assignment of representational content is both necessary and sufficient to make sense of how we come to have scientific knowledge.

12.5 UNDERSTANDING AND TRUTH

I want to close this chapter by conceding that my arguments against fictionalism do not undermine all positions that deny that the understanding of a mathematical scientific representation requires belief in some mathematical claims. This is because there are some positions in the philosophy of mathematics which assign such a low threshold for belief in a mathematical claim that merely understanding the claim is sufficient for belief. According to this family of positions, the claim s: "the number of sheep is three times the number of goats" entails the claim s'': "there are numbers." But there is nothing about the way the world is that could prevent s'' from being true. As a result, we are free to accept these mathematical beliefs as true in the ordinary sense of "true." Against this sort of opponent, my objections based on the fictionalist approach to what is "true in the story" do not even get off the ground.

The most well-articulated approach along these lines is Azzouni's version of nominalism. He is keen to argue that truth is a central concept for our scientific and mathematical beliefs, and so grants that claims like s'' are true in just the same sense that the claim that there are kangaroos is true. But on Azzouni's view we must impose a further distinction between claims like s'', which are quantifier-committed to numbers, and claims about kangaroos, which genuinely ontologically commit the believer to the existence of kangaroos (Azzouni 2004, p. 127). The argument for this conclusion is that scientific criteria for genuine existence require that a scientist meet several stringent epistemic burdens. These include a robust sense of epistemic independence, an ability to refine our beliefs about the subject matter, some minimal sort of tracking or monitoring, and an account of how we are able to know about these entities (Azzouni 2004, p. 129). These scientific criteria are not met by mathematical entities, so Azzouni concludes that we are not committed to their genuine existence. Instead, certain aspects of mathematical practice indicate for Azzouni that mathematical entities are "ultrathin posits":

> Let's first consider those posits that have no epistemic burdens at all. *Ultrathin* posits are what I call such items, and they can be found—although not exclusively—in pure mathematics where sheer postulation reigns: A mathematical subject with its accompanying posits can be created ex nihilo by simply writing down a set of axioms; notice that both individuals and collections of individuals can be posited in this way. (Azzouni 2004, p. 127)[19]

Somewhat confusingly for our purposes, Azzouni adds in a note that "Ultrathin posits, of course, include fictional entities" (Azzouni 2004, p. 127). But this does not mean that Azzouni is a mathematical fictionalist in the sense of Balaguer, Yablo, Bueno, and Leng. Ordinary mathematical claims are true, and we are quantifier-committed to mathematical entities. It is just that these commitments turn out to be much thinner and easier to achieve than we expected. Merely understanding the claims that inaugurate a new domain is sufficient for grasping that these claims are true.

A more recent attempt to preserve belief and truth for ordinary mathematical claims is Rayo's trivialism (Rayo 2008, 2009). Trivialism is the view that

> nothing is required of the world in order for the truth-conditions of a mathematical truth to be satisfied: there is no intelligible possibility that the world would need to steer clear of in order to cooperate with the demands of mathematical truth. This means, in particular, that there is no need to go to the world to check whether any requirements have been met in order to determine whether a given mathematical truth is true. So once one gets clear about the sentence's truth-conditions—clear enough to know that they are trivial—one has done all that needs to be done to establish the sentence's truth. (Rayo 2009, p. 240)

A central innovation of this proposal concerns how one is to think about a semantic theory. On Rayo's approach, a semantic theory's primary aim is to clarify truth-conditions where these are understood as ranging only over circumstances an individual can render intelligible. Rayo grants that it can be a nontrivial task to determine which possibilities are intelligible to an individual. But the basis for our mathematical beliefs limits these possibilities, and so it follows that the falsity of such beliefs is, strictly speaking, unintelligible. As a result, there is no gap between understanding a mathematical belief, where that involves rendering it intelligible or unintelligible, and the truth-value of that belief. As with Azzouni, then, Rayo's position avoids my objections to fictionalism.

---

19. There are thin posits between the ultrathin and the ordinary thick posits that I skip due to considerations of space.

Though my arguments against fictionalism do not affect Azzouni and Rayo, we can see that they escape these arguments only at the cost of closing the gap between understanding and truth. According to the resulting picture of mathematical practice, mathematicians need not debate the truth of some newly proposed axioms because their truth has so little substance. This is an approach to mathematics that seems incorrect. I argue against it by an appeal to features of our mathematical concepts that I develop in the next two chapters.

13

Facades

13.1 PHYSICAL AND MATHEMATICAL CONCEPTS

Back in chapter 2, I articulated a structural account of the content of a mathematical scientific representation. The basic idea was that each representation says that a concrete system S stands in the structural relation M to some mathematical system S^* (section 2.2). To erect representations with these sorts of contents, a scientist must be able to refer to the concrete entities, properties, and relations of the target system as well as the mathematical entities, properties, and relations that make up the relevant mathematical structure. In chapter 2 I assumed an especially simple story about both of these capacities. For physical properties, I just took for granted our abilities to refer, even in the case of phlogiston where the purported stuff turned out not to exist. I also assumed that features of the world beyond what the subject was aware of could be part of the content along the lines of semantic externalism. For mathematical properties, a slightly more nuanced picture was proposed based on the notion of a core conception of a mathematical domain. I suggested that we could limit the mathematical content of these representations to what was determined by these core conceptions. This placed my approach to mathematical content more or less on the semantic internalist side. Both proposals for how the physical and mathematical concepts relate to the contents of our representations have their limitations. In this chapter I aim to determine what options are available to flesh out these assumptions about reference. Unsurprisingly, my aim is to consider which options are available that also preserve the account of the contributions of mathematics to science that I have discussed.

A useful way to probe these assumptions is to engage with Mark Wilson's *Wandering Significance: An Essay on Conceptual Behavior* (Wilson 2006). As I will explain, Wilson's challenge to what he calls the classical conception also seems to pose a challenge to my proposal about reference to physical properties. Furthermore, in the course of considering Wilson's views on physical concepts, associated issues about mathematical concepts come to the fore. This will allow us to sharpen our account of core conceptions and deliver a more nuanced picture of the contents that our scientific representations actually have.

The outline of this chapter is as follows. After summarizing Wilson's challenge to the classical approach to physical concepts, I discuss the central role of mathematics in his account. Wilson's positive proposal for how successful representations work is built around an analogy to some developments in the history of mathematics. To come to grips with Wilson's views, we spend some time considering these sorts of developments and what they tell us about our mathematical concepts. The upshot of our discussion is that the admissible ways to represent a physical property can become clearer in the course of scientific investigation. Analogously, the proper setting for a mathematical domain of investigation can reveal itself as we do more mathematics. These sorts of developments do not undermine a more modest approach to the determination of the content of our representations, but it does lead to a further qualification of what knowledge science can deliver. I end the chapter by surveying the implications of the "patient" realism I find in Wilson's work. As a bonus, our discussion of mathematical concepts allows us to criticize Azzouni's and Rayo's picture of mathematics based on the close association they offer between understanding and truth. The problem for these views will be that our mathematical beliefs develop over time in a way that is highly constrained by what we take to be the preexisting mathematical facts. The significance of this for our understanding of pure mathematics is considered in our concluding chapter 14.

13.2 AGAINST SEMANTIC FINALITY

The central feature of the classical approach to concepts that Wilson objects to is "semantic finality." This is the view that most of our ordinary concepts are grasped in a definitive way early on in our intellectual development so that the property that this concept picks out is fixed:

> with respect to a wide range of basic vocabulary, competent speakers acquire a *complete conceptual mastery* or *grasp* of their word's semantic contents by an early age—no later than 10 or 11, say. This core content then acts as an *invariant* that underwrites many of our characteristic endeavors: "If we don't share common, fixed 'contents,'" it is asked, "how can we possibly understand what others are talking about? For that matter, how can we be sure sure that we are addressing the questions we pose to ourselves?" (Wilson 2006, p. 19)

A philosopher convinced of semantic finality then must search for mental acts that are able to associate our words with concepts that have sufficiently rich contents. With Russell, this is obtained via his notion of direct acquaintance according to which a mental act puts us in contact with a property like being red or being square. But Wilson argues that the problems this approach encountered infect a wide range of philosophers, including Peacocke's more recent attempt to base concept possession on the right sort of acceptance of inferential rules (Peacocke 1992).[1] In both cases, the actual deployment of these concepts typically exceeds the strictures imposed by semantic finality. This leads philosophers into confusion concerning what standards are appropriate for evaluating our conceptual activities and in some

---

1. Peacocke's views are considered in more detail in chapter 14.

cases an overcompensating shift to a form of holism which severs any link between our concepts and genuine properties of objects in the world.

Wilson illustrates the negative consequences of the classical conception with several examples drawn from science. In science and its applications in experiment, engineering, and manufacturing, the failure of semantic finality is most dramatically demonstrated. As an extended example, Wilson considers the concept of hardness. Two advocates of the classical conception here are Descartes and Reid. Descartes insisted that "there is a sensation-type called 'resistance' and **hardness** simply represents the disposition to engender such feelings in the agent under appropriate conditions of testing" (Wilson 2006, p. 335). That is, we grasp the concept of hardness because we have an appreciation of what hardness is. For Descartes, this appreciation could only be delivered by a sensation, so hardness itself must be a dispositional property to produce a certain kind of sensation in the right conditions. Reid pursues semantic finality by different means. Dropping the need for a sensation, Reid identifies hardness with "the cohesion of the parts of a body with more or less force" and insists that it "is perfectly understood, though its cause is not: we know what it is, as well as how it affects the touch" (given at Wilson 2006, p. 336). So the sensations are irrelevant to the physical property. Reid then must struggle to articulate how we obtain the concept of hardness in such a definitive fashion.

Wilson argues that the commitment to semantic finality has blinded Descartes, Reid, and many other philosophers to the "fine-grained structure that we are unlikely to have noticed" that our "usage of the predicate 'is hard' displays" (Wilson 2006, p. 336). In fact, "our everyday usage is built from local patches of evaluation subtly strung together by natural links of prolongation" (Wilson 2006, p. 336). The challenge is to understand what these families of local patches, or "facades," amount to. How does our "everyday usage" relate to the different "patches," and what picture of conceptual content results from abandoning semantic finality and taking these facades seriously? The key piece of evidence that Wilson offers in the hardness case is the wide variety of ways we evaluate the hardness of an object in everyday contexts. These include squeezing, indenting, scratching, and rapping. The methods chosen vary with the material being considered as well as broader features of the context, like the purpose to which we aim to put the object in question. This tacit network of procedures is made more explicit when more sophisticated uses of materials in manufacturing and elsewhere lead to the development of instruments that test the hardness of materials in this or that respect. Each measuring technique develops out of our prior squeezing, indenting, and other practices, but offers further refinements via precision and in a more explicit recognition of the limits of validity for this or that measuring apparatus.

For Wilson's hardness example to have any bearing on Descartes and Reid, it must be the case that these measurement procedures fail to correspond to the means by which either philosopher assigned their conceptual contents. With Descartes this is easy to see because there is no sensation associated both with the squeezing of automobile tires and scratching of minerals. For Descartes, these cannot both be measurements of hardness. Reid's view is more difficult to criticize because it is not immediately clear what his notion of cohesion comes to. As Wilson explains, a crack may result in a plastic when we indent it due to the "pull" from the material as it piles up in the region around the indentation. This sort of cracking is less of a problem for metals, where the results of pushing down on the metal are most salient. So it is fair to ask Reid if the effects of pushing and pulling are relevant to his notion of cohesion or if only pushing matters. Given that Reid "is obviously not considering whether the 'force' comes in the form of a push or pull" (Wilson 2006, p. 340), his attempt to assign a clear conceptual content fails. If Reid were to go on to resolve this issue, he would wind up in the same boat as Descartes: our practices of measuring hardness turn out to not really all be measurements of that physical property.

A determined defender of the classical conception will view this situation in a negative light. It looks as if Wilson is showing that we lack a determinate concept of hardness. Each of Wilson's patches seems to correspond to a different concept that picks out a different physical property. So our original usage deployed an ambiguous concept whose multiplicity should be removed in a sufficiently regimented practice. Alternatively, one may insist that there is a single classical concept that covers all of these patches, but that the concept itself is multiply realized as we move from patch to patch. One problem with these responses is that they assume the classical approach is correct and aim to adjust our scientific practices to conform with this picture of how our concepts must work. The dangers of actually implementing these sorts of reforms are reflected in some of the foundational work carried out by scientists like Duhem and Hertz. These scientists pursued "radical panaceas" (Wilson 2006, p. 356) that sought to rid physics of any confusion once and for all. Their failures thus illustrate the more general failures of trying to force science to adhere to the classical conception.[2]

Wilson's own prescription is to take the patches associated with our scientific concepts as an admirable means to navigate the often overwhelming complexity of the physical world. The solutions to the problems arising within nineteenth-century physics came not from a foundational program based on the commitment to semantic finality but from further developments within physics and mathematics. Wilson argues that we can see, at any given time, that "successful instrumentalities, whether they be of a mechanical or a symbolic nature, always work for reasons, even if we often cannot correctly diagnose the nature of these operations until long after we have learned to work profitably with the instruments themselves" (Wilson 2006, p. 220). But we cannot conclude too much from this sort of success because we typically do not have a clear view of the network of patches we deploy. Getting clearer on this requires the hard work of further scientific investigation and the articulation of "correlational pictures" (Wilson 2006, p. 515) which put our instrumental success in a better light. This approach to science can be called a patient form of scientific realism. Science tells us much about the world, but exactly what it tells us may not be clear at a given time. As we do science more and come to understand the basis of its success, we can be more confident that we are getting this or that part right. This is not consistent with a final, fully rigorous theory (in the classical sense) of some domain, but it is enough to give us knowledge of this or that aspect of a domain of investigation.

---

2. See Wilson (2006, pp. 348–352) for the problems encountered by Samuel Williams in his attempt to remove the problems he found in our concept of hardness by erecting a single hardness scale.

13.3 DEVELOPING AND CONNECTING PATCHES

In my exposition so far, I have been careful to minimize the significance of mathematics for the articulation of the patches of Wilson's facades. In the hardness case, for example, it is fair to see the unanticipated features of materials as primarily responsible for the challenges of measuring hardness. But mathematics enters into Wilson's account in at least two places. First, mathematics itself provides pressures for the formation of new patches and information on the relationship between the patches of a facade. Second, the way our mathematical concepts have developed over time provides Wilson with a "prototype" (Wilson 2006, p. 317) for his own facades. The interaction between mathematical and physical concepts, then, is a central component of Wilson's positive picture of conceptual evaluation and development.

The way mathematics shapes the patches of a facade comes through clearly in Wilson's discussion of two examples, which we have already noted: shock waves and boundary layer theory (sections 4.3 and 5.6). The mathematics of our representation of the density of a compressible fluid allows the development of a discontinuous variation in density. This possibility follows from the sort of partial differential equation deployed in this representation. Part of the value of this representation is that it simplifies the representation of what happens on either side of the shock. This benefit is possible, though, only because what happens within the shock is collapsed down to a surface that we label the shock. As already noted in chapter 4, this is a central instance of "physics avoidance": "if we can examine a situation from several sides and discern that some catastrophe is certain to occur, we needn't describe the complete details of that calamity in order to predict when it will occur and what its likely aftermath might be" (Wilson 2006, p. 189). This approach to representing fluids permits an accurate representation of some class of target systems.

This sort of representation clearly cannot be extended to all cases, even when the fluid itself remains essentially the same physical stuff, for example, ordinary air. For as we saw with the boundary layer approach to the representation of air flow around an object like an airplane wing, a completely different mathematical structure is needed to do justice to this different situation. Now we consider the fluid as incompressible and eventually come to recognize the significance of the viscosity of the air as we approach the object. This leads us to impose another sort of surface between the thin boundary layer, where viscosity is most significant, and a further outer layer, where viscosity is discounted. A different class of mathematical equations results with their own restrictions and peculiarities. This is dictated partly by the features of the system itself, but also partly by the mathematics we have at our disposal. Again, as with the shock wave case, many of the complexities are avoided so that we can accurately represent the main features of concern. As Wilson says in connection with his own discussion of the boundary layer case,

> the macroscopic objects we attempt to treat in classical mechanics are enormously complicated in both structure and behavior. Any *practical vocabulary* must be strategically framed with these limitations firmly in view. To be able to discuss such assemblies with any specificity, our stock of descriptive variables must be radically reduced, from trillions of degrees of freedom down to two

or three (or smoothed out to frame simpler continua). Even systems that are quite simple in their basic composition often need to be partitioned into more manageable subsections, either spatially or temporally. (Wilson 2006, p. 184)

The most important means to carry out this partitioning is mathematical. Once we do this in several different ways, for several different kinds of target systems, we have articulated more of the patches that make up Wilson's facades. This allows us to accurately cover a wider range of target systems. Sometimes the patches overlap to a significant degree, as with the case of the outer layer of the boundary layer theory case and the treatment of ideal fluids via Euler's equations (section 4.2). Still, when we take this network of partially overlapping patches seriously, we recognize that we must give up the goal of unifying the patches into a rigorous theory of the sort envisaged by the classical conception. This is the price we pay for adequate coverage of a wide range of target systems.

What, then, is the relationship between the patches that deploy the concept of hardness or the patches that deploy the concept of fluid if the classical diagnosis of ambiguity or multiple realization is rejected? Wilson often compares the situation of the facade for a concept like hardness or fluid to the results of extending functions defined on the real numbers to the complex numbers. The analogy is subtle and depends on appreciating the details of the mathematics concerned, so we will spend some time unpacking what this extension amounted to. At the same time, this case can serve as a central example of how mathematics itself can develop over time. So it will help us appreciate the picture of pure mathematics that fits best not only with Wilson's account but also with the contributions mathematics makes to science, which have preoccupied us in this book.

As noted in chapter 4, complex numbers can be thought of as numbers of the form $a + bi$ where a and b are real numbers and i behaves like $\sqrt{-1}$. That is, $i^2 = -1$, so $(a + bi)(c + di) = (ac - bd) + (ad + bc)i$. They entered mathematical practice in the sixteenth century when mathematicians like Cardan used complex numbers to find the roots of equations like $x(10 - x) = 40$.[3] Though this algebraic technique proved helpful, it did not convince Cardan that these solutions had any genuine significance. As he put it on one occasion, "So progresses arithmetic subtlety the end of which, as is said, is as refined as it is useless" (given at Kline 1972, p. 253). This attitude seems to have persisted into the beginning of the nineteenth century. Even Euler, who was responsible for some of the most sophisticated uses of complex numbers in the eighteenth century, could contrast complex numbers with real numbers and insist that the former "are impossible numbers ... [that] exist only in the imagination" (given at Kline 1972, p. 594). As late as 1831 De Morgan could claim to have "shown the symbol $\sqrt{-a}$ to be void of meaning, or rather self-contradictory and absurd" (given at Kline 1972, p. 596). All this suggests that the extension of algebraic techniques to complex numbers was not sufficient to render the theory of

---

3. See Kline (1972, p. 253). Beyond Kline (1972), I have drawn on the accessible Nahin (1998) and Nahin (2006).

complex numbers acceptable. To say that they are "impossible" or "self-contradictory" is to say that they are admissible only as formal calculating devices.

What seems to have turned the tide in favor of complex numbers as legitimate mathematical entities was the way functions of real numbers were found to be extendable into functions of complex numbers. One of the most striking instances of this development was the articulation of what is known as Euler's formula:

$$e^{i\theta} = \cos\theta + i\sin\theta \qquad (13.1)$$

Here e is the base of the natural logarithm. It can be introduced as the nontrivial real-valued function which is identical to its own derivative, that is,

$$\frac{d}{dx}e^x = e^x \qquad (13.2)$$

Cosine and sine are the usual functions from trigonometry which concern the ratios of the sides of a right-angled triangle: for a right-angled triangle with angle θ, $\cos\theta$ is the ratio of the side *adjacent* to the angle to the hypotenuse of the triangle, and $\sin\theta$ is the ratio of the side *opposite* to the angle to the hypotenuse. All three functions had been studied extensively prior to the introduction of complex numbers with little hint that it would even be coherent to extend their domains in this radical way. A further remarkable implication of Euler's formula comes when θ is given in terms of radians so that π radians $= 180°$. Then $\cos\pi = 1$ and $\sin\pi = 0$, so

$$e^{i\pi} - 1 = 0 \qquad (13.3)$$

The truth of this claim was eventually widely accepted by mathematicians in the nineteenth century. A contemporary writer says it "reaches down to the very depths of existence" (Keith Devlin, given at Nahin 2006, p. 1). As Nahin relates, it scored first in the 1988 poll in the *Mathematical Intelligencer* of the most beautiful theorems in mathematics (Nahin 2006, p. 2). Among other things, this suggests that nearly all mathematicians take it to be true.[4]

There are two dramatically different ways to view the transformation of complex numbers from absurd to profound. The first involves an application of the classical conception to mathematical concepts. That is, we assume that our mathematical concepts have a fixed content that suffices to motivate their inferential features. So there may just be some very restricted domain of application for these mathematical concepts, as with Descartes's approach to hardness. On this view, the original concepts of e, cos, and sin picked out just the real-valued functions, and any extension must involve a new concept. This means that the acceptance of 13.1 involves a change in concept, perhaps for pragmatic or aesthetic reasons. An alternative

---

4. Although I do not develop the point further, the justification of Euler's formula provides one example of the link between aesthetic judgments and evidence that I considered when responding to Wigner and Steiner in chapter 8.

classical proposal is that our functions applied to complex numbers all along, so mathematicians like Cardan, Euler, and De Morgan simply failed to fully grasp the features of their own mathematical concepts. Finally, one could insist that our concepts were ambiguous or multiply realized. Ambiguity entails that it was merely a process of regimentation and clarification that led to the acceptance of 13.1. A proponent of multiple realizability must say in virtue of what a single concept is realized by both the real-valued e and the complex-valued e.

The second approach, which I wish to defend, rejects all of these options as an implausible reconstruction of this part of the history of mathematics. This forces one to take seriously a nonclassical picture of our mathematical concepts. It is this picture that Wilson deploys in his account of our physical concepts via his patches and facades.[5] The reason to take this route for 13.1 is that its rational acceptance depends on proof. Two popular proofs of 13.1 are sketched in Appendix D. They draw attention to mathematical facts arising out of the widely accepted work on real-valued functions and indicate how these commitments suffice to support 13.1. Beyond this, there are any number of consequences that can be drawn from 13.1 which bear on features of real-valued functions. These include the proof of the Poisson sum deployed in our discussion of the rainbow (11.4) and Gauss's proof that a 17-sided polygon was constructible with ruler and compass.[6]

All this is evidence that no change in our mathematical concepts was involved. At the same time, the earlier skepticism shown by Cardan, Euler, and De Morgan makes it very hard to argue that the validity of 13.1 is somehow implicit in these concepts all along. If it was implicit, then why was it so difficult for the best mathematician of the century, if not of all time, Euler, to appreciate the standing of complex numbers? Ambiguity is hard to motivate as the clarification and regimentation of our concepts was not sufficient to support 13.1. The only remaining classical option is to claim some form of multiple realizability. That is, the very same concepts that apply to the real-valued functions e, cos, and sin also apply to the analogous complex-valued functions. To carry off this proposal, the classicist must say in virtue of what our concept of e applies to both functions. An inspiration here could be similar claims about our concept of pain, which can arguably be multiply realized in organisms with brains and without brains. For pain the strategy is to focus on the functional role of the pain states in the life of the organism. We might argue that the two instances of e have enough in common to render our univocal concept applicable to both functions. The properties we might focus on here include the fact that both functions are identical to their own derivatives. The key question, though, is whether this sort of overlap in properties is the right way to conceive of the relationship between the real-valued e and the complex-valued e. Similarly, can we find properties of the fluids treated via shock waves, boundary layer theory, ideal fluids, and so on, so that it looks like a univocal concept of fluid is multiply realized? We find decisive evidence against

---

5. I should be clear that Wilson does not articulate the details of his nonclassical approach to mathematical concepts, but I believe that what I give here is in the spirit of his discussion.

6. See Nahin (2006, §5.5 and §1.6, respectively). This book is primarily concerned with spelling out the many important applications of 13.1.

the multiple realizability approach by considering how other functions fare when they are extended to the complex numbers.

Wilson rejects all these classical proposals based on the following analogy:

> A facade assembly should be regarded, in analogy to the two-sheeted Riemann surface for $\sqrt{z}$, as a *strategically informed platform* upon which a stable linguistic usage can be settled....As long as a speaker respects the boundary divides marked by δA, she can employ an unevenly founded language to freely express what she wishes locally, while exploiting the boundary restrictions between regions to create an overall employment that may prove more effective and efficient overall. (Wilson 2006, p. 379)

To appreciate the allusion to the complex-valued square root function, it is helpful to recast our approach to the complex numbers in more geometric terms. We can think of $a + bi$ as giving the coordinates for a vector in a two-dimensional plane whose x-axis is the real numbers and whose y-axis is the imaginary numbers, that is, numbers of the form bi for some real number b. Then we can shift from the representation of the number $a + bi$ in rectangular coordinates (a, b) to polar coordinates (r, θ), where r is the length of the vector given by $\sqrt{a^2 + b^2}$ and θ is the angle that the vector makes with the x-axis, measured in the counterclockwise direction. See figure 13.1. Notice that $a = \cos \theta$ and $b = \sin \theta$. Using 13.1 we can then represent any complex number as $re^{i\theta}$. When $r = 1$, $e^{i\theta}$ is a vector lying on the unit circle around the origin. Multiplication and exponentiation of such complex numbers is then handled using the ordinary rules for exponents: $e^{i\theta} e^{i\psi} = e^{i(\theta+\psi)}$ and $(e^{i\theta})^n = e^{i(n\theta)}$. In this approach complex-valued functions take points on one complex plane to points on another complex plane.

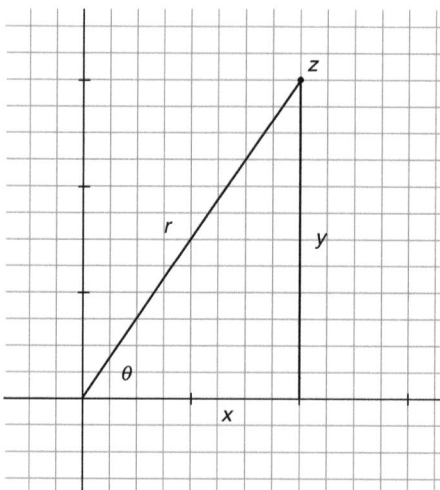

Figure 13.1: Polar Coordinates

So far, so good. But what happens if we consider extending other functions to the complex plane? In some cases the natural choices lead to "multiple-valued" functions, that is, functions with more than one output for the same input. This situation arises even for the square root function. To see this consider $\sqrt{e^{i\theta}} = (e^{i\theta})^{1/2} = e^{i\theta/2}$. We know that $e^{i\pi}$ and $e^{i(3\pi)}$ correspond to the same vector, so it looks like $e^{i\pi} = e^{i(3\pi)}$. But unfortunately $\sqrt{e^{i\pi}} \neq \sqrt{e^{i(3\pi)}}$. The former is $e^{i(\pi/2)}$, and the latter is $e^{i(3\pi/2)}$. The situation is not as bad as it may seem—further increases in the angle do not produce any new values. For example, $\sqrt{e^{i(5\pi)}} = e^{i(5\pi/2)}$ and $\sqrt{e^{i(7\pi)}} = e^{i(7\pi/2)}$. The former corresponds to the same vector as $e^{i(\pi/2)}$, whereas the latter corresponds to the same vector as $e^{i(3\pi/2)}$.[7] So we see that increasing the angle by 4π returns us to the same output.[8]

The simplest way to deal with this situation is to restrict the values of θ somehow and use this to avoid the appearance of multiple values. For example, we can require that $-\pi < \theta \leq \pi$. Then for this *branch* of the square root function, we only get one answer. This function is not defined for $e^{i(3\pi)}$. But we can allow a second branch of the square root function to be defined on $\pi < \theta \leq 3\pi$. We achieved these single-valued branch functions only at the cost of deleting the ray from the origin at the angles $-\pi$ and π, respectively. Our branch functions will not have a derivative on any point on the rays where they are undefined. This motivated Riemann to rethink the situation in terms of his notion of a Riemann surface. A Riemann surface has a more convoluted structure than the flat two-dimensional complex plane. For the square root function, we achieve it by stitching together *two* sheets using the rays where the square root functions just given are not defined. Think of each sheet as the complex plane with a ray removed. Then shift one part of one sheet upward so that one edge is above the other. The second sheet is then attached to the upper and lower edges of the first sheet. See figure 13.2. As Brown and Churchill describe things,

> A Riemann surface for $z^{1/2}$ is obtained by replacing the z plane with a surface made up of two sheets R_0 and R_1, each cut along the [negative] real axis and with R_1 placed in front of R_0. The lower edge of the slit in R_0 is joined to the upper edge of the slit in R_1, and the lower edge of the slit in R_1 is joined to the upper edge of the slit in R_0. (Brown and Churchill 1996, §77)[9]

A path on such a surface might start on R_0 at $e^{i(\pi/2)}$. Increasing θ keeps the vector on R_0 until we reach π where our first branch function becomes undefined. At this stage the vector would "ascend" to R_1 where the second branch function is defined. Continuing to increase θ for 2π radians would sweep through a circle on R_1. But at the angle 3π the second branch function becomes undefined. So it is necessary

---

7. That is, a unit vector pointing straight up and a unit vector pointing straight down.

8. Here we are considering how to extend the *positive* square root function, so the problem is not due to the fact that $n^2 = n \cdot n = (-n) \cdot (-n)$.

9. I have replaced "positive" by "negative" to mesh with our choice of branch functions.

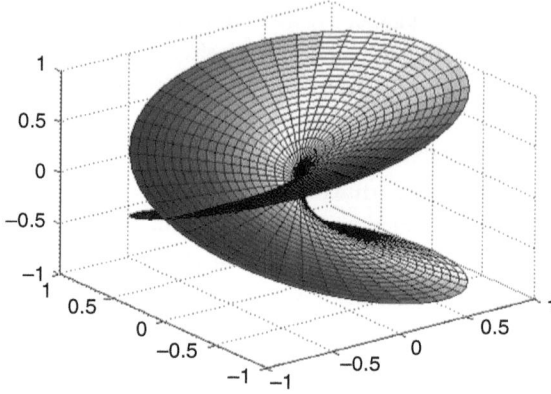

Figure 13.2: A Riemann Surface

to "descend" back to R_0. By employing the Riemann surface, one is able to extend the real-valued square root function to a surface that generalizes the complex plane in a way that achieves a single-valued function whose derivatives exist in as wide a domain as possible.[10]

The main question from a purely mathematical perspective is how we are to think of the square root function once we become aware of this situation. Speaking strictly, we might distinguish four different square root functions: the real-valued function, the first branch function defined on $-\pi < \theta \leq \pi$, the second branch function defined on $\pi < \theta \leq 3\pi$, and the function defined on the Riemann surface just described. My claim is that it is wrong to say that this is a situation where a classically conceived concept of square root is being multiply realized by these four functions. Instead, there is a clear sense in which the proper setting for the study of the square root function is its Riemann surface. This fits with the judgments of contemporary mathematicians like Weyl, who claimed, "Riemann surfaces are not merely a device for visualizing the many-valuedness of analytic functions ... but their native land, the only soil in which the functions grow and thrive" (given at Wilson 2006, p. 319). What this means is that the three other functions are derivative on the function defined on the Riemann surface. The relationship between our original real-valued function and the function defined on the Riemann surface is then not simply that they are two instances of a single concept. It is instead the relationship between a partial account of some domain and a deeper understanding of that domain. It is very hard to make sense of this in terms of the unfolding of something concerning our concepts. It is, rather, a deeper understanding that results from discovering features of things we had not previously been aware of. This realist approach to mathematics is the best way to make sense of these surprising developments in our understanding of functions of complex numbers. It seems that Wilson agrees with this picture, as he endorses Weyl's talk of "native land" in his own presentation of Riemann surfaces

---

10. The derivative is still not defined at the origin.

and says that it "provides a vivid picture of the special personality that the inferential principles natural to the expression '$\sqrt{z}$' display" (Wilson 2006, p. 319).

Finally, then, we can make sense of Wilson's analogy between Riemann surfaces and the domains in which we deploy our physical concepts like hardness and fluid. For fluids, there is every reason to think that there is some complicated set of interlocking target systems where fluids manifest very different features. This is analogous to the Riemann surface of the square root function. At the same time, we do not have a sufficient grip on the features of this complicated set, so we must resort to representations that accurately portray this or that region of the complicated set. These representations are Wilson's patches. They involve the successful treatment of a limited part of our target domain, so in line with Wilson's patient realism, it is reasonable to conclude that they are sufficiently accurate in relevant respects. In Wilson's discussion, he compares these patches to the unproblematic investigation of the square root function in a limited region of the complex plane using series expansions. Unfortunately, the limitations on a given series expansion may not be well understood as "inferential principles which make good sense locally often lose meaning on a global scale" (Wilson 2006, p. 318). Further investigation may locate the problems in the need for the two branch functions of the square root function. Of course, the domains of these branch functions also have boundaries where the branch function is undefined, so even here there are limitations to how much regularity we can impose on the situation. This resistance corresponds to the fact that each patch has its own boundaries, beyond which it can no longer be coherently extended. These limits come from features of both the target systems being represented and the mathematics used to structure the patches. Finally, the dogma of semantic finality closes off the possibility of adding more patches or linking them together in the sophisticated way scientists and engineers actually proceed. To complete the analogy, the classicist closes off the possibility that the complicated set of interlocking target systems has a structure anything like the Riemann surface of the square root function. In doing so, she stands in the way of getting clearer on how the world is by finding the "proper setting" for our scientific concepts.

13.4 A NEW APPROACH TO CONTENT

Let us now consider the potential problems that Wilson's patches and facades pose for our account of the content of a mathematical scientific representation. The key upshot of Wilson's arguments is that we cannot ascribe a substantial and fixed classical grasp of a concept to scientists. As a result, it is not possible for scientists to simply tack on a physical interpretation to a part of a mathematical structure when they devise a new mathematical scientific representation. To the extent that I have assumed this simple picture in my discussion so far, there is a need for a revision. But I want to argue that the revisions necessary to do justice to Wilson's discussion do not fundamentally alter the picture of how mathematics contributes to the success of science, which I have defended in the earlier chapters of this book. It is possible to factor in the complications that scientists encounter when they refer to physical properties. These complications create new epistemic barriers to accurately representing the physical world, and so can be seen as increasing the need for the epistemic contributions from mathematics that I have emphasized.

The simple picture I have deployed allows a scientist to have a substantial and univocal concept of hardness or fluid that she can deploy in the interpretation of widely different mathematical structures. As Wilson emphasizes, our actual concepts of hardness and fluid are much more open-ended than this picture would allow. We must allow for this open-ended character if we are to do justice to how science actually proceeds. That is, many of these inferential practices are not due to features of the concept itself, but the local context in which the concept is deployed. Here the context includes both the features of the target system and the mathematics at work in the representation. In the fluid case, some of the earliest successful treatments of fluids using mathematics involved the Euler equations for ideal fluids. This, then, is a successful patch for deploying the concept of fluid. The content of the representations associated with this patch can be handled using the recipe deployed in chapter 2. That is, once the purely mathematical structure is identified, some parts of it are physically interpreted so that this aspect of the mathematical structure comes to stand for this aspect of the target system. However, the meaning of the claim that the water in the pool is accurately represented by a representation in terms of Euler's equations for an ideal fluid must now be considerably weakened. As the content associated with the concept of fluid is quite minimal and open-ended, the content of the initial representations of fluids is also quite minimal and open-ended. Even this minimal content is consistent with testing the representation and revising it in light of experimental failures. At a certain point, though, the pressures created by experimentation and application in engineering suggest the development of a new patch. One indication that a new patch has been offered is a shift in the purely mathematical structure at the heart of the representation. A clear case of this was the shift from Euler's equations for an ideal fluid to Prandtl's boundary layer theory treatment. The new role for viscosity in the boundary layer was based on an empirical claim about the significance of the viscosity of fluids in these contexts. So the shift to the boundary layer patch involves representations with different contents. Crucially, we need not associate this change in content with a change in the concept of fluid. The change is instead based on additional claims about this class of target system. As the boundary layer theory representations prove empirically successful, these additional claims receive further support. So our interactions with the world and our experimental success contribute to the contents of the well-supported representations we have at our disposal. The concepts that physically interpret the mathematics play a role in this process, but their role is quite minimal. Features of the world play just as important a role as some prior grasp of a concept.

The same world-driven story can be told for the development of our mathematics. Initially we arrive at the representation of the e function based on the calculus and our prior representation of the real numbers. Taking this first step for granted, we have seen how pressures internal to real analysis prompted mathematicians to consider the results of extending e to the complex numbers. Special aspects of the e function itself allowed this extension to go smoothly. But the similar extension in the case of the square root function generated resistance in the form of a multiple-valued function. Rectifying this defect using Riemann surfaces further propelled mathematics into a new patch where mathematical considerations themselves led mathematicians like Weyl to claim that the proper setting for these functions had been isolated. This allows a retrospective reconstruction of what mathematicians were doing prior to the discovery of Riemann surfaces. As with the concept of fluid, there is no need to see the concept of the square root function itself as defective or in need of regimentation. Instead, the interaction between the minimal content delivered by this concept and the features of the domains themselves allow us to enrich our mathematical knowledge.

It is not desirable or possible, then, to factor our representations into contents arising from conceptual resources alone and contents arising from the world independently of these conceptual resources. Our concepts are necessary to get the process started because without them we would not be able to refer to purely mathematical structures or physical features of our target systems. But this minimal role in starting the process of mathematics and science is consistent with features of the mathematical domain and the physical world playing a more significant role in prompting new patches of successful representation. In this respect, Wilson's emphasis on the limited nature of the patches of a facade is reminiscent of Quine's and Putnam's attack on the analytic/synthetic distinction. Our concepts are simply not up to the task of providing a rich domain of conceptually grounded knowledge. Grasping concepts is necessary for science, but grasping a concept is just the first step in the process of developing sophisticated accurate scientific representations. Learning more about the world leads us to assert and test new claims beyond those licensed by our concepts. This flexibility is crucial to progress in mathematics and science.

What this shows is that our epistemic situation is actually worse than many philosophers of science seem to believe. In line with Wilson's patient realism, the successful treatment of a kind of target system using a patch in a facade should not lead to any simple acceptance of the "theory" associated with the patch. This is partly because the basis for the success of the patch may be poorly understood. Again, this was a recurring theme of chapter 7 on the ways premature interpretation and acceptance could lead to scientific failures. In the treatment of the rainbow in chapter 11, we saw how difficult it was to isolate what a successful explanation of supernumerary bows actually tells us about rainbows. So the resulting picture of our scientific and mathematical knowledge is highly restricted. Before we endorse a broader conclusion or theory on the basis of some successful treatment of a target system, we must be careful to check that we have a good understanding of why we have been successful. There is no simple recipe that can be deployed to check how good our understanding is, and the features of one case may not easily transfer over to another. However, in some cases (like our contemporary representation of fluids and rainbows), I believe the case can be made that we have reached that threshold. A central part of the story of how this threshold can be reached is the way mathematics makes the epistemic contributions it does. As Wilson emphasizes, these contributions do not come for free or without risk. But the articulation and extension of our representations is highly dependent on mathematics. Without the mathematics we have at our disposal, it is hard to see how any of the success cases we have discussed would have been possible.

13.5 AZZOUNI AND RAYO

To conclude this chapter, I want to argue that our discussion of complex numbers undermines Azzouni's nominalism and Rayo's trivialism. The structure of this argument is that these views link the understanding and the truth of mathematical claims too closely. As a result, they must place too much of a burden on our grasp of mathematical concepts. Their positions, then, come to resemble the unacceptable classical approach to mathematical concepts that I have argued is inconsistent with the history of this part of mathematics.

Recall that Azzouni insists that purely mathematical claims are settled by "simply writing down a set of axioms" which introduce a new concept (section 12.5). Rayo claims that mathematical truths are true on all intelligible possibilities (section 12.5). Given that these possibilities are fixed conceptually, there is nothing more to the acceptance of a mathematical claim than recognizing the possibilities that our concepts determine, or else shifting to a new concept. How is either proposal consistent with the acceptance of 13.1 or Weyl's conclusion about Riemann surfaces and the square root function? Clearly, writing down axioms for complex numbers and complex-valued functions was not sufficient to motivate the acceptance of complex numbers. For similar reasons, there is a problem for Rayo's view, as mathematicians seem to have taken what was once an unintelligible possibility and deemed it a fully intelligible possibility. If I am right that the shift was based on a deeper appreciation of the facts that obtain in the relevant mathematical domain, then both Azzouni and Rayo are in trouble. The truth of 13.1 is based on features of the domain that far exceed what can be reasonably ascribed to the concepts possessed by Cardan, Euler, and De Morgan. As we saw, all three classified complex numbers as unintelligible. So, as with the classical conception of concepts, Azzouni and Rayo must find a shift in our concepts or fall back on the multiple realizability of some univocal concepts. They offer no real discussion (as far as I am aware) of how a shift in concepts can be made for theoretical reasons, so they will have difficulty addressing the claim that it is right to accept 13.1 because it is true. Similarly, multiple realizability fails to do justice to the relationship between the real-valued square root function and the square root function defined on the Riemann surface. The widely accepted claim that Riemann surfaces are the proper setting for the square root function is not consistent with the view that our concept is just multiply realized. Multiple realization does not allow for a hierarchy of instances where some are more significant than others. Or, if this is the picture of mathematical progress that Azzouni and Rayo wish to defend, then the burden is on them to articulate it and show how it fits with the many different ways mathematics has changed in modern times.

14

Conclusion: Pure Mathematics

14.1 TAKING STOCK

In this concluding chapter, I want to take stock of the main claims I have made throughout this book and see what implications they suggest for the proper interpretation of pure mathematics and its epistemology. As will quickly become clear, I am sympathetic to the interpretation of pure mathematics known as structuralism. But it will prove difficult to come up with an argument that structuralism is mandated by our discussion of the contributions mathematics makes to the success of science. A second potential upshot of our discussion is that some mathematical knowledge is justified by a priori means. This is strongly suggested by the failure of indispensability arguments to support the substantial mathematical knowledge that makes its contributions to the success of science. However, as with structuralism, we have not assembled the materials to show in detail how this a priori justification is possible or what its limits might be. Instead, a more sustained investigation of the history and current practice of mathematics is needed.

To start, I set out the 12 claims given in chapter 1 that I have defended in this book.

1. A promising way to make sense of the way mathematics contributes to the success of science is by distinguishing several different contributions (chapter 1).
2. These contributions can be individuated in terms of the contents of mathematical scientific representations (chapter 2).
3. A list of these contributions should include at least the following five: concrete causal, abstract acausal, abstract varying, scaling, and constitutive (chapters 3–6).
4. For each of these contributions, a case can be made that the contribution to the content provides an epistemic aid to the scientist. These epistemic contributions come in several kinds (chapters 3–6).
5. At the same time, these very contributions can lead to scientific failures, thus complicating any general form of scientific realism for representations that deploy mathematics (chapter 7).
6. Mathematics does not play any mysterious role in the discovery of new scientific theories. This point is consistent with a family of abstract

varying representations having some limited benefits in suggesting new representations that are worthy of testing (chapter 8).

7. The strongest form of indispensability argument considers the contributions I have emphasized and argues for realism of truth-value for mathematical claims (chapter 9).
8. These contributions can be linked to explanatory power, so we can articulate an explanatory indispensability argument for mathematical realism (chapter 10).
9. However, even such an argument based on explanatory contributions faces the challenge of articulating a plausible form of inference to the best explanation (IBE) which can support mathematical claims (chapter 10).
10. This challenge to IBE for mathematical claims is consistent with mathematics contributing to successful IBE for nonmathematical claims, as in the extended example of the rainbow (chapter 11).
11. Fictionalist approaches to mathematics and scientific models face challenges that undermine their main motivations (chapter 12).
12. The way our physical and mathematical concepts relate to their referents suggests that our representations depend for their contents both on our grasp of concepts and our beliefs (chapter 13).

These 12 claims summarize the core of an answer to our initial question of why mathematics is so central to our best contemporary science. Mathematics is there because it is an ideal tool for arriving at well-confirmed and widely applicable scientific representations. Of course, mathematics is not the only input into the process leading to representations that we are right to trust in designing airplanes and building bridges. Experimentation and scientific ingenuity are also crucial ingredients. This fruitful cooperation between mathematics and experimentation is typically ignored in the philosophy of science. But based on what we have seen, I hope it is clear why Daniel Bernoulli's slogan is so appealing: "There is no philosophy which is not founded upon knowledge of the phenomena, but to get any profit from this knowledge it is absolutely necessary to be a mathematician."[1] We have not shown that using mathematics is the only way in principle for science to be done. However, it is hard to see what alternatives a scientist has in practice to learning and deploying a large amount of mathematics.

14.2 METAPHYSICS

What, if anything, does this tell us about the proper interpretation of pure mathematics? In chapter 9 I argued that several different realist interpretations of pure mathematics are consistent with the contributions I have emphasized. These include traditional objects platonism, ante rem structuralism, Hellman's modal structuralism, and Lewis's mereological interpretation. Each of these interpretations delivers a subject matter for mathematics that renders the right claims true and false.

---

1. Given at Truesdell (1968, p. 318) and Lin and Segel (1988, p. 22).

Conclusion: Pure Mathematics

In chapter 12 I explained why fictionalist interpretations run into problems. But it seems that any interpretation that assigns the right truth-values to the claims of pure mathematics is adequate to account for our epistemic contributions. The argument for this can be put briefly. These epistemic contributions have been analyzed in terms of the contents of our scientific representations. However, when mathematics is central to this content, the way we should make sense of the content is in terms of the existence and nonexistence of structural relations. So as long as the subject matter of mathematics is assigned the right structure, these contributions can obtain. As we saw with Hellman and Lewis, delivering the right structures need not involve the existence of abstract structures.

There is an influential argument in favor of objects platonism that I have so far ignored. This argument turns on applications, so it is important that I engage with it and see why it is not successful. This argument can be found in Frege. In its most succinct form, the argument is "it is application alone that elevates arithmetic beyond a game to the rank of a science. So applicability necessarily belongs to it" (given at Wilholt 2006, p. 72). That is, arithmetic, or any branch of mathematics that can be applied, is not just a game as imagined by some early versions of formalism. More to the point, this difference between a game and a genuine mathematical theory is a necessary part of the theory. As what is added to a formalist practice to get a genuine theory is also what is necessary to apply the theory, "applicability necessarily belongs to it." I am prepared to concede this argument, but also believe that its conclusion has little to do with the objects platonism defended by Frege. The key difference between a game and a genuine mathematical theory is that the sentences of a theory express claims. This feature of mathematical sentences is respected by many interpretations of pure mathematics, including Hellman's and Lewis's proposals. So if "applicability" means only expressing claims, then a structuralist can concede that applicability necessarily belongs to mathematical theories. As Frege himself says in this context, "Why can one get applications of arithmetical equations? Only because they express thoughts" (given at Wilholt 2006, p. 72).

Frege deploys a different argument when he engages with Mill's empiricist view that arithmetical claims are about physical regularities. This argument turns not simply on the applicability of a mathematical theory but on the scope of that applicability. With arithmetic, this scope is especially wide because it seems that any objects can be counted. Frege insists that empiricism is not able to account for the wide scope of the applicability of arithmetic:

> The basis of arithmetic lies deeper, it seems, than that of any of the empirical sciences, and even than that of geometry. The truths of arithmetic govern all that is numerable. This is the widest domain of all; for to it belongs not only the actual, not only the intuitable, but everything thinkable. Should not the laws of number, then, be connected very intimately with the laws of thought? (Frege 1980, §14)

For example, we can count the figures of the syllogism. On the assumption that these things are not physical objects, the empiricist is without an explanation of the applicability of numbers to this nonphysical subject matter. Frege's own proposal relates the applicability of numbers in counting to the applicability of a concept: "the content of a statement of number is an assertion about a concept" (Frege 1980, §46). As concepts have all sorts of objects falling under them, including nonphysical objects like the figures of the syllogism, the wide scope of the applicability of arithmetic is accounted for.

Frege takes his argument more seriously than this, though. Later in *Foundations of Arithmetic* he argues that the natural numbers should be identified in terms of the concepts to which those numbers are related in applications, especially counting. If this is right, then ante rem structuralism, Hellman's modal structuralism, and Lewis's mereological interpretation are all ruled out. This is because each interpretation fails to identify the natural numbers this way. Ante rem structuralists have an abstract structure whose positions are the natural numbers. But what it is to be the natural number 2, on the structuralist view, is determined by the relations between positions in that structure. Similarly, a modal structuralist does not identify the 2 of a possible structure in terms of its relations to anything outside the possible structure. Finally, a mereological approach interprets claims about the natural number 2 using mereological theory. Lewis winds up taking a fairly structuralist attitude to this theory, especially when it comes to the relationship between an object and the singleton whose only member is that object. So whatever relationship obtains between the natural number 2 and objects it is used to count is not used to identify the natural number. All three of these interpretations deploy what could be called a two-stage account of applications. In the first stage, the subject matter of the mathematics is identified. This identification turns on features that do not yet involve the way the mathematics is applied. In the second stage, then, a further explanation is needed of how the subject matter of the mathematics relates to the domain of application. Though differing in certain respects, all three interpretations can specify these relations in terms of the structural relations I have emphasized. In a precise sense, applications turn on merely structural relations, on these approaches. The existence of these structural relations does little to pin down what the mathematical entities are "in themselves."

On Frege's one-stage approach, of course, what a natural number is, is related to its application in counting. Does the scope of applicability of arithmetic support this strong conclusion? It does not. For a two-stage account need only ensure that the relevant structural relations have a scope that is up to the task of accounting for the wide range of applicability of arithmetic or whatever mathematical theory is under consideration. It is arguable that Mill failed in this task. He never articulated an account of applicability in terms of structural relations to the subject matter of his arithmetical claims. But this does not disqualify other approaches to mathematics that identify the subject matter of mathematics in physical terms. Lewis, in particular, can deliver structural relations with a wide enough scope to count whatever objects are counted in applications, including cases where we count the numbers themselves. So the second argument we have found in Frege fails to support objects platonism.

A third Fregean argument turns on considerations of meaning and language learning. Even Dummett, who is quite critical of many aspects of Frege's platonism, supports this sort of argument:

Conclusion: Pure Mathematics

> The historical genesis of the theory will furnish an indispensable clue to formulating that general principle governing all possible applications.... Only by following this methodological precept can applications of the theory be prevented from assuming the guise of the miraculous; only so can philosophers of mathematics, and indeed students of the subject, apprehend the real content of the theory. (Dummett 1991, pp. 300–301)

For Dummett, the "real content" of arithmetic involves the link between the natural numbers and counting, so in any philosophical account of arithmetic we must include this link. And, Dummett seems to think, this link should be part of how we identify the numbers themselves. This requirement has come to be known as Frege's constraint. As Wright summarizes it, "a satisfactory foundation for a mathematical theory must somehow build its applications, actual and potential, into its core—into the content it ascribes to the statements of the theory—rather than merely 'patch them on from the outside'" (Wright 2000, p. 324). For Wright, as for Dummett, a central argument in support of Frege's constraint turns on the way we learn arithmetic: "Someone can—and our children typically do—first learn the concepts of elementary arithmetic by a grounding in their simple empirical applications and then, on the basis of the understanding thereby acquired, advance to an a priori recognition of simple arithmetical truths" (Wright 2000, p. 327). Setting aside the point about a priori justification, Dummett and Wright seem to argue that our initial understanding of arithmetic is a crucial guide to what the natural numbers are. When we can see that this initial understanding involves applications of some sort, these applications should be part of our account of what the mathematical entities are. Applying this sort of test, Wright quite honestly concedes that this link between learning and applicability does not extend to all mathematical domains and so concludes that Frege's constraint need only be met in some cases (Wright 2000, p. 329).

Given the example of real-valued functions and Riemann surfaces from the last chapter, however, I think a case can be made that we should never just trust our initial introduction to a mathematical theory when we propose an account of its subject matter. The reason for this is that the concepts we use to get started thinking about a mathematical domain play only a very minimal role in tracking the features of that domain. To think otherwise is to overburden our mathematical concepts and thus block a reconstruction of the history of our increased knowledge of this or that part of mathematics. Perhaps with the natural numbers it is hard to take seriously the possibility of this sort of dramatic rethinking of what is going on. But at least from a structuralist perspective, the eventual characterization of the natural numbers in structural terms is of significant mathematical importance. Just as with the claim that the square root function finds its proper setting on a two-sheeted Riemann surface, the structuralist argues that what the natural numbers are is best approached structurally. True, this involves some change in our initial picture of what we study. But the debate between the Fregean and the structuralist is now carried out primarily in mathematical terms. There should be mathematical considerations which push us away from the Fregean approach and toward a more structuralist definition.

I am not able to pursue this debate here, but I hope it is clear that its resolution should not turn on how we initially learn the relevant mathematics, but on our current best theory of that mathematical domain. Tait defends the structuralist proposal against these Fregean arguments using Dedekind's definition of the natural numbers (Tait 2005). More recently, Parsons has argued that "a structuralist understanding of what the numbers are does not stand in the way of a reasonable account of their cardinal use" (Parsons 2008, p. 74), that is, their use in counting. For Dedekind, Tait, and Parsons, the picture of the use of the natural numbers in counting fits the two-stage approach. First, we identify the natural numbers in wholly structural terms. Then we explain the truth of claims involving counting using the existence of what I have called structural relations. In the simplest case, "There are n F's" is true just in case there is an isomorphism between the F's and the numbers from 1 through n. Some delicate issues arise for a structuralist in distinguishing "what the numbers are" from additional properties and relations that the numbers accrue in application. But there is little reason to think these concerns cannot be addressed for the natural numbers or any of the other domains of mathematics we have discussed.[2]

14.3 STRUCTURALISM

The intended contrast between objects platonism and ante rem structuralism is well summarized by Shapiro:

> The essence of a natural number is its *relations* to other natural numbers. The subject matter of arithmetic is a single abstract structure, the pattern common to any infinite collection of objects that has a successor relation with a unique initial object and satisfies the (second-order) induction principle. The number 2, for example, is no more and no less than the second position in the natural number structure; 6 is the sixth position. (Shapiro 1997, p. 72)

I have just argued that considerations based on applications cannot be used to support objects platonism against ante rem structuralism. But can anything arising out of our discussion of applications be used to support ante rem structuralism over objects platonism? Even if this sort of argument could be found, it would still not be enough to mandate an interpretation of mathematics in terms of abstract structures. There are any number of alternative interpretations that have a broadly structural flavor but dispense with abstract structures. The interpretations I have emphasized are Hellman's modal structuralism and Lewis's mereological theory.

The most influential argument in favor of structuralism is Benacerraf's discussion from "What Numbers Could Not Be" (Benacerraf 1983). Unfortunately, Benacerraf presents his own argument based on considerations of learning. He describes a situation where two children, Ernie and Johnny, learn about the natural numbers

---

2. See Parsons (2008, p. 75) for Parsons's own proposal in terms of the "internal" and "external" features of the natural numbers. See also Linnebo (2008) for a recent discussion of related issues.

Conclusion: Pure Mathematics

in different ways. Ernie comes to identify the natural numbers 1, 2, 3, ... with the sets {∅}, {∅, {∅}}, {∅, {∅}, {∅, {∅}}}, ... whereas Johnny treats the same numbers as {∅}, {{∅}}, {{{∅}}}, ... Ernie and Johnny agree that 1 is identical with the set whose only member is the empty set. But they disagree on the nature of 2. For Johnny, the only member of 2 is the set {∅}, but for Ernie 2 has two members, namely, ∅ and {∅}. Benacerraf's main point in his article is that this disagreement does not block either student from doing mathematics. This suggests that there is no mathematical reason to prefer one identification of the numbers with sets over another. But Benacerraf argues that if the numbers were objects, there would be such a reason as presumably which objects the numbers are would manifest itself in mathematical practice. The conclusion is then that the numbers are not objects, only positions in structures.

It seems like Benacerraf makes the same move from an acceptable introduction to the numbers to what the numbers are that we saw with Dummett and Wright. Analogously, it might seem like a person who learns about the square root function for real numbers is a reliable guide to what this function is. For this sort of argument to go through, there needs to be some additional premise saying that what the numbers are is given by what all the acceptable introductions to the numbers have in common. This sort of premise might come out of a certain view of how we acquire our mathematical concepts and what the concepts have to be like to allow us to think about the numbers. In this view, possessing the concept of natural number requires that our reasoning reflect the central features of the natural numbers. Dropping irrelevant accretions to this concept in an individual by considering what features are held in common across many individuals is then a sensible procedure. If this sort of refinement leads to a purely structural concept of the natural numbers, then this is evidence that there is nothing more to the natural numbers than their structural features.

The problem is that this approach to mathematical concepts is not able to do justice to my reconstruction of our best current theory of the square root function. If we take this sort of case seriously, then mathematical concepts are poor guides to the features of a mathematical domain. The most important features of this domain will reveal themselves only when we learn more about that domain. A structuralist, then, is well advised to emphasize these mathematical considerations in favor of structuralism. But as these sorts of reasons for structuralism turn on pure mathematics and its development, I must set them aside here. The structuralist approach to applications might naturally suggest a structuralist interpretation of pure mathematics, but there is no clear argument from the one to the other. The structuralist approach to applications is too open-ended to rule out any interpretation of pure mathematics that renders the right structural claims true.

14.4 EPISTEMOLOGY

Turning from the interpretation of mathematics to its epistemology, I have repeatedly suggested that some sort of a priori justification for pure mathematics is needed if we are to make sense of how mathematics makes its epistemic contributions to the success of science. The issue first arose in connection with constitutive contributions in chapter 6 and received a more sustained treatment when I discussed explanatory indispensability arguments for realism in truth-value about mathematical claims in chapter 10. In the constitutive case, I argued that the argument that delivers the practical reasons to believe certain constitutive representations requires that the relevant mathematics be justified to a substantial degree. At that point it was hard to see how this mathematics could receive this justification empirically, although in chapter 6 I left open the possibility that some form of indispensability argument could achieve this. However, the discussion of these indispensability arguments in chapters 9 and 10 concluded that there is little hope of using the presence of mathematics in our best scientific theories to empirically support the mathematics. Instead, I claimed that the contributions which mathematics makes to science require that the mathematics be justified prior to its use in science. That is, some central parts of mathematics are justified a priori.

This is a significant cost of my proposal for making sense of how mathematics helps in science. There is no adequate epistemology for mathematics that renders it a priori. But in the space remaining I want to suggest a plan of attack on this daunting problem. I suggest that an a priori epistemology for mathematics must take seriously the now-familiar example from chapter 13 involving the square root function and Riemann surfaces. This example shows how evidence can be produced in support of a mathematical claim even though that evidence is not tied to the concepts involved in the claims. At the same time, it is arguable that this sort of evidence is a priori because it does not depend on perceptual experience of the physical world. This "extension-based" epistemology, as I call it, overcomes the problems that I raise for its main competitors. These are an exclusively concept-based approach insisting that all evidence for a priori claims be conceptual and an empiricist approach demanding that concepts be grounded in perceptual interactions with the physical world.

Peacocke has argued that a theory of our concepts and how they allow us to refer to things in the world is sufficient to explain a substantial body of a priori knowledge.[3] The basic idea of this approach is that possessing a concept requires having certain abilities, and these very abilities are sufficient to provide us with "entitlements," which are a priori. The notion of an entitlement can be thought of as a reason for a belief that an agent may not have access to. Then, deploying Burge's definition, an entitlement is "a priori if its justificational force is in no way constituted or enhanced by reference to or reliance on the specifics of some range of sense experience or perceptual beliefs" (Burge 1993, p. 458).[4] For Peacocke, possessing a concept is having a certain ability, namely, the ability to think about the entity the concept refers to. However, a satisfactory explanation of what this ability amounts to should analyze it into something more basic. Initially, Peacocke seems to have hoped that the abilities at issue could be reduced to a collection of more basic abilities of a certain specific kind, that is, the ability to draw inferences or make transitions between experiences and beliefs. More recently, Peacocke has argued

---

3. My main sources here are Peacocke (1998a, 1998b, 1998c, 2004). See also Peacocke (1992, 1993).

4. See also Burge 1998 for a deployment of this notion of entitlement in a mathematical context.

Conclusion: Pure Mathematics

that a more relaxed conception of possession conditions is required to explain our full range of conceptual abilities. This more relaxed conception does not insist on a reduction of the ability in question to a series of more basic abilities, but seeks to explain it by ascribing to the thinker an implicit conception. An implicit conception of a given concept takes the form of a series of principles involving the core features of the concept. Peacocke's thought seems to be that a thinker's most basic grasp of a concept can be captured by these principles and the full range of the thinker's conceptual abilities can be reconstructed using such implicit conceptions and the thinker's reflection on them.

On both the more restricted conception of possession conditions and the more relaxed notion of implicit conceptions, Peacocke imposes a strong condition of adequacy on any particular proposal concerning a concept. This is that the alleged possession conditions for a concept C be able to uniquely pick out the referent or semantic value of the concept. Among other things, this requires that the possession conditions be able to distinguish the intended referent of C from all other candidates:

POSS. An adequate possession condition for concept C must be sufficient to distinguish the semantic value of C from other entities of that kind.[5]

POSS is fairly easy to motivate if we are trying to explain all a priori knowledge in terms of concept possession. Our goal in articulating the possession conditions of a given concept is to account for the thinker's ability to think about the semantic value of that concept. If a possession condition did not include features sufficient to distinguish this semantic value, then it seems that an examination of the possession condition would not settle whether the thinker was thinking about the intended entity E or some other entity E' that shares some but not all of the features of E. But if the possession conditions do not settle this issue, they are thereby apparently inadequate to explain how the thinker can think of E.

It is hard to know how demanding POSS is really intended to be. For example, does POSS require that the possession conditions distinguish the intended semantic value from all logically possible entities or merely from all the actual entities the thinker could be thinking about? In an effort to be fair to Peacocke, I read POSS in the weaker sense: for a possession condition of a concept to be adequate, it must be sufficient to distinguish the semantic value of that concept from all other actual entities.

Hill has argued that Peacocke's proposal in *A Study of Concepts* fails to meet POSS (Hill 1998). Imagine that we have completed a specification of the concepts pertaining to the natural numbers such as 0, successor, addition, and multiplication. Presumably the possession conditions can do no better than the Peano axioms for arithmetic and the standard recursive definitions of addition and multiplication. But if we think of these axioms as part of a theory given in the language of first-order logic, we can prove that the theory admits of a number of different models including

---

5. See, for example, Peacocke (2004, p. 186).

so-called nonstandard models that are different in structure from the intended standard model whose domain includes just the natural numbers. The problem is now that there seems to be no way to use the possession conditions of these concepts to fix their semantic values. There will be a variety of potential referents for our mathematical concepts, each of which fits with the possession conditions.

Peacocke responds to Hill by presenting an implicit conception of the concept of natural number that he believes can meet POSS. Following Dummett and Wright, Peacocke tries to pin down arithmetical concepts by linking them essentially to the canonical application of counting, endorsing Frege's claim that "it is applicability alone which elevates arithmetic from a game to the rank of a science. So applicability necessarily belongs to it" (Peacocke 1998a, p. 107). First, he links 0 and successor to their canonical applications using principles P1 and P2:

P1. $Nx(Fx) = 0 \leftrightarrow \neg \exists x(Fx)$
P2. $Nx(Fx) = sa \leftrightarrow \exists x(Fx \& Ny(y \neq x \& Fy) = a)$

Second, he claims that this pins down successor by noting that P1 and P2 imply S1 and S2, two core features of the successor function:

S1. $0 \neq sa$
S2. $sa = sb \rightarrow a = b$

Finally, Peacocke presents three principles that, together with the previous principles about successor, are meant to ground our ability to think about the natural numbers:

i. 0 is a natural number.
ii. The successor of any natural number is also a natural number.
iii. Only what can be determined to be a natural number by clauses i and ii is a natural number.

He goes on to claim that reflection on the "can" contained in iii is sufficient to rule out nonstandard models.

Although one could question the particular choices made by Peacocke here, especially in light of our earlier rejection of Frege's constraint, the basic strategy should be clear. An agent who possesses the concept of natural number is able to think about natural numbers because she accepts the appropriate implicit conception. This acceptance may occur on a subpersonal level that she is not aware of, but as long as her thinking is guided by this implicit conception she can come to have justified a priori beliefs about natural numbers. This is because the way she arrives at these beliefs tracks the real features of the natural numbers themselves, which are captured by the implicit conception. The resulting picture of a priori knowledge is highly externalist. An agent can come to know that there are infinitely many natural numbers without having a full awareness of what justifies this belief. At the same time, Peacocke's approach to a priori knowledge bypasses the implausible attempt to find some causal connection between our mathematical beliefs and its subject matter.

To illustrate in more detail how Peacocke's proposal is supposed to work, let's consider a theorem from group theory: every group of prime order is cyclic. A group

Conclusion: Pure Mathematics

is a set of objects with a binary operation $*$ defined on the set that meets three conditions:

G1. $(a * b) * c = a * (b * c)$ for all a, b, c in the set (associativity),
G2. There is an element e in the set such that for all a in the set,
$e * a = a = a * e$ (identity),
G3. For each a in the set there is an a' such that $a * a' = a' * a = e$ (inverse). (Fraleigh 1999, p. 52)

A prime order group is a group whose set is the size of some prime number. Finally, a cyclic group is a group where for some a, every element of the group is identical to $a * \ldots * a$ (n times), for some n. In Peacocke's view, learning these definitions tells you all you need to know to refer to groups, if they exist. It exhausts the implicit conceptions of group, prime order group, and cyclic group. Just on the basis of possessing these mathematical concepts, along with some logical concepts, an agent becomes entitled to believe the axioms of group theory and their trivial consequences. For example, an agent becomes entitled to believe that there is no element in any group that lacks an inverse. Reflection on G3 and logical concepts like negation grounds this entitlement.

It should be clear in what sense this entitlement is a priori. We distinguish between the experiences necessary to acquire a concept and the experiences (if any), that figure in the justification of a belief involving that concept. Even though, as Frege said, "without sense impressions we should be as stupid as stones" (Frege 1980, §105), no perceptual experiences or perceptual beliefs figure in the justification of the claim that there is no element in any group that lacks an inverse. The justification begins with the axioms of group theory and involves a simple application of logical inference rules. No appeal is made in the proof to what the agent perceives.

The hope, of course, is that the same sort of thing can ground our entitlement to the theorems of group theory even when their justification involves a long series of steps. For the claim that every group of prime order is cyclic, the series of steps is not that long, but long enough to make the result somewhat surprising and even interesting.[6] Peacocke must believe that all those who possess the concept of group and the required logical concepts are in fact entitled to believe this theorem, even if they do not realize this and even if they are in some sense psychologically precluded from ever realizing it. It might be, for example, that I am prone to run from the room whenever I hear or read the word *cyclic*. A patient instructor might eventually get me to acquire all the relevant concepts, but I am not psychologically able to learn the proof of our theorem. This deprives me of an awareness of what I am entitled to, but the entitlement nevertheless remains.

Focusing on possession conditions of concepts seems to allow us to explain how a priori entitlements are possible. Assume first that anyone with the implicit conception of group is in a position to refer to groups if they exist. Second, groups do actually exist. So anyone with the implicit conception is referring to groups. Furthermore, their implicit conception picks out genuine features of these entities,

---

6. See Fraleigh (1999, p. 124).

and so the proof that begins with these genuine features and follows valid inference rules will terminate in a true claim about groups. That is, the agent winds up with an entitlement to a belief about groups, namely, that all groups of prime order are cyclic.

14.5 PEACOCKE AND JENKINS

Unfortunately, there are major problems with Peacocke's whole approach to mathematical knowledge. The most obvious is the claim that groups exist. Suppose we take for granted our current mathematical beliefs, and so we can conclude from a purely metaphysical perspective that groups really do exist. Still, we can wonder if the existence of groups is sufficient to confer the entitlements in question on someone, even if that agent has not done anything to establish their existence. In ordinary scientific cases we do not allow a scientist to conclude anything from merely reflecting on their concepts. For example, to establish the existence of dark matter, we require not only that scientists acquire the concept of dark matter, but that they go out and develop some experimental evidence supporting the claim that dark matter exists. The worry, then, is that we are in a similar sort of situation with respect to groups.

An inadequate response to this objection is to insist that whichever mathematical entities exist, exist necessarily. This might seem like a promising strategy, as one can then argue that an agent who possesses the implicit conception of group and whose thinking adheres to this conception will by necessity arrive at true claims about groups. From a purely reliabilist perspective, there is nothing more that could be asked from the agent. In all possible worlds, once she has acquired the implicit conception of groups, her group-directed beliefs are true.

Jenkins has recently offered an objection to Peacocke that undermines this response. Her basic concern is that even a perfectly reliable concept is not conducive to knowledge if our possession of that concept came about in the wrong way. She compares two cases where we wind up with what is in fact a highly accurate map. In the first case, which is analogous to a Gettier case, a trustworthy friend gives you a map that, as originally drawn up by some third party, was deliberately inaccurate. The fact that the map has become highly accurate and you have some justification to trust it is not sufficient to conclude that your true, justified beliefs based on the map are cases of knowledge. In the second case, simply finding a map that happens to be accurate and trusting it "blindly and irrationally all the same" will block any beliefs you form from being cases of knowledge. Jenkins extends these points about maps to our mathematical concepts: "A concept could be such that its possession conditions are tied to the very conditions which individuate the reference of that concept...but not in the way that is conducive to our obtaining knowledge by examining the concept" (Jenkins 2008, p. 61). The first sort of problem could arise if the concept came to us via some defective chain of testimony. For example, a "crank" mathematician develops a new mathematical theory with his own foolishly devised concepts and passes them off to our credulous mathematics teacher, who teaches the theory to us. The mere fact that the crank mathematician has happened to pick out the right features of some mathematical domain is insufficient to confer

Conclusion: Pure Mathematics

knowledge on us. The second kind of problem would come up if I was, based on a failure of self-knowledge, like the crank mathematician myself, coming up with new mathematical concepts based on my peculiar reactions to sudoku puzzles. Again, this approach to mathematics would not lead to knowledge even if my concepts happened to reflect genuine features of the mathematical world.[7]

It is not clear to me how Peacocke can respond to Jenkins's challenge. Her diagnosis of the problems with Peacocke's approach to a priori knowledge is compelling:

> The lacuna in Peacocke's project is that, even assuming everything he says is true, he has offered us no reason to think that concepts are anything better than an in-fact-accurate guide. To do more than that he would need to describe an input step into the process of concept examination—some kind of impact on us by the world, in virtue of which we come, non-accidentally, to possess and be guided by the right sorts of concepts. (Jenkins 2008, p. 62)

Here Jenkins's empiricism comes to the fore. She claims that someone already committed to empiricism is able to do better than someone like Peacocke. An empiricist believes that perceptual experience must play a role in our thinking for the relevant true beliefs to be cases of knowledge. But she points out that an empiricist need not claim that these perceptual experiences are part of the *evidence* for what we know. Instead, the perceptual experience may serve merely to *ground* the concepts we deploy in our thinking. On this approach, one can maintain that our mathematical knowledge is a priori because our evidence is purely conceptual. Jenkins thus retains Peacocke's assumption that our mathematical concepts are substantial enough to exhaust the evidence for our mathematical beliefs. At the same time, she departs from Peacocke by arguing that our mathematical knowledge is empirical because perceptual experiences are needed to ground our mathematical concepts (Jenkins 2008, p. 147).

The central question for Jenkins, then, is how perceptual experience can ground our mathematical concepts so that the problems Peacocke runs into are avoided. She restricts her focus to arithmetic, especially the much-discussed $7 + 5 = 12$. The notion of grounding is explained in terms of what is needed to bridge the gap between a concept that leads to accurate beliefs and a concept that leads to cases of knowledge:

> Fitting concepts are concepts which either refer themselves or else are correct compounds of referring concepts [etc.]....
>
> Given some purported a priori knowable proposition p, we can say that a concept C is *relevantly accurate* ... iff C is fitting and neither C nor any concept from which C is composed misrepresents its referent in any respect relevant to our purported a priori way of knowing that p....
>
> A concept is *grounded* just in case it is relevantly accurate and there is nothing lucky or accidental about its being so. (Jenkins 2008, pp. 127–128)

---

7. The mathematical versions of these problems are my own, not Jenkins's. See Jenkins (2008, §5.6) for her discussion of concepts of this sort.

Setting aside some of the complications associated with concept composition, we can see how a claim involving grounded concepts has a better chance of being a case of knowledge. In ruling out the sort of accident or luck that arose in our crank mathematician cases, we are assured that the true, justified beliefs which we form are justified in the right kind of way to ensure knowledge.

Still, for one who is not yet an empiricist, it is important to consider how perceptual experience can perform the grounding task for our mathematical concepts. Here Jenkins admits that she does not have a detailed, positive story for how this works, although she does offer two potential models that could be vindicated by scientific investigation (Jenkins 2008, §4.5). In lieu of these accounts of how experience grounds the relevant concepts, she points to a feature of these concepts that is evidence they are grounded. This feature is that these concepts are indispensable to our best, current science:

> the structure of our sensory input is our best guide to the structure of the independent world. I furthermore suggest that concepts which are indispensably useful for categorizing, understanding, explaining, and predicting our sensory input are likely to be ones which map the structure of that input well. These two suppositions together mean that concepts which are indispensable for understanding our sensory input are probably ones which our best available data suggest correspond to the structure of the independent world; that is, ones it is rationally respectable to rely upon as accurate representations of—and guides to—the independent world. (Jenkins 2008, pp. 144–145)

This point needs to be carefully distinguished from an indispensability argument for the truth of the mathematical claims that appear in our best science of the sort criticized in chapters 9 and 10. Colyvan, for example, argues that the role of these mathematical claims in our best science is the evidence that these claims are true. I criticized this argument based on the difficulty in seeing how this role was sensitive enough to support the truth of the strong mathematical claims appearing in scientific explanations. Recall the difficulty in justifying the claim that prime periods minimize intersections rather than the weaker claim that prime periods up to 100 minimize intersections. But Jenkins does not use this role as her evidence that the claims are true. Instead, this role is evidence only that the concepts are grounded. Given that they are grounded, purely conceptual reasoning of the sort reflected in a mathematical proof can lead to knowledge that prime periods minimize intersections.

Our specific objection to indispensability arguments does not affect Jenkins's position. Still, the more general lessons learned from that discussion should make us wary of accepting her epistemology for mathematics. Again and again we have seen how an appeal to mathematical structures has helped in science without any connection to perceptual experience. This raises the prima facie worry about whether our perceptual experiences are up to the task of grounding our mathematical concepts. The more sophisticated uses of mathematics that we have considered seem to have nothing to do with whatever "structure" we might find in our perceptual experiences. To be fair, Jenkins restricts her focus to arithmetic, so perhaps it is less clear what

Conclusion: Pure Mathematics

the problems might be for grounding these concepts in our experience. And it is plausible to imagine that concepts grounded by perceptual experience could lead an agent to know $7 + 5 = 12$ through an examination of these concepts. But there is more to our arithmetical knowledge than claims about sums and products of small natural numbers. As far as I am aware, Jenkins nowhere provides any positive case for how we know that there are infinitely many primes or that Fermat's last theorem is true. For example, she does not articulate anything like Peacocke's implicit conception for the natural numbers reviewed earlier.[8] The main positive argument for her view is the claim that our arithmetical concepts must be grounded if they are to lead to knowledge and the empiricist assumption that grounding is possible only via perceptual experiences. But for one who is not yet a committed empiricist, it is hard to see why we cannot both accept the former claim and reject the latter. Indeed, Jenkins may herself allow this more flexible option to account for our knowledge of more advanced mathematics. From this perspective, the indispensable role of our arithmetical concepts is not evidence that they are grounded in experience. There may be other means to ground our concepts. More generally, there may be nonconceptual sources for the justification of our mathematical beliefs that do not require Jenkins's sort of grounding.

14.6 HISTORICAL EXTENSIONS

I turn now to a second major problem with Peacocke's approach to a priori knowledge, based on the assumption that all justification for our mathematical beliefs must be traced back to the features of our concepts. To the extent that Jenkins agrees with Peacocke that the evidence for all mathematical knowledge is conceptual, her position is also undermined by the problem to be considered in this section. This paves the way for an alternative approach to mathematical knowledge.

The problem is that Peacocke is forced into an implausible reconstruction of episodes in the history of mathematics like the square root function and Riemann surface case. On Peacocke's view, all justification in these mathematical cases is tied to the implicit conception of the domain. So if a mathematical discovery occurs, and we take the beliefs of the mathematicians to be justified, then we must ascribe to them an implicit conception that is substantial enough to justify their beliefs. In the square root function case, this proposal requires that it is part of the implicit conception of this function that it is defined for the complex numbers and even that its proper setting is the two-sheeted Riemann surface. That is, mathematicians like Cardan, Euler, and De Morgan had this implicit conception of the square root function but were unable to articulate it, and so mistakenly rejected these sorts of claims about that function. Only with Riemann or perhaps Weyl could what was implicit all along be made explicit.

Peacocke does not make these claims for the history of complex analysis, but he makes the equally implausible point that Newton and Leibniz had our implicit conception of limit. It was just difficult for them to appreciate this. Indeed, according

---

8. See Jenkins (2008, pp. 260–261) for an honest assessment of what remains to be shown.

to Peacocke, many mathematicians had this very implicit conception prior to its explicit articulation by Cauchy and Weierstrass:

> There are some cases in which a thinker has an implicit conception, but is unable to make its content explicit. The thinker may even be unable to formulate principles distinctive of the concept his possession of which consists in his possession of that implicit conception.
>
> One of the most spectacular illustrations of this is given by the famous case of Leibniz's and Newton's grappling with the notion of the limit of a series, a notion crucial in the explanation of the differential calculus. It would be a huge injustice to Leibniz and Newton to deny that they had the concept of the limit of a series ... (Peacocke 1998b, p. 49).

This is because they could arrive at limits for particular cases. Their inability to articulate the appropriate general account of limits, then, is merely a failure to grasp something about their own implicit conception. This implicit conception of limit is shared by the mathematicians in the mathematical tradition leading up to Weierstrass. On the basis of this common implicit conception, all these mathematicians were talking about the limit of a series. Furthermore, the widespread acceptance of Weierstrass's definition is best explained by appeal to this common implicit conception.

The problem with this proposal, of course, is that nothing Leibniz or Newton did shows that they had this implicit conception. The general assumption that the only source of justification is a shared implicit conception drives Peacocke's reconstruction, not anything about the history of this particular case. Independently of this assumption, it is much more plausible to think that the basis for Weierstrass's definition and its acceptance was the substantial increase in mathematical knowledge between the seventeenth century and the nineteenth century. Similarly, the acceptance of a particular Riemann surface as the proper setting for the square root function was not part of or the result of some shared implicit conception stretching back from Riemann to Cardan. The role for our mathematical concepts is quite minimal when it comes to the justification of our mathematical beliefs. These concepts are necessary to get mathematics going. But they are not substantial enough to meet Peacocke's POSS or guide the future development of that mathematical domain into the indefinite future. What is crucial, instead, are new mathematical discoveries that put a previous subject matter in a new light. Ignoring the significance of these discoveries and the special epistemic principles that allow them to occur leads Peacocke to impose an unmotivated picture on the history of mathematics.

Does this sort of situation become any more manageable if we adopt Jenkins's empiricist picture of grounding? She does not discuss this sort of case, so it is not clear if she would endorse Peacocke's reconstruction of the history of our implicit conception of limit. However, if Jenkins insists that all of our evidence for our mathematical beliefs is conceptual, then she faces the same dilemma as Peacocke.[9]

---

9. There is an inconclusive discussion of Euclidean geometry at Jenkins (2008, pp. 238–241) that seems to accept this dilemma.

One option is to insist that our concepts are fixed throughout this sort of historical episode. This allows the acceptance of the definition of Weierstrass to be justified based on the stubbornly implicit concept tracing back to Newton and Leibniz. Another option is to claim that our concepts have changed over this period. This is similar to the proposal that our concept of square root function has changed from Euler to Riemann. This second option has problems explaining what is better about Riemann's concept when compared to Euler's concept if we simply have a change in concept. More generally, a shift in concepts is not something that is justified by the evidence tied to those concepts. So the sort of mathematical progress we find in these cases is hard to locate using this second option. It is, of course, unfair to speculate on how Jenkins would handle these cases because they involve considerations quite far from her own focus on our contemporary knowledge of arithmetic. My main reason for considering them here is to explain why I do not opt for her empiricism to address my own project of articulating an account of our knowledge of the mathematics deployed in science.

14.7 NONCONCEPTUAL JUSTIFICATION

The upshot of our discussion is that an a priori epistemology for mathematics cannot be based entirely on features of our mathematical concepts. This is partly because these concepts are not substantial enough to motivate the sorts of historical changes in mathematics we have considered. A further consideration is Jenkins's reasonable demand that our concepts be grounded by some "input," which rules out the sort of accidental or lucky formation of true, justified beliefs that fall short of knowledge. But I have suggested that it is hard to see how an empiricist project of grounding our mathematical concepts in perceptual experience will succeed. The natural alternative, then, is to insist that these concepts can also be grounded by considerations internal to mathematics. What distinguishes the crank mathematician from the professional mathematician is that the professional presents new conjectures, proof techniques, and even new mathematical concepts in response to earlier mathematical developments. The context of the mathematician, then, can be sufficient to ground mathematical innovations in Jenkins's sense of "ground." This context blocks worries about the accidental or lucky character of the innovation. As a result, when these sorts of innovations do lead to justified, true beliefs about some mathematical subject matter, these beliefs have a good chance of being cases of knowledge.

It is obviously an enormous task to make this proposal plausible. My aim in these last two sections is merely to present a strategy that could be employed in future work to make sense of the possibility that our mathematical knowledge is largely a priori.[10] The extension-based strategy I have in mind starts with the assumption that mathematicians have achieved some degree of mathematical knowledge at some

---

10. I have been influenced by Manders (1989), Wilson (1992), and Tappenden (2008), but I take it they do not endorse this approach.

stage in the history of mathematics. For example, returning again to the example from the last chapter, we can take for granted that mathematicians around 1800 knew a great deal about real-valued functions. At this point, it is also fair to attribute to many mathematicians like Euler and Gauss the concept of a complex number. But unlike Peacocke and Jenkins, I do not think we should assign too much substance to this concept beyond merely allowing these mathematicians to think about complex numbers. The concept of complex number is clearly grounded in Jenkins's sense if we take for granted the prior knowledge of real analysis. We can explain how this concept arose out of reflection on real numbers in a nonaccidental way. Typically, these sorts of significant mathematical concepts have several sources, so a review of the way these sources mutually reinforce the articulation of the concept is an important epistemological task. However, given the minimal way I am thinking about these concepts, it is not reasonable to assume that the concepts themselves are sufficient to provide the evidence in favor of a substantial mathematical claim like Euler's formula 13.1. The evidence instead comes in the form of the proofs provided in Appendix D. Crucially, these proofs rely on nothing more than previously known features of the real numbers along with the minimal concept of complex number which I have ascribed to mathematicians like Euler. On this approach, then, the knowledge available in a mathematical context provides a crucial part of the evidence in favor of new mathematical claims. When everything goes well, the grounded concepts and the presentation of the right known theorems can extend our mathematical knowledge.

This is the easiest sort of case for my extension-based epistemology of mathematics. But even with the square root function, we saw that such extensions can go far from smoothly. When mathematicians tried to isolate the complex-valued square root function, they encountered a certain amount of resistance. This resistance came in form of a multiple-valued function. What an extension-based epistemology must confront are the steps that mathematicians like Riemann and Weyl took to argue that, in fact, the best way to approach this function is as a single-valued function on a Riemann surface. This argument is ampliative in a way that the argument for Euler's formula was not. It involves a conclusion that goes beyond previously known theorems and the relevant concepts. Nothing that mathematicians believed prior to Riemann compelled them to accept his proposal for Riemann surfaces. But Riemann still was able to give a reason for this dramatic change in our beliefs about the square root function.

My suggestion is that the sort of argument used here was a purely mathematical version of inference to the best explanation.[11] Like IBE in the scientific case, an appeal to the right Riemann surface helps explain the peculiar multiple-valued appearance of the square root function. In particular, it makes clear why the branch cut appears and why continuing through 4π radians returns us to our original output. These explanatory considerations are evidence that the new collection of claims

---

11. Mancosu has written extensively on the importance of explanation in pure mathematics. See Mancosu (2008a, 2008b) for recent surveys. In Mancosu (2008b, p. 137), Mancosu presents an IBE for pure mathematics. However, he does not endorse this inference.

about the square root function are correct. Riemann was justified in presenting these claims, and arguably he knew them to be true. This sort of purely mathematical IBE is not based on an appeal to perceptual experience. An extension-based epistemology for mathematics that puts this sort of reasoning at the center is consistent with the conclusion that mathematical knowledge is a priori.

I mention two other sorts of cases to convey what needs to be accomplished in a particular case to assimilate it to this approach to mathematical knowledge. After complex analysis, perhaps the most significant mathematical innovation in the nineteenth century was the development of group theory. The implicit conception of group presented earlier is a genuine mathematical achievement that is not best seen as merely making explicit what mathematicians implicitly had all along. Instead, the concept of group arose out of purely mathematical work in several different areas of mathematics, including geometry, number theory, and the study of the solutions of algebraic equations. For example, pioneers like Lagrange and Galois considered permutations of the solutions to these equations and used the features of the group of permutations to investigate the equations themselves.[12] However the definition of group and the study of groups in their own right had to wait for the late nineteenth century. This paved the way for the further development of group theory so that now it is one of the most important areas of mathematics. It is possible to see some of the transitions in this history as involving appeals to IBE. In particular, as with the Riemann surface, peculiar features of this or that mathematical domain become easier to appreciate once these domains are approached in group-theoretic terms.

A second case to note is Peacocke's own case of the development of a theory of limits from Newton and Leibniz through Cauchy and Weierstrass. The definition arrived at by Weierstrass is not merely an articulation of an earlier, tacit implicit conception. It is instead a genuine mathematical advance motivated by the investigation of series and functions far beyond anything considered in the time of Newton and Leibniz. An extension-based epistemology must present these developments and show how they provided reasons for the acceptance of Weierstrass's proposal. Again, these reasons will not be completely based on features of the concepts involved but on the way the new proposals serve to explain and extend the previously known results.

14.8 PAST AND FUTURE

An empiricist like Jenkins may object at this point that I have not given any real grounding to mathematics. What makes it appropriate to assume, for example, that Euler knew a great deal about real-valued functions? This is a fair point, and to address it one would have to return to the earliest stages in the history of mathematics and explain the justification for these mathematical beliefs. Here I think it is fair to bring in an important role for perceptual experience. I agree with Shapiro's claim that perception of small collections of things is an important part of the justification

---

12. See Stewart (2007a, ch. 7). An accessible textbook treatment is Stewart (2004).

for some of our mathematical beliefs.[13] Experience manipulating 5 stones and 7 stones may serve to ground the appropriate concepts here. This may be because the concepts involve simple abstract structures, say, finite progressions up to some threshold established by our experiences. So there is room for a crucial role for experience in getting mathematics going. This does not mean that the subject matter of mathematics is empirical regularities. Furthermore, even this limited role for experience in getting mathematics going is consistent with that role decreasing in significance as mathematics comes to study richer and richer structures whose tie to experience is quite minimal.

When we consider specific claims about addition and multiplication, we are in the domain of mathematics where perceptual experience can play an important epistemic role. But realistic philosophies of mathematics have faced a difficulty in getting beyond this stage because it is not clear how our experience can provide evidence for the existence of an infinite progression. This was the challenge we faced in trying to see how to use the role of a claim like "prime periods minimize intersection" in science to support it. I argued that IBE based on scientific considerations was not sufficient. But now we can see how an extension-based epistemology can provide what was missing in the case of applications. Suppose we notice a certain pattern in our specific claims about addition. For example, $1 + 2 = 2 + 1, 2 + 3 = 3 + 2$ and so on. An explanation of this pattern is that the natural numbers continue indefinitely, so that the unrestricted claim $a + b = b + a$ holds. Based on a prior knowledge of a wide array of claims involving finite sums, we can develop evidence that the relevant structure here is an infinite progression, and not a disjointed collection of finite progressions.

This approach is consistent with the ante rem structuralist interpretation of pure mathematics, but departs from the epistemology articulated by Shapiro in a crucial respect. Shapiro claims that the existence of structures is easy to prove as "any coherent theory characterizes a structure, or a class of structures" (Shapiro 1997, p. 95). With my approach, by contrast, agents must obtain evidence for the existence of the extensions they propose. This evidence may come from reflection on properly grounded concepts, or it may come from proofs whose premises are not all based on the features of the relevant concepts. In particular, we allow explanatory considerations to provide evidence for mathematical claims.

From this perspective, the acceptance of set theory early in the twentieth century marks a decisive turning point in the history of mathematics. Set theory delivers so many structures that it is possible to locate nearly any domain of study in a domain made up of sets. Prior to the acceptance of set theory, each new extension was challenged by the mathematical community and its acceptance came about only after a hard-fought struggle. After set theory was accepted, these debates about mathematical existence became less and less a part of mainstream mathematics. If an extension-based epistemology of mathematics is to be vindicated, then it must provide a compelling reconstruction of this turning point. It must be the case that mathematical considerations were sufficient to convince mathematicians that set

---

13. Shapiro (1997, §4.2). The origin of geometry is obviously another important part of the story.

Conclusion: Pure Mathematics

theory had earned its keep. This should be due to the way the commitments of set theory arise naturally out of earlier commitments, along with the explanatory benefits that the appeal to an intricate domain of sets brings with it.

A consequence of this extension-based epistemology of mathematics is that many of the controversies associated with extensions find their new home in debates about extensions of set theory itself. Koellner's recent "Truth in Mathematics: The Question of Pluralism" (Koellner 2009) provides a compelling discussion of the current state of these debates. We see there that the considerations he appeals to are just the sort of reasons I have pointed to in earlier transitions in the history of mathematics. What is different now is that we may be reaching the limits of what these sorts of transitions can accomplish. This is the pluralism that Koellner concedes may result from future foundational investigations.

Whatever the outcome of this debate, it remains imperative for philosophers of mathematics to return their focus to the history of mathematics and the debates about its development. Only in this fashion can we provide an epistemology of mathematics that renders it a priori. This is the key to understanding how mathematics contributes what it does to the success of science.

APPENDIX A

Method of Characteristics

Suppose we have

$$\rho_t + j'(\rho)\rho_x = 0 \quad (A.1)$$

as our conservation law and $\rho(x, 0) = \rho_0(x)$ as some initial distribution of density. Our goal is to find $\rho(x, t)$ for the later times. Such problems can sometimes be explicitly solved using the method of characteristics. The first step is to consider a new family of functions $x(t)$. The chain rule tells us that

$$\frac{d}{dt}\rho(x(t), t) = \frac{\partial}{\partial t}\rho(x(t), t) + \frac{\partial}{\partial x}\rho(x(t), t)x'(t) \quad (A.2)$$

But the condition that $\frac{d}{dt}\rho(x(t), t) = 0$ is just the claim that ρ is constant. If we impose this condition, then we can identify the left-hand side of A.1 with the right-hand side of A.2 to yield

$$\rho_t + j'(\rho)\rho_x = \rho_t + x'(t)\rho_x \quad (A.3)$$

That is, $j'(\rho) = x'(t)$. Integrating with respect to t while holding ρ fixed yields

$$x(t) = j'(\rho)t + k_1 \quad (A.4)$$

where k_1 is some constant. We can find k_1 by letting $t = 0$. Then

$$x(0) = k_1 = \rho_0(x) = x_0 \quad (A.5)$$

So we obtain the $x(t) = j'(\rho)t + x_0$ curves along which the density is constant. These are the characteristic base curves.

To see the method of characteristics in action, consider the case where $j(\rho) = 4\rho(2 - \rho)$ and imagine the initial density ρ_0 to be 1 for all $x \leq 1$, $1/2$ for $1 < x \leq 3$, and $3/2$ for $x > 3$. Clearly $j'(\rho) = 8 - 8\rho$, so our conservation law is

$$\rho_t + (8 - 8\rho)\rho_x = 0 \quad (A.6)$$

Then, along the different parts of the x-axis, we have $j'(\rho) = 8 - 8\rho$ coming out as 0 when $x \leq 1$, 4 when $1 < x \leq 3$, and -4 for $x > 3$. These are the slopes of the $x(t) = j'(\rho)t + x_0$ base curves along which the density is constant.

To move from the base curves to the solution of ρ, first notice that ρ is conserved along the base curves so $\rho(x(t), t) = \rho(j'(\rho)t + x_0, t) = k_2$. Again letting $t = 0$, we see that $k_2 = \rho_0(x_0)$. So $\rho(x(t), t) = \rho_0(x_0)$. But rearranging $x(t) = j'(\rho)t + x_0$ tells us that $x_0 = x(t) - j'(\rho)t$. Making the final substitution gives us $\rho(x, t) = \rho_0(x(t) - j'(\rho)t)$. Several examples can be found in Illner, et. al. (2005, ch. 9). See also Haberman (1977, §§71–82).

APPENDIX B

Black-Scholes Model

Given that V is a function of both S and t, we can approximate a change in V for a small time-step δt using a series expansion known as a Taylor series

$$\delta V = V_t \delta t + V_s \delta S + \frac{1}{2} V_{ss} \delta S^2 \qquad (B.1)$$

where additional higher-order terms are dropped. Given an interest rate of r for the assets held as cash, the corresponding change in the value of the *replicating portfolio* $\Pi = DS + C$ of D stocks and C in cash is

$$\delta \Pi = D \delta S + r C \delta t \qquad (B.2)$$

The last two equations allow us to represent the change in the value of a *difference portfolio*, which buys the option and offers the replicating portfolio for sale. The change in value is

$$\delta(V - \Pi) = (V_t - rC)\delta t + (V_S - D)\delta S + \frac{1}{2} V_{ss} \delta S^2 \qquad (B.3)$$

The δS term reflects the random fluctuations of the stock price and if it could not be dealt with we could not derive a useful equation for V. But fortunately the δS term can be eliminated if we assume that at each time-step the investor can adjust the number of shares held so that

$$D = V_S \qquad (B.4)$$

Then we get

$$\delta(V - \Pi) = (V_t - rC)\delta t + \frac{1}{2} V_{ss} \delta S^2 \qquad (B.5)$$

The δS^2 remains problematic for a given time-step, but we can find it for the sum of all the time-steps using our lognormal model. This permits us to simplify B.5 so that over the whole time interval Δt,

$$\Delta(V - \Pi) = \left(V_t - rC + \frac{1}{2}\sigma^2 S^2 V_{SS}\right)\Delta t \qquad (B.6)$$

Strictly speaking, we are applying a result known as Itô's lemma. This is glossed over somewhat in Almgren (2002), but discussed more fully in Wilmott (2007, ch. 5 and 6). What is somewhat surprising is that we have found the net change in the value of the difference portfolio in a way that has dropped any reference to the random fluctuations of the stock price S. This allows us to deploy the efficient market hypothesis and assume that $\Delta(V - \Pi)$ is identical to the result of investing $V - \Pi$ in a risk-free bank account with interest rate r. That is,

$$\Delta(V - \Pi) = r(V - \Pi)\Delta t \qquad (B.7)$$

But given that $V - \Pi = V - DS - C$ and $D = V_S$, we can simplify the right-hand side of this equation to

$$(rV - rV_S S - rC)\Delta t \qquad (B.8)$$

Given our previous equation for the left-hand side, we get 7.2 after all terms are brought to the left-hand side.

APPENDIX C

Speed of Sound

We consider a gas at rest with uniform pressure p and initial density ρ_0 and imagine a plane wave that arises due to the compression of the air.[1] The change in density associated with the wave can be given by the condensation σ where

$$\frac{\rho}{\rho_0} = 1 + \sigma \qquad (C.1)$$

If we can calculate the speed at which this disturbance σ moves through the gas, then we will have the speed of sound for that medium. Beginning with equations for the conservation of mass and the balance of momentum, it is possible to derive the plane wave equation

$$\frac{\partial^2 \sigma}{\partial t^2} = c^2 \frac{\partial^2 \sigma}{\partial x^2} \qquad (C.2)$$

A crucial step in this derivation is to treat the pressure p as a function of the density ρ and to expand p into the truncated series

$$p(\rho) = p(\rho_0) + c^2(\rho - \rho_0) \qquad (C.3)$$

where

$$c^2 = \frac{\partial}{\partial \rho} p(\rho_0) \qquad (C.4)$$

But it is known that for this plane wave equation, the velocity of the wave is given by c, so the problem reduces to finding $\frac{\partial}{\partial \rho} p(\rho_0)$.

Here, assumptions about the relationship between density, pressure, and heat come into play. In the Newton-Euler derivation, the temperature θ was assumed to be constant, so the equation of state $p = R\rho\theta$ or $\frac{p}{\rho} = K_0$ for some constant K_0 yielded

---

[1]. Lin and Segel (1988, §15.4). Historical issues are pursued in Fox (1971, pp. 161–162), drawing on an 1822 clarification of an 1816 paper by Laplace, and unpublished work by Bryan Roberts.

$$c^2 = \frac{\partial}{\partial \rho} p(\rho_0) = \frac{p}{\rho} \tag{C.5}$$

If we allow the temperature to vary but assume that there is no net heat flow into the area in question, then we must adjust the equation of state to $p = K_1 \rho^\gamma$ where γ is the ratio of specific heat at constant pressure to specific heat at constant volume:

$$\gamma = \frac{C_p}{C_V} \tag{C.6}$$

Making this change results in a different value for c^2:

$$c^2 = \frac{\partial}{\partial \rho}(K_1 \rho^\gamma) \tag{C.7}$$

$$= \gamma K_1 \rho^{\gamma - 1} \tag{C.8}$$

$$= \frac{\gamma K_1 \rho^\gamma}{\rho} \tag{C.9}$$

$$= \frac{\gamma p}{\rho} \tag{C.10}$$

The basic idea of this adjustment is that an increase in density will correspond to an even greater increase in pressure because of the associated increase in temperature. As the plane wave equation shows, this brings about a faster propagation of the compression wave.

APPENDIX D

Two Proofs of Euler's Formula

Perhaps the most elementary proof of 13.1 comes from the proof that there is only one function $f(x + iy) = u(x, y) + iv(x, y)$ defined on the complex numbers that is defined on the whole complex plane and satisfies both (i) $f(x + i0) = e^x$ and (ii) $f'(z) = f(z)$. My exposition follows the sketch given at Brown and Churchill (1996, p. 68, exercise 15). The proof assumes a natural definition of differentiation for functions in the complex plane, and two theorems quickly follow from this definition. The first theorem is that when $f'(z)$ exists at point $z_0 = x_0 + iy_0$, then $u_x = v_y$, $u_y = -v_x$, and $f'(z_0) = u_x + iv_x$ (Brown and Churchill 1996, p. 50). The second theorem is that the existence of a derivative throughout a domain D requires that u and v be harmonic on that domain, that is, $u_{xx} + u_{yy} = 0$ and $v_{xx} + v_{yy} = 0$ (Brown and Churchill 1996, p. 60). The proof of 13.1 comes in three parts:

a. By ii, $f'(z) = f(z)$. But $f'(z) = u_x + iv_x$. So $u + iv = u_x + iv_x$. This requires that $u = u_x$ and $v = v_x$. We have then shown that the existence of our function f requires that, for some ϕ and ψ, $u(x, y) = e^x \phi(y)$ and $v(x, y) = e^x \psi(y)$.

b. As $f'(z)$ exists for all z, then u is harmonic for all inputs (x, y): $u_{xx} + u_{yy} = 0$. Appealing to a, $u_{yy} = e^x \phi''(y)$ and $u_{xx} = u_x = u = e^x \phi(y)$. Making the substitution, $e^x(\phi''(y) + \phi(y)) = 0$. This requires that $\phi''(y) + \phi(y) = 0$. This ordinary differential equation yields the solution $\phi(y) = A \cos y + B \sin y$, where A and B are as yet unknown.

c. We know that $u_y = -v_x$, but also that $v = v_x$. So $u_y = -v$. As $u_y = e^x(-A \sin y + B \cos y)$, this means that $v = e^x(A \sin y - B \cos y)$. Using the expression for v from a tells us that $\psi(y) = A \sin y - B \cos y$. Putting all the pieces together, our function takes the form

$$e^x(A \cos y + B \sin y + iA \sin y - iB \cos y) \tag{D.1}$$

Given i, when $y = 0$, $(A \cos y + B \sin y + iA \sin y - iB \cos y) = 1$. Solving this equation yields $A - iB = 1$, or $A = 1$ and $B = 0$. This makes $f(z) = e^x(\cos y + i \sin y)$.

A second proof makes use of Taylor series. These series provide a means to represent the real-valued functions e, $\cos$, and $\sin$ in terms of infinite series. If we suppose that each of these functions can be extended uniquely into complex-valued functions, then we can prove 13.1.

To see how, notice that the Taylor series for e^x is

$$e^x = 1 + x + \frac{x^2}{2!} + \frac{x^3}{3!} + \ldots \tag{D.2}$$

whereas the Taylor series for $\cos x$ and $\sin x$ are

$$\cos x = 1 - \frac{x^2}{2!} + \frac{x^4}{4!} - \ldots \tag{D.3}$$

and

$$\sin x = x - \frac{x^3}{3!} + \frac{x^5}{5!} - \ldots \tag{D.4}$$

We have already assumed that $i^2 = -1$, so $i^3 = -i$, $i^4 = 1$, $i^5 = i$, and so on. Consider the result of replacing x by iy in the series for e^x. This yields

$$e^{iy} = 1 + iy - \frac{y^2}{2!} - \frac{iy^3}{3!} + \frac{y^4}{4!} + \frac{iy^5}{5!} - \ldots \tag{D.5}$$

Now consider what happens if we represent $\cos y + i \sin y$ using the Taylor series for $\cos$ and $\sin$. The $\cos$ series remains the same:

$$\cos y = 1 - \frac{y^2}{2!} + \frac{y^4}{4!} - \ldots \tag{D.6}$$

whereas the sin series is altered to include i:

$$i \sin y = iy - \frac{iy^3}{3!} + \frac{iy^5}{5!} - \ldots \tag{D.7}$$

Adding these series together shows that $e^{iy} = \cos y + i \sin y$.

BIBLIOGRAPHY

Adam, J. A. 2002, "The mathematical physics of rainbows and glories", *Physical Reports* 356, 229–365.

Adam, J. A. 2003, *Mathematics in Nature: Modeling Patterns in the Natural World*, Prentice Hall.

Allan, R., Lindesay, J., and Parker, D. (1996), *El Niño Southern Oscillation and Climactic Variability*, CSIRO.

Almgren, R. 2002, "Financial derivatives and partial differential equations," *American Mathematical Monthly* 109, 1–12.

Aspray, W., and Kitcher, P., eds. 1988, *History and philosophy of modern mathematics*, vol. 11 of *Minnesota Studies in the Philosophy of Science*, University of Minnesota Press.

Aubin, D. 2004, "Forms of explanation in the catastrophe theory of René Thom: Topology, morphogenesis, and structuralism", in M. N. Wise, ed., *Growing Explanations: Historical Perspectives on Recent Science*, Duke University Press, pp. 95–130.

Azzouni, J. 2004, *Deflating Existential Consequence: A Case for Nominalism*, Oxford University Press.

Baker, A. 2005, "Are there genuine mathematical explanations of physical phenomena?," *Mind* 114, 223–238.

Baker, A. 2009, "Mathematical explanation in science", *British Journal for the Philosophy of Science* 60, 611–633.

Balaguer, M. 1998, *Platonism and Anti-Platonism in Mathematics*, Oxford University Press.

Balaguer, M. 2001, "A theory of mathematical correctness and mathematical truth", *Pacific Philosophical Quarterly* 82, 87–114.

Bangu, S. 2006, "Steiner on the applicability of mathematics and naturalism," *Philosophia Mathematica* 14, 26–43.

Batterman, R. 1997, " 'Into a mist': Asymptotic theories on a caustic," *Studies in the History and Philosophy of Modern Physics* 28, 395–413.

Batterman, R. 2002a, 'Asymptotics and the role of minimal models', *British Journal for the Philosophy of Science* 53, 21–38.

Batterman, R. 2002b, *The Devil in the Details: Asymptotic Reasoning in Explanation, Reduction, and Emergence*, Oxford University Press.

Batterman, R. 2005, "Response to Belot's 'Whose devil? Which details?' ", *Philosophy of Science* 72, 154–163.

Batterman, R. 2007, "On the specialness of the special functions (the nonrandom effusions of the divine mathematician)," *British Journal for the Philosophy of Science* 58, 263–286.

Batterman, R. 2010, "On the explanatory role of mathematics in empirical science," *British Journal for the Philosophy of Science* 61, 1–25.
Belot, G. 2005, "Whose devil? Which details?," *Philosophy of Science* 72, 128–153.
Benacerraf, P. 1983, "What numbers could not be," in Benacerraf and Putnam (1983).
Benacerraf, P. and Putnam, H., eds. 1983, *Philosophy of Mathematics: Selected Readings*, 2nd ed., Cambridge University Press.
Bishop, R. C. 2008, "Downward causation in fluid convection," *Synthese* 160, 229–248.
Boolos, G. 1998, *Logic, logic and logic*, Harvard University Press.
Brown, J. W. and Churchill, R. V. 1996, *Complex Variables and Applications*, 6th ed., McGraw-Hill.
Bruneau, J., ed. 1973, *Correspondance par Flaubert*, Gallimard.
Bueno, O. 2009, "Mathematical fictionalism," in O. Bueno and Ø. Linnebo, eds., *New Waves in the Philosophy of Mathematics*, Palgrave Macmillan, pp. 59–79.
Bueno, O., and Colyvan, M. 2011, "An inferential conception of the application of mathematics," *Nous*. 45: 345–374.
Burge, T. 1993, "Content preservation", *Philosophical Review* 102, 457–488.
Burge, T. 1998, "Computer proof, a priori knowledge, and other minds," *Philosophical Perspectives* 12, 1–37.
Burgess, J., and Rosen, G. 1997, *A Subject with No Object: Strategies for Nominalistic Interpretation of Mathematics*, Clarendon Press.
Busch, J. 2010, "Can the new indispensability argument be saved from eucliden rescues?," *Synthese*, forthcoming.
Camp, E. 2009, "Two varieties of literary imagination: Metaphor, fiction, and thought experiments", *Midwest Studies in Philosophy* 33, 107–130.
Carlson, S. C. 2001, *Topology of Surfaces, Knots, and Manifolds: A First Undergraduate Course*, Wiley.
Carnap, R. 1956, "Empiricism, semantics, and ontology", in *Meaning and Necessity*, University of Chicago Press, pp. 205–221.
Cartwright, N. 1999, *The Dappled World: A Study of the Boundaries of Science*, Cambridge University Press.
Casullo, A. 2003, *A Priori Justification*, Oxford University Press.
Chandrasekhar, S. 1943, "Stochastic problems in physics and astronomy," *Reviews of Modern Physics* 15, 1–89.
Chang, H. 2003, "Preservative realism and its discontents: Revisiting caloric," *Philosophy of Science (Proceedings)* 70, 902–912.
Chang, H. 2004, *Inventing Temperature: Measurement and Scientific Progress*, Oxford University Press.
Chihara, C. 2004, *A Structural Account of Mathematics*, Oxford University Press.
Colyvan, M. 2001, *The Indispensability of Mathematics*, Oxford University Press.
Colyvan, M. 2002, "Mathematics and aesthetic considerations in science," *Mind* 111, 69–74.
Colyvan, M. 2008, "The ontological commitments of inconsistent theories", *Philosophical Studies* 141, 115–123.
Colyvan, M. 2010, "There is no easy road to nominalism", *Mind* 119, 285–306.
Contessa, G. 2007, "Scientific representation, interpretation and surrogative reasoning", *Philosophy of Science* 74, 48–68.
Contessa, G. 2010, "Scientific models and fictional objects", *Synthese* 172, 215–229.
Daly, C. and Langford, S. 2009, "Mathematical explanation and indispensability arguments", *Philosophical Quarterly* 59, 641–658.

Darrigol, O. 2005, *Worlds of Flow: A History of Hydrodynamics from the Bernoullis to Prandtl*, Oxford University Press.
DiSalle, R. 2002, "Space and time: Inertial frames," in *Stanford Encyclopedia of Philosophy*. E.N. Zalta (ed.), http://plato.stanford.edu/entries/spacetime-iframes/
Dowe, P. 2007, "Causal processes," *Stanford Encyclopedia of Philosophy*. E.N. Zalta (ed.), http://plato.stanford.edu/entries/causation-process/
Dummett, M. 1991, *Frege: Philosophy of Mathematics*, Harvard University Press.
Dummett, M. 1993, "What is mathematics about?," in *The Seas of Language*, Clarendon Press, pp. 429–445.
Earman, J. 1992, *Bayes or Bust?: A Critical Examination of Bayesian Confirmation Theory*, MIT Press.
Elgin, C. Z. 2009, "Exemplification, idealization and scientific understanding", in M. Suárez, ed., *Fictions in Science: Philosophical Essays on Modeling and Idealization*, Routledge, pp. 77–90.
Field, H. 1980, *Science without Numbers: A Defence of Nominalism*, Princeton University Press.
Field, H. 1989, *Realism, Mathematics and Modality*, Blackwell.
Fowler, A. C. 1997, *Mathematical Models in the Applied Sciences*, Cambridge University Press.
Fox, R. 1971, *The Caloric Theory of Gases: From Lavoisier to Regnault*, Clarendon Press.
Fox, R. 1974, "The rise and fall of Laplacian physics," *Historical Studies in the Physical Sciences* 4, 89–136.
Fraleigh, J. B. 1999, *A First Course in Abstract Algebra*, 6th ed., Addison-Wesley.
Franklin, J. 1989, "Mathematical necessity and reality," *Australasian Journal of Philosophy* 67, 11–17.
Franklin, J. 2008, "Aristotelian realism," in A. Irvine, ed., *Handbook of the Philosophy of Science. Philosophy of Mathematics*, Elsevier, pp. 101–153.
Frege, G. 1980, *The Foundations of Arithmetic*, J. L. Austin (Trans.), 2nd ed., Northwestern University Press.
French, S. 2000, "The reasonable effectiveness of mathematics: Partial structures and the applicability of group theory to physics," *Synthese* 125, 103–120.
Friedman, M. 2001, *Dynamics of Reason: The 1999 Kant Lectures at Stanford University*, CSLI Publications.
Friend, S. 2007, "Fictional characters," *Philosophy Compass* 2, 141–156.
Frigg, R. 2010a, "Fiction in science," in J. Woods, ed., *Fictions and Models: New Essays*, Philosophia Verlag, pp. 247–287.
Frigg, R. 2010b, "Models and fiction," *Synthese* 172, 251–268.
Galileo 1974, *Two New Sciences*, S. Drake (Trans.), University of Wisconsin Press.
Gendler, T. S. 2006, "Imaginative resistance revisited," in S. Nichols, ed., *The Architecture of Imagination: New Essays on Pretence, Possibility, and Fiction*, Oxford University Press, pp. 149–175.
Glaskin, M. 2008, "Shockwave traffic jam recreated for first time," NewScientist.com news service, March 4.
Godfrey-Smith, P. 2006, "The strategy of model-based science," *Biology and Philosophy* 21, 725–740.
Godfrey-Smith, P. 2009, "Models and fictions in science," *Philosophical Studies* 143, 101–116.
Haberman, R. 1977, *Mathematical Models: Mechanical Vibrations, Population Dynamics and Traffic Flow*, Prentice Hall.

Hales, T. 2000, "Cannonballs and honeycombs," *Notices of the American Mathematical Society* 47, 440–449.
Hellman, G. 1989, *Mathematics without Numbers: Towards a Modal-Structural Interpretation*, Clarendon Press.
Heyman, J. 1998, *Structural Analysis: A Historical Approach*, Cambridge University Press.
Hill, C. S. 1998, "Peacocke on semantic values," *Australasian Journal of Philosophy* 76, 97–104.
Hillerbrand, R. (n.d.), "Scale separation as a condition for quantitative modelling," in preparation.
Hitchcock, C. 2008, "Causation," in S. Psillos and M. Curd, eds., *Routledge Companion to Philosophy of Science*, Routledge, pp. 317–326.
Howson, C., and Urbach, P. 1989, *Scientific Reasoning: The Bayesian Approach*, Open Court.
Humphreys, P. 2004, *Extending Ourselves: Computational Science, Empiricism, and Scientific Method*, Oxford University Press.
Illner, R., Bohun, C. S., McCollum, S., and van Roode, T. 2005, *Mathematical Modelling: A Case Studies Approach*, American Mathematical Society.
Jenkins, C. S. 2008, *Grounding Concepts: An Empirical Basis for Arithmetical Knowledge*, Oxford University Press.
Kevorkian, J., and Cole, J. D. 1981, *Perturbation Methods in Applied Mathematics*, Springer-Verlag.
Kitcher, P. 1984, *The Nature of Mathematical Knowledge*, Oxford University Press.
Kitcher, P. 1989, "Explanatory unification and the causal structure of the world", in P. Kitcher and W. Salmon, eds., *Scientific Explanation*, University of Minnesota Press, pp. 410–505.
Kline, M. 1972, *Mathematical Thought from Ancient to Modern Times*, Oxford University Press.
Koellner, P. 2009, "Truth in mathematics: The question of pluralism," in O. Bueno and Ø. Linnebo, eds., *New Waves in the Philosophy of Mathematics*, Palgrave, pp. 80–116.
Kolman, J. 1999, "LTCM speaks," *Derivatives Strategy*, April.
Krantz, D. H., Luce, R. D., Suppes, P., and Tversky, A. 1971, *Foundations of Measurement: Additive and Polynomial Representations*, vol. 1, Academic Press.
Kuhn, T. S. 1970, *The Structure of Scientific Revolutions*, 2nd ed., University of Chicago Press.
Kuhn, T. S. 1979, "Second thoughts on paradigms," in *The Essential Tension*, University of Chicago Press, pp. 293–319.
Kundu, P. K., and Cohen, I. M. 2008, *Fluid Mechanics*, 4th ed., Academic Press.
Lange, M. 2009, "Dimensional explanations," *Nous* 43, 742–775.
Leng, M. 2007, "What's there to know?," in M. Leng, A. Paseau, and M. Potter, eds., *Mathematical Knowledge*, Oxford University Press, pp. 84–108.
Leng, M. 2009, " 'Algebraic' approaches to mathematics," in O. Bueno and Ø. Linnebo, eds. *New Waves in the Philosophy of Mathematics*, Palgrave Macmillan, pp. 117–134.
Leng, M. 2010, *Mathematics and Reality*, Oxford University Press.
Lewis, D. 1991, *Parts of Classes*, Blackwell.
Lewis, D. 1993, "Mathematics is megethology," *Philosophia Mathematica* 1, 3–23.
Lewis, D. 1999a, *Papers in Metaphysics and Epistemology*, Cambridge University Press.
Lewis, M. 1999b, "How the eggheads cracked," *New York Times Magazine*, January 24.

Bibliography

Liggins, D. 2008, "Quine, Putnam and the 'Quine-Putnam' indispensability argument," *Erkenntnis* 68, 113–127.

Lin, C. C., and Segel, L. A. 1988, *Mathematics Applied to Deterministic Problems in the Natural Sciences*, SIAM.

Linnebo, Ø. 2008, "Structuralism and the notion of dependence," *Philosophical Quarterly* 58, 59–79.

Lowenstein, R. 2000, *When Genius Failed: The Rise and Fall of Long-Term Capital Management*, Random House.

Lyon, A., and Colyvan, M. 2008, "The explanatory power of phase spaces," *Philosophia Mathematica* 16, 227–243.

MacBride, F. 1999, "Listening to fictions: A study of Fieldian nominalism," *British Journal for the Philosophy of Science* 50, 431–455.

Machamer, P., Darden, L., and Craver, C. F. 2000, "Thinking about mechanisms", *Philosophy of Science* 67, 1–25.

Maddy, P. 1998, *Naturalism in Mathematics*, Clarendon Press.

Maddy, P. 2008, "How applied mathematics became pure," *Review of Symbolic Logic* 1, 16–40.

Mancosu, P. 1997, *The Philosophy of Mathematics and Mathematical Practice in the Seventeenth Century*, Oxford University Press.

Mancosu, P. 2008a, "Explanation in mathematics," in *Stanford Encyclopedia of Philosophy*. E. N. Zalta (ed.), http://plato.stanford.edu/entries/mathematics-explanation/

Mancosu, P. 2008b, "Mathematical explanation: Why it matters," in P. Mancosu, ed., *The Philosophy of Mathematical Practice*, Oxford University Press, pp. 134–150.

Mancosu, P., ed. 2008c, *The Philosophy of Mathematical Practice*, Oxford University Press.

Manders, K. 1989, "Domain extension and the philosophy of mathematics," *Journal of Philosophy* 86, 553–562.

Marion, J., and Thornton, S. 1995, *Classical Dynamics of Particles and Systems*, 4th ed., Saunders College Publishing.

Mayo, D. 1996, *Error and the Growth of Experimental Knowledge*, University of Chicago Press.

McGivern, P. 2008, "Reductive levels and multi-scale structure," *Synthese* 165, 53–75.

Melia, J. 2000, "Weaseling away the indispensability argument," *Mind* 109, 455–479.

Melia, J. 2002, "Response to Colyvan," *Mind* 111, 75–79.

Melnyk, A. 2003, *A Physicalist Manifesto: Thoroughly Modern Materialism*, Cambridge University Press.

Morrison, M. 2000, *Unifying Scientific Theories: Physical Concepts and Mathematical Structures*, Cambridge University Press.

Nahin, P. J. 1998, *An Imaginary Tale: The Story of $\sqrt{-1}$*, Princeton University Press.

Nahin, P. J. 2004, *When Least Is Best: How Mathematicians Discovered Many Clever Ways to Make Things as Small (or as Large) as Possible*, Princeton University Press.

Nahin, P. J. 2006, *Dr. Euler's Fabulous Formula: Cures Many Mathematical Ills*, Princeton University Press.

Narasimhan, T. N. 1999, "Fourier's heat conduction equation: History, influence, and connections," *Reviews of Geophysics* 37, 151–172.

Nickel, K. 1973, "Prandtl's boundary-layer theory from the viewpoint of a mathematician," *Annual Review of Fluid Mechanics* 5, 405–428.

Norton, J. 2003, "A material theory of induction," *Philosophy of Science* 70, 647–670.

Nussenzveig, H. M. 1977, "The theory of the rainbow," *Scientific American* 236(4), 116–127.

Nussenzveig, H. M. 1992, *Diffraction Effects in Semiclassical Scattering*, Cambridge University Press.

Parker, W. 2006, "Understanding pluralism in climate modeling," *Foundations of Science* 11, 349–368.

Parsons, C. 1983, *Mathematics in Philosophy: Selected Essays*, Cornell University Press.

Parsons, C. 2008, *Mathematical Thought and Its Objects*, Cambridge University Press.

Peacocke, C. 1992, *A Study of Concepts*, MIT Press.

Peacocke, C. 1993, "How are a priori truths possible," *European Journal of Philosophy* 1, 175–199.

Peacocke, C. 1998a, "The concept of a natural number," *Australasian Journal of Philosophy* 76, 105–109.

Peacocke, C. 1998b, "Implicit conceptions, the a priori, and the identity of concepts," *Philosophical Issues* 9, 121–148.

Peacocke, C. 1998c, "Implicit conceptions, understanding and rationality," *Philosophical Issues* 9, 43–88.

Peacocke, C. 2004, *The Realm of Reason*, Oxford University Press.

Petroski, H. 1994, *Design Paradigms: Case Histories of Error and Judgment in Engineering*, Cambridge University Press.

Pinchover, Y., and Rubinstein, J. 2005, *An Introduction to Partial Differential Equations*, Cambridge University Press.

Pincock, C. 2004, "A revealing flaw in Colyvan's indispensability argument," *Philosophy of Science* 71, 61–79.

Pincock, C. 2005, "Critical notice of Torsten Wilholt, *Zahl und Wirklichkeit* [Number and Reality]," *Philosophia Mathematica* 13, 329–337.

Pincock, C. 2007, "A role for mathematics in the physical sciences," *Nous* 41, 253–275.

Pincock, C. 2009, "From sunspots to the Southern Oscillation: Confirming models of large-scale phenomena in meteorology," *Studies in the History and Philosophy of Science* 40, 45–56.

Pincock, C. 2011, "Modeling reality," *Synthese* 180: 19–32.

Porter, T. 1995, *Trust in Numbers: The Pursuit of Objectivity in Science and Public Life*, Princeton University Press.

Prestini, E. 2004, *The Evolution of Applied Harmonic Analysis*, Birkhäuser.

Psillos, S. 1999, *Scientific Realism: How Science Tracks Truth*, Routledge.

Psillos, S. 2010, "Scientific realism: Beyond platonism and nominalism," *Philosophy of Science (Proceedings)*. 77: 947–958.

Psillos, S. 2011, "Living with the abstract: Realism and models," *Synthese* 180: 3–17.

Putnam, H. 1979a, *Mathematics, Matter and Method: Philosophical Papers*, vol. 1, 2nd ed., Cambridge University Press.

Putnam, H. 1979b, "Philosophy of logic," in *Mathematics, Matter and Method: Philosophical Papers*, vol. 1, 2nd ed., Cambridge University Press, pp. 323–358.

Quine, W. V. 2004, *Quintessence: Basic Readings from the Philosophy of W. V. Quine*, Roger F. Gibson Jr. (Ed.), Harvard University Press.

Rayo, A. 2008, "On specifying truth-conditions," *Philosophical Review* 117, 385–443.

Rayo, A. 2009, "Toward a trivialist account of mathematics," in O. Bueno and Ø. Linnebo, eds., *New Waves in the Philosophy of Mathematics*, Palgrave Macmillan, pp. 239–260.

Redhead, M. 2004, "Discussion note: Asymptotic reasoning," *Studies in the History and Philosophy of Modern Physics* 35, 527–530.
Resnik, M. 1997, "Holistic mathematics," in M. Schirn, ed., *Philosophy of Mathematics Today*, Oxford University Press, pp. 227–246.
Rosen, G. 2001, "Nominalism, naturalism, epistemic relativism," *Philosophical Perspectives* 15, 69–91.
Saatsi, J. 2011, "The enhanced indispensability argument: Representational versus Explanatory Role of Mathematics in Science," *British Journal for the Philosophy of Science.* 62: 143–154.
Segel, L. A. 2007, *Mathematics Applied to Continuum Mechanics*, SIAM.
Shapiro, S. 1997, *Philosophy of Mathematics: Structure and Ontology*, Oxford University Press.
Steiner, M. 1998, *The Applicability of Mathematics as a Philosophical Problem*, Harvard University Press.
Sterrett, S. (n.d.), "What makes good models good?," in preparation.
Stewart, I. 2004, *Galois Theory*, 3rd ed., Chapman and Hall.
Stewart, I. 2007a, *Why Beauty Is Truth: A History of Symmetry*, Basic Books.
Stewart, J. 2007b, *Calculus*, 6th ed., Brooks Cole.
Suárez, M. 2009a, "Scientific fictions as rules of inference," in M. Suárez, ed., *Fictions in Science: Philosophical Essays on Modeling and Idealization*, Routledge, pp. 158–178.
Suárez, M., ed. 2009b, *Fictions in Science: Philosophical Essays on Modeling and Idealization*, Routledge.
Suárez, M. 2010a, "Fictions, inference and realism," in J. Woods (ed.), *Fictions and Models: New Essays*, Philosophia Verlag, pp. 225–245.
Suárez, M. 2010b, "Scientific representation," *Philosophy Compass* 5, 91–101.
Sugiyama, Y., Fukui, M., Kikuchi, M., et al. 2008, "Traffic jams without bottlenecks—experimental evidence for the physical mechanism of the formation of a jam," *New Journal of Physics* 10, 033001.
Sussman, H. J., and Zahler, R. S. 1978, "Catastrophe theory as applied to the social and biological sciences: A critique," *Synthese* 37, 117–216.
Tait, W. 2005, "Frege versus Cantor and Dedekind: On the concept of number," in *The Provenance of Pure Reason: Essays in the Philosophy of Mathematics and Its History*, Oxford University Press, pp. 212–251.
Tappenden, J. 2008, "Mathematical concepts: Fruitfulness and naturalness," in P. Mancosu, ed., *The Philosophy of Mathematical Practice*, Oxford University Press, pp. 276–301.
Thomson-Jones, M. 2010, "Missing systems and the face value practice," *Synthese* 172, 283–299.
Toon, A. 2010, "The ontology of theoretical modelling: Models as make-believe," *Synthese* 172, 301–315.
Truesdell, C. 1968, *Essays in the History of Mechanics*, Springer.
Truesdell, C. 1984, "The role of mathematics in science as exemplified by the work of the Bernoullis and Euler," in *An Idiot's Fugitive Essays on Science*, Springer-Verlag, pp. 97–132.
van de Hulst, H. C. 1957, *Light Scattering by Small Particles*, Wiley.
van Fraassen, B. 1980, *The Scientific Image*, Clarendon Press.
van Fraassen, B. 2008, *Scientific Representation: Paradoxes of Perspective*, Oxford University Press.

Vincenti, W. G., and Bloor, D. 2003, "Boundaries, contingencies and rigor: Thoughts on mathematics prompted by a case study in transonic aerodynamics," *Social Studies of Science* 33, 469–507.

Weisberg, M. 2006, "Robustness analysis," *Philosophy of Science* 73, 730–742.

Weisberg, M. 2007, "Who is a modeler?," *British Journal for the Philosophy of Science* 58, 207–233.

Weisberg, M. 2008, "The robust Volterra principle," *Philosophy of Science* 75, 106–131.

Weisberg, M. n.d., *Simulation and Similarity*, in preparation.

Weslake, B. 2010, "Explanatory depth," *Philosophy of Science* 77, 273–294.

Wigner, E. P. 1960, "The unreasonable effectiveness of mathematics in the natural sciences," *Communications of Pure and Applied Mathematics* 13, 1–14.

Wilholt, T. 2004, *Zahl und Wirklichkeit [Number and Reality]*, Mentis.

Wilholt, T. 2006, "Lost on the way from Frege to Carnap: How the philosophy of science forgot the applicability problem," *Grazer Philosophische Studien* 73, 69–82.

Wilmott, P. 2007, *Paul Wilmott Introduces Quantitative Finance*, 2nd ed., Wiley.

Wilson, M. 1992, "Frege: The royal road from geometry," *Nous* 26, 149–180.

Wilson, M. 2006, *Wandering Significance: An Essay on Conceptual Behavior*, Oxford University Press.

Wimsatt, W. 2007, *Re-engineering Philosophy for Limited Beings: Piecewise Approximations to Reality*, Harvard University Press.

Winsberg, E. 2006, "Handshaking your way to the top: Inconsistency and falsification in intertheoretic reduction," *Philosophy of Science (Proceedings)* 73, 582–594.

Winsberg, E. 2009, "A function for fictions: Expanding the scope of science," in M. Suárez, ed., *Fictions in Science*, Routledge, pp. 179–192.

Woodward, J. 2003, *Making Things Happen: A Theory of Causal Explanation*, Oxford University Press.

Wright, C. 1980, *Frege's Conception of Numbers as Objects*, Aberdeen University Press.

Wright, C. 2000, "Neo-Fregean foundations for real analysis: Some reflections on Frege's constraint," *Notre Dame Journal of Formal Logic* 41, 317–334.

Yablo, S. 1998, "Does ontology rest on a mistake?," *Proceedings of the Aristotelian Society, Supplement* 72, 229–261.

Yablo, S. 2001, "Go figure: A path through fictionalism," *Midwest Studies in Philosophy* 25, 72–102.

Yablo, S. 2005, "The myth of the seven," in M. E. Kalderon, ed., *Fictionalism in Metaphysics*, Oxford University Press, pp. 88–115.

Zeeman, E. C. 1976, "Catastrophe theory", *Scientific American* 234, 65–83.

INDEX

Page numbers for case studies are in bold face.

a priori, 22, 138
   absolute, 138–140
   relative, 131–137
A Tale of Two Cities, 247
absolute space, 132
abstract acausal representation, 6, 9, 121, 145
   and Batterman, 63
   capture features of scientific interest, 53
   and column storage, 142–144
   and change in time and counterfactuals, 51
   and meteorological modeling, 89
   not just precausal, 51
   Psillos on, 164
   reflects formal structure, 53
   sets threshold, 64
abstract and concrete, 29
abstract structures, 257
abstract varying representation, 6, 66–68, 78–80, 82
   epistemic aspects of, 10
   and explanation, 210
   and failure, 146, 151–152
   reviews of, 121, 187
   and scale similarity, 95
abstraction and mathematics, 228
account of applications
   one-stage, 282
   two-stage, 282
accuracy, failure of, 115
accurate representation, 3

adequate, set of scales, 101, 104
aesthetics and mathematics, 180, 182
agnosticism, about metaphysics, 120
air, viscosity of, 108
Airy representation, 236, 237, 239
algorithm, 36
almost always, mathematical sense of, 159
Ampère's law, 186
amplitude of a wave, 97
analogical reasoning and discovery, 18
analogy, mathematical, 221
analytic functions, 71, 183
analytic statements in logical empiricism, 45
angle of deflection, 225
angular displacement versus length, 66
angular momentum, 232
anomalies of a paradigm, 127
anthropocentrism, 183, 187
antirealism about fictional characters, 244
antirealism about mathematics, 198
applicability as expressing claims, 281
applied mathematics and mathematics, 68, 82–86, 182
approximations and scaling, 96
arithmetic, basis of for Frege, 281
asymptotic representation versus acausal, 64
asymptotic expansions, 104, 109, 111, 112, 114
asymptotic explanation, 221–223
asymptotic limits, 64

asymptotic reasoning, 223
asymptotically emergent structures, 222
atoms, 211, 215–216
Aubin, David, 156
auxiliary assumptions, 57
axioms, new, 251
Azzouni, Jody, 100n13, 261–263, 265, 278

Baker, Alan, 203, 205, 206, 208n5
Balaguer, Mark, 16n6, 26n1, 251–252, 255
Bangu, Sorin, 184n6
basic contents, 27–29
Batterman, Robert, 63–64, 93–94, 239–241
  on asymptotic explanation, 210
  on size parameter, 227n5
  and strut, 142n1
  on representations without details, 63
Bayesianism, 39–40
  objective, 39
  subjective, 38–39
beauty and mathematics, 180
begs the question, indispensability argument, 211
beliefs and ray theory, 241
Belot, Gordon, 239–241
Benacerraf, Paul, 169, 174, 284–285
Bénard cells, 116–118
Bernoulli, Daniel, 280
Bessel function, 231
best explanation and belief, 237
bifurcation set, 156
Bishop, Robert C., 115–117, 119, 120
Bjerknes, Jacob, 91
Black-Scholes model, 141, **146–148**, 303–304
  and boundary layer theory, 155
  and Long-Term Capital Management, 150–151
  and parameter illusion, 165
Boolos, George, 173
boundaries, artificial, 65
boundary conditions, 240–241
boundary layer, 7, **106–112**, 120, 227
  equations, 111
  theory, 102n15, 154–155, 276, 268–269
  and viscosity, 178
  width of, 110
Bovary, Emma, 249, 256
branch functions, 273, 275, 296
bridges of Königsberg, **51–54**, 56, 59, 61
  and abstract explanation, 228
  ignores details, 223
  and mathematical explanation, 206
  not a causal explanation, 208
  and weaker alternatives, 212
Brownian particle, 216
Bueno, Otàvio, 28n4, 251
Burge, Tyler, 286
Busch, Jacob, 219n17

caloric, 74, **162–163**, 254–255
  and failure, 179
  free versus latent, 162
CAM. *See* complex angular momentum representation
Camp, Elisabeth, 246n2
Cardan, Girolamo, 269, 271, 278, 293
Carnap, Rudolf, 8, 122–126, 137
  and Friedman, 131
  and Kuhn, 127, 129–130
Carton, Sidney, 249
Cartwright, Nancy, 3
case study of
  Black-Scholes model, **146–148**
  boundary layer theory, **106–112**
  bridges of Königsberg, **51–54**
  caloric, **162–163**
  catastrophe theory, **156–161**
  column storage, **142–145**
  concrete scale model, **94–95**
  El Niño Southern Oscillation, **89–91**
  electrostatics, **73–74**
  fluid, irrotational, **68–73**
  fluid, shock wave, **77–78**
  fluid, wave dispersion, **96–103**
  harmonic oscillator, damped, **114–115**
  heat equation, **93–94**
  Lotka-Volterra equations, **58–60**
  rainbow, angle, **224–226**
  rainbow, color, **226–228**

Index

rainbow, supernumerary bows, **229–236**
traffic, cars, **48–50**
traffic, density, **54–55**
traffic, shock waves, **74–77**
Casullo, Albert, 138
catastrophe set, 156
catastrophe theory, **156–161**, 166
   and rainbows, 239
catastrophe, elementary, 158
Cauchy, Augustin-Louis, 294, 297
Cauchy-Riemann equations, 70, 176, 178, 187
causal representation, 48–51, 58, 67, 91–93
   and acausal associates, 61
   and caloric, 163
   of effects, 163
   and explanation, 13
   and relevance, 145
   and structure, 5
   and testing, 253
cause, 4, 6, 45–48, 209
Chang, Hasok, 162–164
characteristic base curves, 75
charged particle, 73–74
chess versus mathematics, 181
cicada, 205, 208, 212, 223, 228
circular frequency of a wave, 99
claims, reason to believe from fiction, 246
classical mechanics and quantum mechanics, 221
climate change, 60, 91–92
clouds, parameters for, 91–92
coarse-grained level of representation, 31
column storage, **142–145**, 163
Colyvan, Mark
   on content, 28n4
   on honeycomb, 205
   on inconsistency, 100n13
   and indispensability arguments, 15, 190, 200–202, 292
   on mathematical explanation, 203, 208n5
   on Melia, 209n6
   on unification, 208n5
common features and self-similarity, 94
comparison test for mathematical explanations, 204
complex analysis, 70, 182
complex angular momentum representation, 230, 236–239
complex function, 71, 232
complex numbers, 17, 202, 269
   and extensions of mathematics, 183
   as vectors, 37, 272
complex potential, 177, 179, 187
   for electrostatics, 74
   for irrotational fluid, 71
computer simulation, 80
concept, 25–27
   change in, 270
   classical conception of, 264
   grounded, 291
   open-ended, 276
   possession, 26
conceptual scheme, 191
concrete causal representation, 5, 9, 121
concrete core, 253
concrete description, 257
concrete fixed representation, 6
concrete scale model, 7, **94–95**
conditional knowledge and applied mathematics, 83
conditionalization, Bayesian, 39
confirmation, 18, 20, 38–41
   and abstract varying representations, 66
   and acausal representations, 61
   and causal representations, 50
   and confirmational holism, 136
   and constitutive representations, 135
   direct, 79
   versus heuristic use, 199
   illusion of, 161
   indirect, 57, 81
   and inference to the best explanation, 238
   and testing, 33
   of Volterra principle, 60
confirmational holism, 62, 136, 218–220
   Friedman against, 135–137
   Quine on, 190–191
connected graph, 51
conservation law, 75
conservation of charge, 186

conservative extension, 198
constitution, physical, 54
constitutive frameworks, 12, 129n3, 131
constitutive representation, 121–122, 137–138
　and failure, 155
　and Friedman, 133, 135
　and Kuhn, 129
　and mathematics, 8, 139, 218
constructive empiricism, 3, 164
content, 25–27, 32–34, 36, 40, 275–277
　assigning, 257
　causal, 46
　conceptual, 267
　empirical, 131, 134
　fictional, 252–253, 255, 256, 258–260
　and Goodman's challenge, 188
　inferential approach to, 28
　and mathematics, 16
　and mystery, 178
　real, 283
　representational, 250, 252–255
　and semantic problem, 171–172
　and structural relation, 27, 197, 217, 264
Contessa, Gabriele, 28n4, 256n14
context, 25, 276, 295
contingent propositions, 39
continuity equations, 69
continuous trading, 148
contributions of mathematics to science, 5–8, 21, 197
　via constitutive representation, 138–139
　descriptive, 177
　different kinds of, 67
　explanatory, 203–210
　and failure, 161–163
　and indispensability arguments, 196, 198
　versus metaphysical problem, 173
　and pure mathematics, 190
　via scaling, 87
convection, fluid, 116
conversion, 131
coordination, principles of, 131, 139
Copericans, 128
core conception, 35–37

correlation between losses, 152
correlational pictures, 267
cosine, 270
counterfactuals, 48, 54, 57
Craver, Carl, 46
critical density, 55
critical points, 239
critical set, 156
cusp catastrophe, 157, 159

d'Alembert's paradox, 154
Dalton's law of proportions, 216
Daly, Chris, 204n2, 207
Darden, Lindley, 46
Darrigol, Olivier, 154–155
data model, 50
De Morgan, Augustus, 269, 271, 278, 293
Debye expansion, 231–232
decomposition, 117
decoupling and idealization, 32, 97, 101
Dedekind, Richard, 284
deep-water waves, 99
degree of belief, 38
delayed action oscillator model, 91, 118n25
density, 54, 77–78, 97
　in boundary layer theory, 107
　decreased by heating, 116
depth, 97, 103
derivative representation, 139
derived elements, 29
Descartes, René, 266–267
descriptive policies, 65
descriptive problem
　general, 176, 179–183
　particular, 175–179
details and understanding, 104, 145
determinacy worry, 253
determinate concept, 267
determinate truth-value, 125
Devlin, Keith, 270
Dickens, Charles, 247, 255
difference-making, 53
difference portfolio, 147, 152
differential equations, 68, 86, 172–173
　and asymptotic expansion, 105
　and dynamic similarity, 94
diffusion, heat, 116

Index

dimensional analysis, 84
dimensionless variables, 101
direct confirmation, 61–62
DiSalle, Robert, 133n4
disanalogies, mathematical, 81
disciplinary matrix, 127
disconfirmation, 57
discontinuous jump, 159
discovery, 21, 80, 82, 187
   and mathematics, 179–183
displacement current, 185–186
displacement function, 97
distance, 87
division of labor, 193
dog attacks, 159
domain
   overlapping, 113–114
   understanding of, 274
Dowe, Phil, 46
downward causation, 118
drag, 94, 106
Duhem, Pierre, 267
Dummett, Michael, 172, 175, 282–283, 285, 288
dynamic boundary condition, 98
dynamic representation, 5, 49, 208
dynamic similarity, 94, 95, 102n14
   versus geometric similarity, 153

e, 55, 71, 270, 276
Earman, John, 38n10
Earth-sun representation, 88
edge of boundary layer, 112
efficient market, 147, 148
Einstein, Albert, 128
Einstein's field equations, 8
El Niño Southern Oscillation, **89–91**, 100n12
Eleatic principle, 211
electrodynamics, 73, 221
electrons, 211, 215–216
electrostatics, **73–74**, 78, 177, 210
   and irrotational fluid, 68–74
Elgin, Catherine Z., 249n4
elliptic, partial differential equation, 83
emergence, 118–119
empiricism, 291

end of science, 199
enriched content, for heat equation, 30–31, 55
ENSO. *See* El Niño Southern Oscillation
entities, existence of, 165
entitlement, 286, 289–290
entrenched predicates, 188
epiphenomenal properties, 117
epistemic benefits of mathematics, 8, 12
   and abstract varying representations, 80
   for concrete scale models, 94
   for constitutive representations, 138
   for multiscale representations, 119
   and problem of applications, 8–12
   and scale similarity, 96
   for single scale representations, 92
epistemic possibility, 214
epistemology for pure mathematics, 285–290
equations, rescaling of, 101
equilibrium equation, 156
equilibrium representation, 42
equinumerous concepts, 172
equity vol position, 150
equivalent constructions, 193, 201
errors due to mathematics, 81
estimate of size of terms, 108
ether theories, 130
Euclidean geometry, 131
Euler equations, 69, 108, 139, 154
   and boundary layer theory, 269
   and concept of fluid, 276
   and linguistic frameworks, 123
Euler, Leonhard, 269, 271, 278, 293, 295, 296
Eulerian graph, 51–52
Euler's conjecture, 213
Euler's formula, 270, 296, 307–308
Euler's theorem for graphs, 214
evidence, 10, 47
   and fiction, 250, 255, 260
   versus ground, 291
evolutionary biology, 206
exemplar, 127–129
exemplification in fiction, 249
existence of groups, 290
experience, contents of, 129

explanation
  asymptotic, 210
  intrinsic and extrinsic, 172, 174
  mathematical, 204, 209, 216–217, 226
  minimal, 164
  noncausal, 46, 209
  power of, 228–229, 236
  using light rays, 241
export problem for fictionalism, 244, 252–256, 258–260
  for historical fiction, 247
  for literary fiction, 248
expository work, 249
extensions, historical, 293–295
external questions, 122–123
externalism, semantic, 27, 125–126, 264
extrapolating, 19, 143
extrinsic mathematics of a representation, 5, 38, 49–50, 64, 177

facades, 65, 266
failure, 19, 141–142, 179
  and holism, 218–219
feedback loop, 118n26
Fermat's last theorem, 36, 182, 293
fictional characters, 244–245
fictional models, 257–258
fictionalism, 15–16, 217, 281, 242–244, 256–261
  objections to, 244, 255–256
Field, Hartry, 16, 196, 198–200, 251
  Colyvan on, 201–202
  on intrinsic explanations, 215n13
  and mathematical truth, 243
  and Melia, 207
  and metaphysical problem, 172, 174
  objections to, 199, 209
film of Tacoma Narrows Bridge, 154
financial economics, 146
fixed interpretation in robustness analysis, 67
Flaubert, Gustave, 247–249, 255
flexibility in applying Thom's theorem, 160
flow net, 70
flow, 72, 106
fluid, 125, 271, 276
  and abstract acausal representation, 10
  dynamics, 176, 177
  and explanation, 210
  ideal, 69
  irrotational, **68–73**, 68–74, 78, 178, 261
  linguistic framework for, 123
  shock wave, **77–78**, 268
  source, 70
  wave dispersion, **96–103**, 105, 120
flux, 55
flux function, 75
flux lines, 74
Fock, V. A., 237n17
fold catastrophe, 157
force, 73, 130, 132
forecasting, 91
formal network, 27
formalism versus platonism, 281
formalist analogy, 183
Foundations of Arithmetic, 282
Fourier, Jean-Baptiste Joseph, 164
Fowler, A. C., 84
Fr. See Froude number
frames of reference, 132–133
Franklin, James, 4n1, 51n7
Frege, Gottlob, 170, 172, 281, 288–289
Frege's constraint, 283, 288
French Revolution, 247
French, Steven, 186–187
Friedman, Michael, 8, 12, 122, 131–137
  objections to, 134–135
Friend, Stacie, 244
Frigg, Roman, 256–261
Froude number, 95
functional property, 118
functions
  analytic, 176
  of complex numbers, 270
  multiple-valued, 273

Galileo, 142–144, 152
Galois, Évariste, 297
game and pretense, 245, 252
gauge freedom, 68
Gauss, Carl Friedrich, 271, 296
Gendler, Tamar Szabo, 248
general relativity theory, 8, 139
genuine content, 32, 101

geometric similarity versus dynamic similarity, 95
Gettier case for concepts, 290
glory and rainbows, 237
goals of representation, 32
Godfrey-Smith, Peter, 256, 259n17
Golden Gate Bridge, 154
Goldman Sachs, 153
good mathematics, 182, 184
Goodman, Nelson, 187–188
graph, 51
gravitation, 74, 132
greater than, 35
ground versus evidence, 291
grounding, 294–295
group theory, 126, 288–289, 297
grue, 188

Hales, Thomas C., 213
Hankel function, 231
hardness, 266–267
Hardy, G. H., 181–182, 251
harmonic functions, 69
harmonic oscillator, 6, 10, **114–115**
    and discovery, 18
    as exemplar, 128
    and explanation, 14, 66
    and overlap of mathematics, 78
    risks in applying, 19
heat, 74, 79, 163
    equation, 30, 34, **93–94**, 146
    and Fourier, 165
    and partial differential equation, 29
    and representation, 259
hedge funds, 164
hedge, risk-free, 146, 148
Heisenberg, Werner, 186
Hellman, Geoffrey, 195, 197–198, 207, 280, 284
    versus fictionalism, 217
    and modal logic, 194
    and semantic problem, 171
Hertz, Heinrich, 267
heuristics, 81, 113
Hill, Christopher S., 287–288
Hillerbrand, Rafaela, 88–89
historical fiction, 247–248
history of mathematics, 182

Hitchcock, Christopher, 45
holism and semantic finality, 266
Holmes, Sherlock, 246
homogeneous fluid, 69
homomorphism, 27
honeycomb, 205–206, 208, 212, 223, 228
Howson, Colin, 40n11
Hume, David, 45
Hume's principle, 173
Humphreys, Paul, 36, 80
hurricanes, 91–92
hyperbolic, partial differential equation, 83

IBE. *See* inference to the best explanation
idealization, 4, 97, 151, 257
    and Black-Scholes model, 146, 150–151
    and continuous medium, 81
    deep water, 96, 100–101
    and derived elements, 29
    and explanation, 226–227
    and heat equation, 150–151
    and isospin, 186
    and limits, 229
    and perturbation theory, 88
    and scale, 87, 96–104, 107n19, 120
    sequence of, 223, 238, 241
    small amplitude, 96–97, 102–103, 108
    and water waves, 163
Illner, Reinhard, 55
illusion, 165
    causal, 161–163
    completeness, 142–145
    parameter, 146–153
    scale, 153–155
    segmentation, 142–145, 154
    traction, 155–161
imaginative resistance, 248
impact parameter, 232
implicit conception, 287–288, 294
implied volatility, 150, 152
import practices for fiction, 247
impossibility arguments, 155
impredicativity, 175
incommensurable paradigms, 128
incompressible, 69
inconsistent beliefs, 100
inconsistent estimates of boundary layer, 113

index of refraction, 226
indexing and applications, 204, 208–209
indirect confirmation, 61–63
indispensability, 15, 190, 200, 280
   argument, explanatory, 21, 203, 206–208, 286
   argument, objections to, 16, 196–200
   argument for platonism, 12, 191–193, 195
   argument for realism, 21, 194–196, 202
   and empirical justification, 140
   and grounding, 292
indispensability$_C$, Colyvan, 200, 205
indispensability$_P$, Putnam, 194–195
indispensability$_Q$, Quine, 195
induction, material theory of, 42
inertial frame, 136
inference, 28, 33–35
inference to the best explanation, 13–14, 142, 165, 221
   and caloric, 163
   and indispensability arguments, 203, 210–217
   for pure mathematics, 280, 296–297
   and rainbows, 237
   restrictions on, 21, 212, 238
infinite progression, 298
infinitely deep ocean, 258
instantiation, 29
instrumentalism, 3, 119, 165
intelligibility, 262
intensity, 229
intermediate asymptotics, 222
internal questions, 122–123
internalism, semantic, 26, 264
interpretive conjecture, 234
interpretive flexibility, 19
interpretive puzzles, 100, 113, 178–179
intervention, 57
intrinsic mathematical interest, 112
intrinsic mathematics of a representation, 5, 37
   and descriptive problem, 177
   versus extrinsic, 22, 36–38
   overlap of, 78
   for traffic, cars, 49
intrinsic version of a theory, 198
inverse method, 70, 177

inviscid, 77
iron bar, 31
irrelevant feature, 102, 210, 228
irrotational, 69
isomorphism and counting, 27, 53, 284
isospin, 186

Jenkins, C. S., 290–293, 296
Julius, inventor of the zipper, 130
justification, 136
   a priori, 138, 140, 211, 279, 285
   empirical, and mathematics, 192, 211
   nonconceptual, 295–297
   practical, 137

Kelvin, William Thomson, 155
key, 260
key claims, list of, 21, 279–280
kinematic versus dynamic boundary condition, 98
kinetic energy, 162–163
Kitcher, Philip, 14n4, 182
Koellner, Peter, 299
Kuhn, Thomas, 8, 12, 122, 126–133, 155
   and Friedman, 131, 134
   objections to, 129–130

Lagrange, Joseph-Louis, 297
Lange, Marc, 211n8
Langford, Simon, 204n2, 207
Laplace equation, 69, 83, 98, 177–178, 210
Laplace, Pierre-Simon, 141, 162, 164–165, 179, 254–255
Laplacian school, 165
Lavoisier, Antoine, 128
law of universal gravitation, 132, 218
laws of motion, Newton's, 131
Legendre function, 231
Leibniz, Gottfried Wilhelm, 293–295, 297
Leng, Mary, 251
less content, 43, 57–58, 62, 92
levels, 119
Lewis, David
   versus fictionalism, 217
   on laws, 46n1
   and mathematics, 172, 195, 197, 207, 280
   and structuralism, 207, 284

Index

lift, 106, 155
Liggins, David, 193n2
light and electromagnetic waves, 221
light dispersion, 226
light rays, 222, 224
limit, 293–295, 297
　and representation of light, 222
Lin, C. C., 78, 84, 88, 111
linear equation, 70, 176–177
linguistic framework, 122–126, 137
Linnebo, Øystein, 284n2
literary fiction, 244–250
load, calculation of, 153
localization principle, 232
logic, classical, 175
logic, first-order, 191
logical empiricists, 45
logical omniscience, 20, 39
logical paraphrase, 33
lognormal model, 147, 152
Long-Term Capital Management, 141, 148, 149
Lotka-Volterra equations, **58–60**
Lowenstein, Roger, 148
LTCM. *See* Long-Term Capital Management
Lyon, Aidan, 203n1, 205

Machamer, Peter, 46
macroscale, 7
Madame Bovary: A Tale of Provincial Life, 247
Maddy, Penelope, 100n13, 183, 216n16, 256n15
magnitude and perturbation theory, 105
Mancosu, Paolo, 183, 296n11
Manders, Kenneth, 295n10
map and concepts, 290
mapping account. *See* content and structural relation
market participants, 152–153
mass, 130, 132
materials, strength of, 144
mathematics
　and acausal representations, 61
　and analogy, 185, 210
　belief in, 217–220
　Carnap on, 123

　and causal relations, 50
　and causal representations, 48, 60
　confirmation of, 219–220
　constructive, 175
　created ex nihilo, 262
　and descriptive success, 190
　and explanation, 12–15, 207
　extrapolating in, 214
　and failures, 141
　fictionalism about, 250–256
　flexibility of, 161
　and fruitful concepts, 178
　generality of, 139
　history of, 265, 271, 293
　important, 185
　language, 34
　and linguistic frameworks, 125
　versus logic, 45
　models, 84
　and new representations, 93
　nonempirical justification for, 41
　and patches, 268
　precision of, 139
　presupposed, 140
　prior justification of, 219, 222, 238
　similarities, 68, 80
　story of, 243
　structure, 68, 83, 256
　as symbolic generalizations, 128
　versus tautologies, 40
　terms of, 126
　as true in all possible worlds, 33
　understanding of, 217–220
maxim of minimal mutilation, 219
Maxwell, James Clerk, 185
Maxwell's equations, 73
Mayo, Deborah, 50n6, 216
McGivern, Patrick, 114, 117, 119–120
meaning
　and Carnap, 125–126
　and Friedman, 131, 134
　and Kuhn, 128–130
　and Putnam, 193
measurement of hardness, 266
measurement theory, 176, 193, 198–199
Melia, Joseph, 204, 207, 215
Melnyk, Andrew, 118
Merton, Robert C., 148, 153

metaphysical conception of applied mathematics, 4, 13
metaphysically necessary, 214
metaphysics of mathematics, 280–284
metaphysics, reform of, 119
meteorological modeling, 89
method of characteristics, 75, 301–302
microphysical, 63–64
microscale, 7
Mie representation, 230–231, 236
Mill, John Stuart, 281
Miller, Merton H., 149
Millikan, Robert, 216
miracle of applied mathematics, 180
modal-structuralism, 171, 194
model, 11, 25–26
   and fictionalism, 243–244, 256
   nonstandard, 288
Morrison, Margaret, 14, 186n7
multiply realized concept, 267, 271, 274
multiscale representation, 7, 88, 106, 113–120
mystery of discovery, 177

Nahin, Paul J., 270
Narasimhan, T. N., 146
natural kinds, 187–189
natural numbers, 214, 251
naturalism, 17, 169, 183–187
Navier-Stokes equations, 69, 96, 98, 261
   and boundary layer theory, 107
   and fluid convection, 117
   limitations of, 104
   in a linguistic framework, 125
   relevant parameters of, 94–95
   simplifications of, 112
   and Steiner, 178
Newton, Isaac, 128, 179, 254, 293–295, 297
Newtonian mechanics, 130
Newtonian paradigm, 129
Newton's law of gravitation, 192
Newton's laws of motion, 161, 218
nominalism, 15
nominalistic adequacy, 253
nonfundamental representations, 63
nonlinear equation, 70
nonmathematical description, 175

normal science, 126–127
Norton, John, 42
no-slip condition, 109
numeral, 170
Nussenzveig, H. M., 230, 234

objective chance, 39
objectivist account of explanation, 13
objectivity of aesthetics, 184–185
one-one correspondence, 173
ontological commitment, 191, 201, 254, 261
operation, limit of, 228–229
operator approach to fiction, 245
option, call, 146
order of a function, 104
ordinal, largest, 136
oscillations of bridge, 154
overview of book by chapter, 21–22

P1, 43–44
P2, 43–44
pain, 271
Pappus, 213
paradigm, 126–131
   articulation of, 127, 129
parameters, 87, 91, 151
Parker, Wendy, 60, 91–92
Parsons, Charles, 173n3, 284
partial differential equations, 83, 146, 268
partial overlap of mathematics, 81
partial representation and accuracy, 104
patches, 268–275
path of integration, 232–233
Peacocke, Christopher, 265, 286, 288, 296
   and implicit conceptions, 35n9, 293
   objections to, 290–293
Peano axioms, 36, 194, 287
Peirce, C. S., 183
perception and grounding, 292
period of a wave, 98
perturbation theory, 84, 88, 104–105
   and idealization, 96, 101
   and power series, 178
   regular, 105, 112
   and scale, 101–104
   singular, 106, 112, 227

Index

pessimistic meta-induction, 163–164, 165
Petroski, Henry, 142–144, 145, 154
phase velocity, 99, 100
phlogiston, 26, 130
physical insight and explanation, 238
physical interpretation, 79, 227–229
physicalism, 117–118
physically impossible, 100
physically related, 96
physics avoidance, 268
physics of the rainbow, 230–231
planetary orbits, 215
platonism, 15, 26, 190, 281
   indispensability argument for. *See* indispensability argument for platonism
   and mathematical importance, 182
   and the metaphysical problem, 172
   and objects, 280
   versus realism, 197
   and the semantic problem, 170
Poincaré, Henri, 88
Poisson sum formula, 232, 238
Poisson's equation, 83
polar coordinates, 71, 272
poles, 233, 235–236
population biology, 58
Porter, Theodore, 4
positivism, 134, 165
possession condition for a concept, 287, 289
potential. *See* scalar potential
potential function, 70
potential theory, 74, 84
power series, 176, 178
practical vocabulary, 268
pragmatic account of explanation, 13
Prandtl, Ludwig, 106, 110, 154–155
precausal representation, 49, 60
predator and prey, 58, 259
pressure, 97
presupposition, 124, 133
pretense approach to fiction, 245, 247
Priestly, Joseph, 128
primary bow, 224
prime numbers, 205

prior probabilities, 41–44
probability calculus, 38
probing for error, 81
problem of applications of mathematics, 3, 280
problem of old evidence, 40
problems of applicability in Steiner, 169
   descriptive, 175–179
   discovery, 169, 181
   metaphysical, 169–175
   semantic, 169–175
process and two time scales, 115
process theory, 53
projectible predicates, 188
proof, 271
prop in a game, 245
proper setting, 265, 275, 277
pseudo-statement, 124
Psillos, Stathis, 164–165, 203n1, 256n15
Putnam, Hilary, 15, 190–196, 277
Pythagorean analogy, 17, 183, 185–186

quantifier-committed, 261
quantum mechanics, 180, 221
quarks, 215
Quine, W. v. O., 277
   and confirmational holism, 135–137
   on indispensability, 15, 139–140, 190–197
Quinean holism, 137

rainbow, 210, 221, 223, 277
   angle, **224–226**
   color, **226–228**
   nuclear, 237
   supernumerary bows, **229–236**
Ramanujuan, Srinivasa, 252
rarefaction wave, 76
rational belief and linguistic frameworks, 123
rational justification for constitutive representation, 137
ray theory of light, 223
Rayo, Agustín, 262, 265, 278
Re. See Reynolds number
reaction time, 48
real numbers, 174

realism, 15, 194, 197
   and fictional characters, 244
   indispensability argument for. *See* indispensability argument for realism
   patient, 265, 267
realistic novel, 248
real-valued functions, 296
red, 226
Redhead, Michael, 240n19
reduction sentences, 123
reduction, 119, 179, 221
reference, 26
regime, 88
regimentation, 191, 198
regular polygons, 213
Reichenbach, Hans, 136
Reid, Thomas, 266–267
relativized a priori, 136
relevance and explanation, 209
relevant structural features, 239
reliabilism, 290
replacement test for mathematical explanations, 204
replicating portfolio, 147
representation, 26
   constitutive, 121–122
   content of, 28
   derivative, 121
   theorems, 199
   versus theory, 222, 241
rescaling, 101
residues, 235
Resnik, Michael, 200n6
Reynolds number, 87, 95, 109, 117
Riemann hypothesis, 182
Riemann surface, 272–274, 296
   and epistemology, 286
   and patches, 276–277
   and Peacocke, 293
   and Rayo, 278
   and Wright, 283
Riemann, Bernhard, 295–297
Riemannian manifolds, 131
rigor, lack of, 111–112
risk, 151–152
Roberts, Bryan W., 305n1
robust result, 59
robustness analysis, 10, 58, 60, 67, 92n5

Rosen, Gideon, 251, 253, 255
Rossby waves, 118
rules, semantic, 126
Russell, Bertrand, 265
Russell class, 136

Saatsi, Juha, 204n2, 205n3
saddle points, 233, 234
Salmon, Wesley, 46
scalar potential, 68, 73, 78, 97
scale, 6–7, 69, 87–88, 120–121
   chosen for empirical reasons, 111
   effects and engineering, 153
   for Euler equations, 108
   for fluid, wave dispersion, 102
   and harmonic oscillator, 114–115
   for heat equation, 93
   and idealization, 96–104
   multiple, 105–113
   separation, 87–93, 120
   similarity, 11, 87, 93–96
   and Tacoma Narrows Bridge, 154
scattering amplitudes, 231
schematic content, 26, 32, 55, 97, 101
Scholes, Myron, 148
science versus metaphysics, 119
scientific failures, 141
scientific knowledge, 163
scientific realism, 3, 65, 142, 166, 203
   and caloric, 163–164
   and constitutive representation, 135
scientific revolutions, 126
scope, 151, 163–166, 213
   of applicability, 155, 166, 282
   of a representation, 101, 212
sea surface temperature, 89
Segel, L. A., 78, 84, 88, 111
self-similar phenomena, 7, 93
semantic finality, 265–267, 275
semantic theory, 262
semantic value, 287
sensitivity and explanation, 212, 214–215
series expansions, 275
set theory, 173, 298
shallow-water waves, 99–100
Shapiro, Stewart, 29n5, 284, 297–298
ship design, 124
shock wave, 74–78

Index

similarity and content, 258–260
simplicity, 160
sine, 270
single scale representation, 11, 88
singularities, 64, 112, 223
sink, 72
size parameter, 227, 231
smooth, curve, 176–178
Snel's law, 224
social conception of science, 3
source, 72
space-time points, 199
speed of sound, 77, 305–306
    and caloric, 179, 254
    Newton and Laplace on, 162
spinning top model, 260
spring, 66, 78
square root function, 272, 276, 285, 296
    branches of, 273
    and Riemann surface, 274
SST. *See* sea surface temperature
stability analysis, 56, 84, 86
standards
    in applied mathematics, 113
    of a paradigm, 128, 130
statement of number, 282
steady flow, 69
steady-state representation, 54
Steiner, Mark, 17–18, 169, 183, 270n4
    on aesthetics in mathematics, 181
    and descriptive problem, 176
    on Goodman, 187
    and metaphysical problem, 172
    and semantic problem, 170–171
    on Wigner, 17, 180–181
Sterrett, Susan, 94
Stokes, Sir George G., 239
strain, 142
streamline function, 70
structural engineering, 153
structural relation, 27, 31, 199
    and applications, 282, 284
    and content, 197, 217, 257
    versus keys, 260
    and structuralism, 174
structuralism, 279–280, 282, 284–285
    modal, 282
structure, abstract, 174

strut, 64, 142n1
$SU(2)$, 186
Suárez, Mauricio, 28n4, 29, 256
substitution test, 27–28
supernumerary bows, 224, 229–236
superposition, 98, 177
surface tension, 95
surface waves, 235
surfaces, classification of, 156
Sussman, Héctor, 160
symmetry principles, 186
systematic failures, 164

Tacoma Narrows Bridge, 154
Tait, William, 284
Tappenden, Jamie, 295n10
temperature, 30, 93
Tennant, Neil, 57
terminal value problem, 147
testimony, 246, 250, 290
testing and the relative a priori, 131
theism, 184
theory, 25, 222, 241
theory choice, criteria for, 200
thermocline, 89, 100n12
Thom, René, 156
Thom's theorem, 156–159
Thomson-Jones, Martin, 256n15
Thomson, Joseph John, 216
three-support system, 144
time, 48, 49, 87
time dependence, 75
toolkit of applied mathematics, 68
Toon, Adam, 256
tractable, mathematics, 104, 111
trade-off
    accuracy and interpretative debate, 113
    completeness and accuracy, 104
    mathematics and other assumptions, 213
traffic jam, 77
traffic
    cars, 5, **48–50**, 51, 56, 208
    density, 6, **54–55**, 56, 58–59, 61–62
    shock waves, **74–77**
    steady state. *See* traffic, density
transfer from physical representation, 80–81

translated facts, 260
translation, 128
t-represents, 260
trivialism, 262
true in the story, 217
truth, 129, 137, 208
   conditions, 262
   and inference, 172
   values, 197
turbulence, 117

ultrathin posits, 262
understanding and truth, 261–263
underdetermination, 135–136
unification and explanation, 13, 95, 208, 237
uniform flow, 70
unreasonable versus unscientific, 128
unreasonable effectiveness, 17, 63, 179–180
unspecified parameter, 32
Urbach, Peter, 40n11

validity, 105, 109
van de Hulst, Hendrik C., 230
van Fraassen, Bas, 13, 29, 211
variable reduction, 65
variously interpreted mathematics, 177
varying interpretation, 67
varying representation. *See* abstract varying representation
vector field, 68
velocity, 48, 55, 87, 97
verificationism, 123
viscosity, 69, 97, 155, 178, 261
   and boundary layer theory, 107
volatility, 146
Volterra principle, 59, 61
von Neumann, John, 181

Walker, Sir Gilbert T., 91
Walton, Kendall, 245, 252, 257

war model, 159
water waves and sound waves, 163
water, viscosity of, 108
wave dispersion. *See* fluid, wave dispersion
wave equation, 29–30, 162
wave number, 99, 227
wave theory of light, 226
wavelength, 97
weaker alternatives, problem of, 212–213
web of belief, 219
Weierstrass, Karl, 294, 297
Weisberg, Michael, 11n3, 59, 94
   and robustness analysis, 10n2, 58, 67
Weslake, Brad, 213n11
Weyl, Hermann, 274, 277, 293, 296–297
whole numbers, 174
Wigner, Eugene P., 17, 179–180, 270n4
Wilholt, Torsten, 51n7, 174
Williams, Samuel, 267n2
Wilson, Mark, 22, 166, 183
   on concepts, 27, 264
   on costs of mathematics, 242
   on fluid and traffic, 78
   on Fourier, 165
   and the impossibility of flight, 154
   and mathematical extensions, 295n10
   on Quine and Putnam, 277
   and Riemann surface, 275
   on variable reduction, 65
Wimsatt, William C., 119n27, 260n18
Winsberg, Eric, 114n22
Woodward, James, 47, 50n5, 54n9, 57, 215
Wright, Crispin, 173n3, 283, 285, 288

Yablo, Stephen, 26n1, 251–252, 255

Zahler, Raphael, 160
Zeeman, E. C., 156, 159
ZFC, 173
zombie sheep, 254

www.ingramcontent.com/pod-product-compliance
Ingram Content Group UK Ltd.
Pitfield, Milton Keynes, MK11 3LW, UK
UKHW041307180426